CATCH EFFORT SAMPLING STRATEGIES

CATCH EFFORT SAMPLING STRATEGIES

THEIR APPLICATION IN FRESHWATER FISHERIES MANAGEMENT

EDITED BY

I.G. COWX

Humberside International Fisheries Institute
University of Hull

FISHING NEWS BOOKS

Copyright © I.G. Cowx 1991

Fishing News Books
A division of Blackwell Scientific
  Publications Ltd
Editorial offices:
Osney Mead, Oxford OX2 0EL
25 John Street, London WC1N 2BL
23 Ainslie Place, Edinburgh EH3 6AJ
3 Cambridge Center, Cambridge
  MA 02142, USA
54 University Street, Carlton,
  Victoria 3053, Australia

First published 1991

Set by Setrite Typesetters Ltd
Printed and bound in Great Britain by
Hartnolls Ltd, Bodmin, Cornwall

DISTRIBUTORS

Marston Book Services Ltd
PO Box 87
Oxford OX2 0DT
(*Orders*: Tel: 0865 240201
       Fax: 0865 721205
       Telex: 83355 MEDBOK G)

USA
  Blackwell Scientific Publications Inc
  3 Cambridge Center
  Cambridge, MA 02142
  (*Orders*: Tel: (800) 759-6102)

Canada
  Oxford University Press
  70 Wynford Drive
  Don Mills
  Ontario M3C 1J9
  (*Orders*: Tel: (416) 441−2941)

Australia
  Blackwell Scientific Publications
  (Australia) Pty Ltd
  54 University Street
  Carlton, Victoria 3053
  (*Orders*: Tel: (03)347−0300)
British Library
Cataloguing in Publication Data
Catch effort sampling strategies: their
  application in
  freshwater fisheries management.
  1. Angling waters. Management
  I. Cowx, I.G. (Ian G)
  333.956110151952

ISBN 0−85238−177−8

Contents

Preface

At the Fifteenth Session of the European Inland Fisheries Advisory Commission (EIFAC) in Goteborg, Sweden, 31 May–7 June 1988, the need to improve sampling methodology for studying fish stocks in large fresh waters as an aid to management was identified. In particular, it was felt that catch effort sampling strategies were extremely important techniques that warranted further investigation and discussion.

To this end, a symposium was organized and hosted by the Humberside International Fisheries Institute at the University of Hull, UK from 2–6 April 1990. The symposium was a joint venture with EIFAC but with the co-operation of the Fisheries Society of the British Isles and the Institute of Fisheries Management. The Steering Committee for the symposium comprised I.G. Cowx (HIFI, UK) Convener, B. Steinmetz (The Netherlands) Chairman, P. Hickley (UK), W.L.T. van Densen (The Netherlands), S.P. Malvestuto (USA), F.L. Orach-Meza (Uganda) and M. Pawson (UK).

The objectives of the symposium were to advance the scientific and management basis of catch effort sampling strategies, and provide a medium for the dissemination and exchange of ideas. The symposium was attended by 116 delegates from 31 countries representing all continents. Forty-eight papers and posters, covering various aspects of catch effort sampling in relation to both commercial and recreational fisheries, were presented. From these presentations, and the associated discussions, a number of aspects of catch effort sampling strategies were recognized as common to both commercial and recreational inland fisheries, and the different types of water bodies, i.e. lakes, reservoirs and rivers. There was seen to be a need for a simple common mechanism which could be applied to locally specific situations by the responsible organizations which would enhance management decision-making to the benefit of the fishery.

These proceedings, which contain selected papers from the symposium, illustrate these points and, it is hoped, will stimulate fisheries scientists, managers and academics to collaborate in further research to improve our understanding of catch effort sampling strategies.

The production of these proceedings has involved considerable effort by a number of people. In particular, thanks must go to members of the Steering Committee for their assistance in the successful running of the symposium. I am indebted to S. Axford, K. O'Hara, C. Machena, M. Aprahemian, S. Welton, P. Hickley, P. Gerard, J.A. Timmermans and E.H.H.R. Lammens for critically refereeing the papers and making useful suggestions for their improvements. I would like to thank Julia Cowx and Debbie Leake for their considerable assistance in the running of the symposium and production of these proceedings. Finally, I would like to thank the many international funding agencies, especially British Council, Commonwealth Secretariat, NORAD, and the FAO for their financial support, thus ensuring a truly international symposium.

Ian G. Cowx

Chapter 1
The Atlantic salmon fishery of the River Severn (UK)

A.S. CHURCHWARD and P. HICKLEY *National Rivers Authority, Severn-Trent Region, 550 Streetsbrook Road, Solihull, B91 1QT, UK*

Catch statistics for Atlantic salmon taken by rod fishermen and commercial instruments during the period 1940–89 have been analysed and trends identified. Catches from fishing weirs have remained remarkably stable whereas those from commercial netting methods have decreased slightly. Rod catches appear to be increasing.

There have been two noticeable long-term changes in composition of the catch. First is the decreased contribution of 3 and 4 sea-winter fish, which may be the result of overexploitation by rod fishermen. Second is that in recent years there has been a relatively dramatic increase in the percentage number of grilse (1 sea-winter fish).

Management strategies to redress the age group balance of the stock are discussed in the context of fishing seasons and their relationship to the principal months in which salmon of different sea ages are taken.

It is concluded that statutory returns can produce a very cost-effective and useful supply of data, provided that limitations in accuracy are recognized.

1.1 Introduction

1.1.1 *The river*

The River Severn rises in mid-Wales. At first it flows north-east to Shrewsbury and then turns southward to Gloucester and its ultimate union with the Bristol Channel. At 354 km in length with an annual discharge of $62.70 \, \mathrm{m^3 S^{-1}}$, it is ranked first and seventh respectively for British rivers (Ward, 1981). Downstream of Shrewsbury the river system supports extensive cyprinid recreational fisheries. The upper reaches contain resident brown trout (*Salmo trutta* L.) populations and also serve to provide spawning and nursery grounds for Atlantic salmon (*Salmo salar* L.). Migrating adult salmon are taken both by rod anglers and commercial fishermen.

1.1.2 *Historical perspective*

The River Severn catchment has been managed for fisheries purposes since 1865 when the Board of Conservators for the Severn Fishery District was established.

1

The Severn River Board was formed and took responsibility in 1950, the Severn River Authority in 1963, the Severn-Trent Water Authority in 1974 and, finally, the Severn-Trent region of the National Rivers Authority on 1 September 1989.

There is a long history of information gathering and details of salmon catches have been reported since 1869. The statutory requirement to submit catch returns was introduced for the 1938 fishing season with the need to include nil returns commencing in 1940. The voluntary supply of scale samples by fishermen started in 1951.

1.1.3   *Fishing methods*

There are four methods used to take salmon, namely rod and line, draft net, lave net and fishing weir. Rod fishermen generally use artificial fly, spinner or worm and fish most of the river length upstream of the tidal limit. The three commercial methods are used in the estuarial zone.

The draft net is a single sheet of netting up to 185 m in length. One end of the net is held fast on the shoreline and the other is paid out from a boat which takes it across the channel. Once the net has been taken three-quarters of the way across, the free end is then towed downstream until the end of the fishing zone is reached. Finally, the boat is brought rapidly back to the shoreline landing area and the net is drawn in. In tidal waters draft netting only takes place on the ebb tide. On a neap tide it is possible to get nearly six hours fishing whereas on a spring tide only three hours may be available.

The lave net is hand-held to dip salmon out of the water during low tide in the estuary. The frame of the lave net is shaped like a large 'Y' with a bag net suspended between the 1.9 m-long arms. The fisherman sees the mark of a salmon as it rushes upstream through the shallows and runs to intercept it.

The principal type of fishing weir, described in legal parlance as a fixed engine, consists of a framework of stakes on which is mounted a series of traps known as putchers. Each putcher is a loosely woven basket-work cone up to 1.8 m in length and many hundreds are placed in three of four tiers. These putcher ranks are staked out across the flow of the river with the mouths of the traps facing upstream. They catch salmon which try to drop back into the safety of deeper water when the tide is on the ebb. At the start of each close season these fixed engines must have their putchers removed. Also, the location of the ranks are controlled by law.

Fishing seasons for the various methods are as follows: rod and line – 2 February to 30 September inclusive; draft net and lave net – 2 February to 31 August inclusive; and putcher ranks – 16 April to 15 August inclusive.

1.2   Materials and methods

Data on salmon catches have been drawn from both Severn Fishery District estimates (pre 1938) and statutory returns (post 1938). All fishermen are legally obliged to declare how many fish they caught during the season. The return forms require information on location, date of capture and a breakdown into weight categories of <7 lb (3.2 kg), 7–15 lb (3.2–6.8 kg) and >15 lb (6.8 kg). Generally there is a 100% return from commercial fishermen but that for rods only reaches 80%, even after a second reminder. It is believed, however, that the majority of defaulters are those with a nil catch.

Fishermen are encouraged to send in scales, together with details of lengths and weights of the fish, and about 10%, mostly commercial, actually do so. Scales are read on a back-projection microscope and fish are assigned both freshwater and sea ages.

Since 1987, fisheries staff have had extremely good access to salmon taken from one particular putcher rank. These fish are gutted prior to smoking and have provided details on length, weight, age and sex ratio.

There is also one fish counter on the River Severn which is located on the weir at Shrewsbury at the arbitrary division between the upper and middle reaches of the river. Counting started in 1979.

A visual assessment of the number of spawning redds cut in the river is attempted each winter. Although redd counting has been ongoing for many seasons, only the last ten years of data can be treated with any degree of confidence.

1.3 Results

1.3.1 *Total catches*

Total salmon catches 1869–1989

Since 1936 there has been a substantial decrease in salmon catches, although catches in 1930 and 1931 were also low (Fig. 1.1). There is a decreasing trend since 1911 which may have been contributed to by the building of the weir, albeit with a fish pass, at Shrewsbury in 1910. The statutory returns for 1938 onwards probably under-represent catches but there is also little double that the pre-return figures could be overestimates. The situation elsewhere in the country, whereby some riparian owners believe that they can increase income to private fisheries by inflating their catch declarations, does not apply to the River Severn.

Catches taken by rods and commercial instruments 1940–89

Changes in annual catch since the introduction of statutory returns are presented in Fig. 1.2. There is an upward trend evident for the rods which is probably related to the increase in licence sales (see also Fig. 1.5).

Draft net catches show several changes. The first bye-law controlling fishing in the estuary was introduced in 1953 and this brought about a reduction in the number of salmon drafts which could be fished. By the mid-1970s, however, the

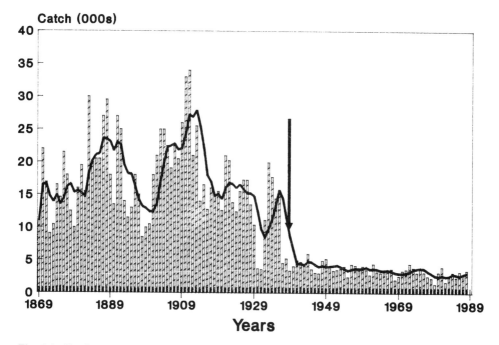

Fig. 1.1 Total annual catches of salmon reported for the River Severn for the period 1869–1989 inclusive. The vertical bars show actual catch figures and the solid line the five-year moving average. The vertical arrow indicates 1938, which is the year that statutory returns were introduced.

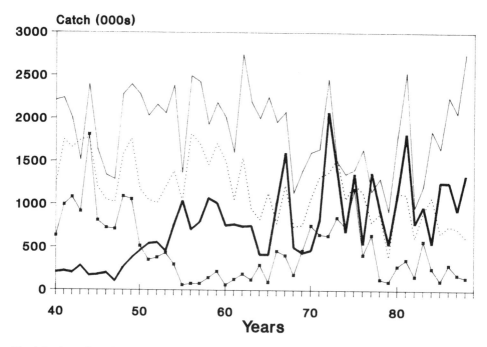

Fig. 1.2 Annual catches of salmon taken by rods and commercial instruments for the period 1940–89 inclusive. (———— rod; · · · · · lave net; ■——■ draft net; ———— fixed engine)

number of draft nets in use and fish taken had increased again. Since that time, with the number of nets remaining fairly constant, landings have declined (see also Fig. 1.6).

Lave net catches show a slight, but steady, declining trend. The number of salmon taken by the fixed engines is remarkably consistent within cycles of variability. Average annual catches for the long-term compared with those for the last ten years are given in Table 1.1.

Ratio of rod catch to commercial catch 1940−89

Figure 1.3 clearly shows the steady and considerable increase in the proportion of the total salmon catch which is attributable to rods. Note that the unexpectedly

Table 1.1 Average annual salmon catches by different instruments for the long term compared with the last ten years

Method	Average no salmon caught 1940−89	Average no salmon caught 1980−89
Rod and line	759	1112
Draft net	474	262
Lave net	1170	846
Putcher rank	1878	1897

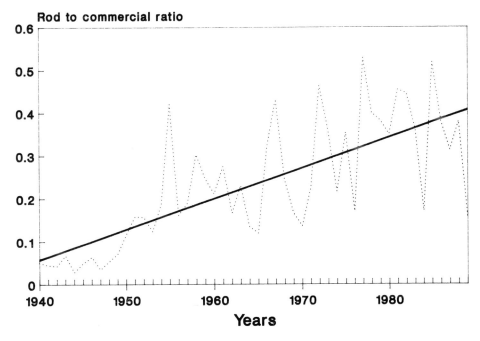

Fig. 1.3 Ratio of rod catch to commercial catch for the period 1940−89 inclusive. The dotted line follows actual values and the solid line shows the overall trend

low value for 1989 is undoubtedly due to river conditions where unusually low flows were followed by exceptional floods.

1.3.2   *Catches according to age*

Sea age composition of rod and fixed engine catch 1970–88

There is a considerable difference in the way the rods and fixed engines exploit salmon of various sea ages (Table 1.2). The putcher ranks take far more grilse and 2+ sea-winter fish, whereas rods may be slightly biased toward 3 sea-winter specimens.

Monthly catches of different sea ages 1970–88

Some of the features of Table 1.2 can be explained by the spread of salmon migration times (Fig. 1.4). Three sea-winter fish move into the river first, reaching a peak in April, followed by 2 sea-winter salmon, predominating during May. These two age classes coincide with prime rod fishing time in a river which is traditionally regarded by anglers as being 'spring run'. More susceptible to commercial methods are 2+ sea-winter fish, which peak in June, and the grilse which are the latest running category and make their maximum contribution to catches during July. Numbers of 3+ sea-winter stock remain low but also peak in July.

1.3.3   *Sex ratio*

The monthly sex compositions of salmon taken by fixed engine during the period 1987–89 are given in Table 1.3. It can be seen how different timescales for the migration of male and female fish affect the sex ratios of captured salmon on a seasonal basis. The grilse run is dominated by males with a F/M ratio of 1:2.9. For 2 sea-winter salmon, females are the principal component of the catch with a high F/M ratio of 3.5:1. Females again dominate 2+ sea-winter catches with an even higher F/M ratio of 5.85:1. The interrelationship between the contrasting sex compositions for fish of different age classes and the actual numbers of fish taken

Table 1.2 Total catches and percentages of salmon of different sea ages taken during the period 1970–88 by rods and fixed engines (PS = previous spawner)

Method		1	2	2+	3	3+	4	PS
				Sea age				
Rods	No	3 130	7 898	3 887	4 525	80	273	310
	%	15.6	39.3	19.3	22.5	0.4	1.4	1.5
Fixed engine	No	8 260	8 010	10 965	3 357	616	429	322
	%	25.8	25.1	34.4	10.5	1.9	1.3	1.0

Total monthly catches (000s)

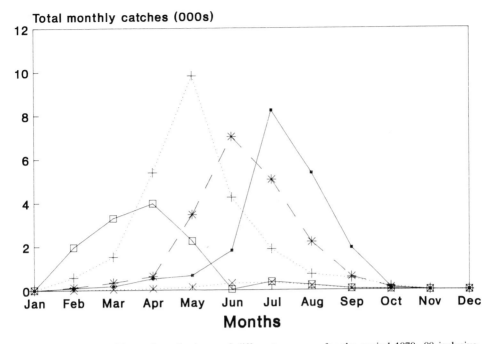

Fig. 1.4 Total monthly catches of salmon of different sea ages for the period 1970–88 inclusive. (■————■ 1 sea-winter; +······+ 2 sea-winter; *------* 2+ sea-winter; □————□ 3 sea-winter; x······x 3+ sea-winter)

Table 1.3 Total number of salmon of different sea age and sex taken in different months from a fixed engine (putcher rank) during the period 1987–9. Also given is the total % contribution made by females of all age groups

Descriptor	Apr	May	Jun	Jul	Aug
No 1 sea winter females	0	1	15	115	57
No 1 sea winter males	0	5	69	298	176
No 2 sea winter females	13	69	30	13	8
No 2 sea winter males	4	12	7	9	6
No 2+ sea winter females	0	47	81	95	11
No 2+ sea winter males	0	9	12	15	4
% contribution by females	77.7	74.3	58.7	55.2	29.3

results in a reversal in sex ratio as the season progresses. Females comprise almost 80% of the catch in April, falling to about 30% in August.

1.3.4   *Catch per unit effort*

Catch per rod licence and number of licences 1940–89

Figure 1.5 shows that 20 years of decline in catch per licence, albeit during a period when total numbers of fish increased (Fig. 1.2), was accompanied by a

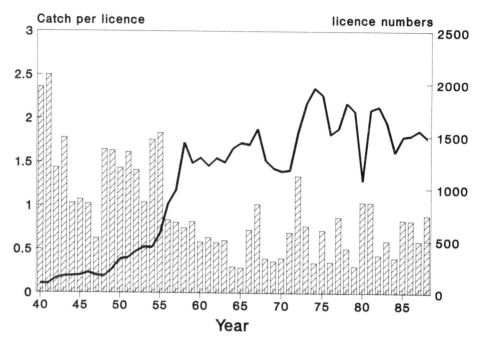

Fig. 1.5 Annual catches of salmon per rod licence (vertical bars) and number of rod licences (solid line) for the period 1940–89 inclusive

corresponding build up in licence sales. Since the early 1960s both catch and effort trends have levelled off.

Catch per draft net licence and number of licences 1940–89

Figure 1.6 shows an apparent trough in catch per licence for the late 1950s but this is not so great as was seen for catches (Fig. 1.2). There is a noticeable and possibly serious reduction in catch per licence from 1977. However, it is unknown how closely the number of licences reflects actual effort.

Catch per lave net licence and number of licences 1940–89

There is definite evidence (Fig. 1.7) of a decline in catches but, as for draft nets, the laves cannot take fish if the fishermen do not go out. Nonetheless, the number of licences has remained relatively constant so the downward trend in catch per licence may indicate a genuine reduction in the number of fish available.

Catch per 50 putchers and number of putchers licensed 1940–89

With the exception of a slight trough in the late 1970s, the salmon catch has been remarkably constant (Fig. 1.8). Unlike other methods, if a putcher rank is licensed then, in general terms, it can be relied upon to be actually fishing. Fixed engine

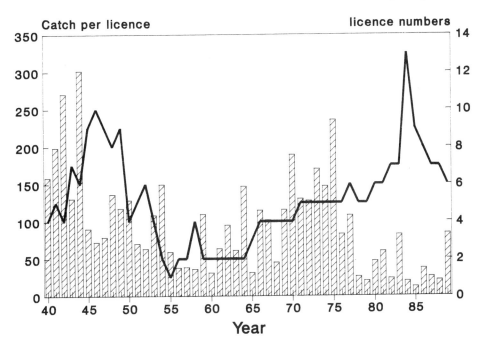

Fig. 1.6 Annual catches of salmon per draft net licence (vertical bars) and number of draft net licences (solid line) for the period 1940–89 inclusive

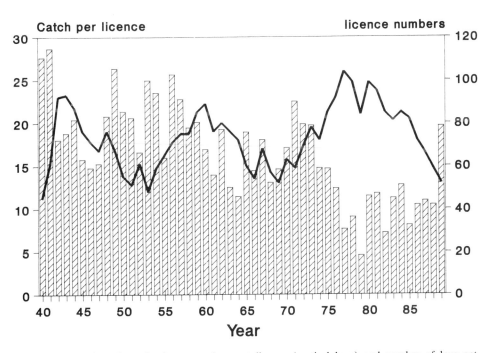

Fig. 1.7 Annual catches of salmon per lave net licence (vertical bars) and number of lave net licences (solid line) for the period 1940–89 inclusive

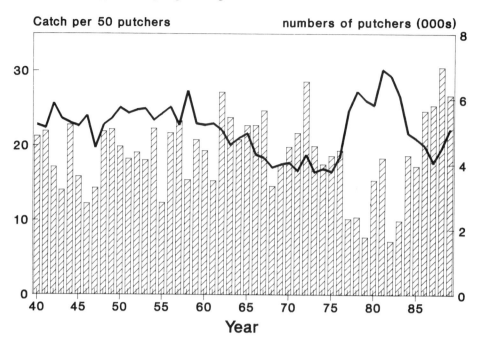

Fig. 1.8 Annual catches of salmon per licensed group of 50 putchers (vertical bars) and total number of putchers (solid line) for the period 1940–89 inclusive

results, therefore, come closest to providing a genuine CPUE. The situation in the last few years, however, may not be representative as significant changes to one particular putcher rank may have increased its efficiency sufficiently to effect overall figures.

1.3.4   *Year class strength*

Figure 1.9 shows the catch of each year class, as derived from scale samples, grouped according to sea age for the period 1950–81. Year class strength is a good indicator of change, being more representative than the simple consideration of sea age of the catch. The contribution of grilse (1 sea-winter fish) has shown an overall increase during the period whilst the percentage of 3 sea-winter fish has declined. Indeed, old annual reports often noted salmon >30 lb (13.6 kg) where such fish are now almost unheard of. Two points about the year class graph should be noted, however. Firstly, almost every River Severn salmon has a freshwater age of 2+ years. Secondly, of those fish hatched in the early months of 1985, only the grilse component has so far returned with the older sea ages yet to follow.

1.3.5   *Catch as an indication of stock*

Catches of salmon taken above Shrewsbury, counts of fish through the fish pass and the number of identifiable redds are compared in Fig. 1.10. It is evident that

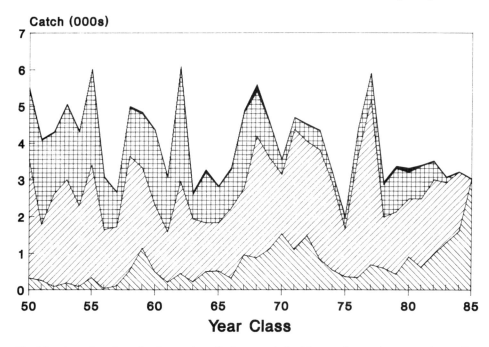

Fig. 1.9 Annual catches of each year class of salmon, as derived from scale samples, grouped according to sea age for year classes 1950–81 inclusive. (From the bottom upwards, the graph shows 1 sea-winter; 2/2+ sea-winter; 3/3+ sea-winter; 4 sea-winter.)

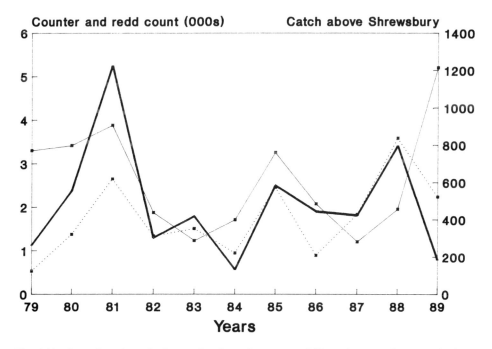

Fig. 1.10 Annual catches of salmon taken by rod upstream of Shrewsbury, total counts of salmon recorded for the fish pass at Shrewsbury and total number of redds counted by observers upstream of Shrewsbury for the period 1979–89 inclusive. (———— rod catch; ———— count through fish pass; ······ redd count)

the patterns of the three plots are similar. This supports the view that, in general terms, catch declarations do reflect trends of change in stock. Nonetheless, exceptional years will always cause problems and render statistics less reliable. For example, in 1989, 95% of salmon using the fish pass did so after the close of the angling season.

1.4   Discussion

1.4.1   *Usefulness of the catch statistics*

When considering information collected from fishermen it is important not to be misled by data sets which suggest a level of accuracy that does not really exist. All possible errors must be identified and acknowledged when interpreting the catch statistics. In the case of the River Severn salmon, likely errors comprise:

(1) The returns may not be honest declarations because of unwarranted concerns about income tax liability.
(2) The scale sub-samples may not be representative as there is no control over who is motivated to send them in.
(3) It is difficult for fisheries staff to count redds unless water conditions are ideal.
(4) Automatic salmon counters are prone to failure.
(5) The only measure of effort currently available is the number of licences which may not relate to time actually spent fishing.

Provided that these limitations are recognized, the results presented show that statutory returns can provide a very cost-effective and usable supply of data which is capable of indicating any major trend. In practice, however, knowledge of such trends may be of limited value with respect to any attempt to formulate management strategies.

Another drawback of the present style of catch statistics is that management has to be reactive rather than proactive. This situation would not necessarily be of major consequence if the legal framework did not put a long timescale on the speed of any reaction.

If sophisticated management is required then accurate, reliable data becomes vital and much greater expenditure by way of manpower and monetary resources would be necessary. Nevertheless, there are times when the possession of even such low-key information as can be derived from catch returns may be important. For example, in the Severn-Trent region, evidence from the salmon catch statistics has enabled the continued renewal of certain bye-laws which restrict the use of nets in the lower reaches of the River Severn.

1.4.2   *The future*

The long-term aim should be to work steadily towards developing a proven and reliable model for stock/recruitment relationships.

A significant increase in the value of the salmon catch data could be brought about by the collection of effort data as indicated by Bunt (chapter 3) for rivers in Wales. During the 1989 fishing season rod licence holders were asked, for the first time, to report the number of days they had spent fishing. It is intended to put a similar request to the commercial fishermen, namely the number of tides fished by nets and the dates of installation for fixed engines. In addition, it really is necessary to try to devise some methodology for validation of the catch returns so that, at the very least, the order of magnitude for under-reporting can be determined.

If funding and resources could be made available then further work should include the installation of high quality fish counters, the collection of long-term information on juvenile stocks and the adoption of a programme to microtag migrating smolts.

1.4.3 *Management proposals*

There is a perception held by the rod anglers that the commercial fishery is operating to their detriment. If there is a problem, however, it is much more likely to be overexploitation by the rods of the 3 sea-winter, large spring fish. There appears to be a need to encourage the capture of grilse whilst taking steps to reduce the fishing pressure on larger salmon.

In a multi-sea-winter fishery, which uses four different methods to exploit all age groups, management strategies designed to protect any one group are bound to be limited in option. Nonetheless, two possibilities exist. Firstly, the fixed engine season could be altered to both start and finish later so as to coincide more fully with the timing of the grilse run. Secondly, anglers could be encouraged to return any fish taken above a certain size. This would be a less severe measure than the curtailment of the season and there is anecdotal evidence (Anon, *pers. comm.*) that salmon, if played carefully, will survive when returned. A limit on the number of rod licences would be almost impossible to administer and such an approach would not direct the benefit towards specific age groups.

Whatever control measures are attempted, true management of the River Severn fishery, or indeed any other salmon resource, will only become feasible when monitoring can be upgraded to give instantaneous knowledge of stock and recruitment. It should be noted, however, that any such achievement would have to be supported by a more flexible legislation in order to provide the ability to adjust close seasons as required.

Acknowledgements

The authors wish to thank the many fishermen who voluntarily sent in salmon scales and/or data for analysis and also colleagues within the fisheries department who are or have been involved with scale reading and the collation of statutory returns. Please note that any views expressed are those of the authors and not necessarily those of the National Rivers Authority.

References

Bunt, D.A. (1990) Analysis of migratory rod catch effort data in Wales. In *Catch Effort Sampling Strategies: Their application in freshwater fisheries management*, (Ed. by I.G. Cowx). Oxford: Fishing News Books, Blackwell Scientific Publications.

Ward, R.C. (1981) River Systems and River Regimes. In *British Rivers* (Ed. by J. Lewin). London: George Allen & Unwin.

Chapter 2
Use of rod catch effort data to monitor migratory salmonids in Wales

D.A. BUNT *National Rivers Authority (Welsh Region), St Mellons, Cardiff, CF3 0LT, Wales*

Rod catch and effort data for salmon and sea trout were analysed from over 11 600 returns for 57 Welsh rivers for the 1988 season. Total reported effort was 159 113 visits with a median of seven visits per angler. Larger rivers (measured as average daily flow, ADF) received greater fishing effort and yielded larger catches than smaller rivers. CPUE (fish per visit) for the whole region was 0.0582 for salmon and 0.1894 for sea trout, with considerable variation between rivers. As it was not possible to separate fishing effort between salmon and sea trout these data underestimate the true CPUE.

An index of fish abundance (n) is proposed which reduces variability in catch (C) due to flow index (q) and fishing effort (f); $n = C/qf$. Calculation of n allows comparison between rivers in any one year and inter-annual trends in individual rivers, though more refined use of flow data is recommended to determine q more accurately. Changes to the method of reporting effort are proposed to allow separation of effort between species for future years, yet maintain comparability with earlier seasons. The issue of logbooks to season anglers is recommended to increase the quality of catch data.

2.1   Introduction

In the Welsh Region of the National Rivers Authority (NRA), over 16 000 licences are sold annually entitling anglers to fish for migratory salmonids, i.e. salmon (*Salmo salar* L.) and sea trout (*Salmo trutta* L.) in over 57 rivers. Catch records have been collected for most of these rivers since 1952, though it is only since 1976 that a standardized approach has been adopted.

It has been recognized, however, that catch data alone are not a reliable measure of the rivers' stocks, and may not even be suitable to indicate trends in catch over the long term (Harris, 1988). Catch per unit effort (CPUE) is widely accepted as a more accurate index of the stock size (Ricker, 1975; Prouzet & Dumas, 1988). In addition, effort data alone can provide useful information to the Fisheries Manager on the patterns of use of the fisheries resources. In 1988, in its publication 'Information on the Status of Salmon Stocks', the Salmon Advisory Committee stated that information on fishing effort was essential as one of the minimum requirements to monitor salmon stocks, representing a significant advance in the interpretation of catch information.

In Wales, information on fishing effort was not collected on a routine and regional basis until 1984. These data have been archived together with catch data and whilst the latter have been published annually, information on fishing effort has not been analysed. Catch and effort data received from anglers statutory returns for the 1988 season were therefore analysed with the following objectives:

(1) To present baseline data on effort and CPUE for 1988.
(2) To recommend how archived effort data should be analysed and presented.
(3) To recommend how future effort data should be collected, analysed and presented.

2.2   Method and materials

It is mandatory for every holder of a salmon and sea trout licence to make a catch return. The information is recorded on a form provided with the licence. Returns were received voluntarily at the end of the fishing season in late October, and following a postal reminder in mid-February. The following information was abstracted from each return:

(1) Number visits (days or part days fished) to each specified river.
(2) Number salmon caught where effort data were specified.
(3) Number salmon caught where no effort data were specified.
(4) Number sea trout caught where effort were specified.
(5) Number sea trout caught where no effort data were specified.

A record was also kept of:

(6) The number of returns received for each river where no effort data were specified, whether fish were caught or not.
(7) The number of returns where no river was specified.
(8) The number of returns that indicated no fishing had taken place.

After collation, data were analysed using the 'Minitab' statistical computer software package. The Hydrology Department of the NRA supplied river flow data.

2.3   Results

2.3.1   *Returns*

Information concerning the number of catch returns made voluntarily and following the postal reminder are given in Table 2.1.

Table 2.1 Summary of returns received

	No. Sold	No. Received	% Return	No. with unknown rivers	No. with no fishing
Voluntary		6095	32.9	318	135
Reminder		5519	29.7	561	285
Total	18594	11614	62.5	876	420

2.3.2 *Effort*

Total reported visits to rivers ranged from a minimum of eight on the Soch to 20110 on the Wye, with the total number of visitors to each river ranging from two on several rivers to 1825 on the Wye (Table 2.2). Fishing effort was highly positively skewed; mean visits per angler = 13.87, median = 7. The average river received a mean of 2785 reported visits, median = 1507.

2.3.3 *Catch*

Total reported catch in 1988 was the highest for salmon since the late 1960s, and the second highest on record for sea trout. Catch was found to be proportional to effort when data for all rivers were plotted (Fig. 2.1) (rivers for which there were less than 20 sets of usable data were excluded). This was a function of the size of the river as those with higher average daily flow (ADF) received greater fishing effort (Fig. 2.2) and yielded higher catches for both salmon and sea trout (Fig. 2.3). For sea trout, the Wye, Usk and Dee are excluded as it can be reasonably assumed that the vast majority of effort on these rivers is for salmon only.

2.3.4 *Catch per unit effort*

The CPUE for salmon and sea trout for each river (Table 2.3) has been calculated as:

$$\frac{\text{No. fish caught (where effort is specified)}}{\text{No. visits}}$$

CPUE for salmon ranged from zero on some of the smaller rivers to 0.1361 on the Lledr (a tributary of the Conwy) and for sea trout from zero on the Kenfig and Rhydhir to 0.6154 on the Gwendraeth Fawr (the latter represents returns from only 11 anglers with one angler being particularly successful with 46 fish; CPUE = 1.92). CPUE for the whole region was 0.0582 for salmon (= 17.2 visits per fish) and for sea trout, omitting the Wye, Usk and Dee, 0.1894 (= 5.3 visits per fish).

Table 2.2 Summary of effort data for all rivers

River	Abbreviation	Visitors with effort	No. Visits	Median visits per angler	Mean visits per angler	Visitors no effort	Total visitors	% No effort
ABER	AB	15	65	3.0	4.33	2	17	11.8
AERON	AE	79	1 507	10.0	19.08	20	99	20.2
AFAN	AF	60	1 372	15.0	22.78	12	72	16.7
ALWEN	AL	19	91	2.0	4.78	6	25	24.0
ARTH	AR	15	225	10.0	15.00	5	20	25.0
ARTRO	AT	60	785	6.0	13.08	5	65	7.7
E. CLEDDAU	EC	148	1 908	10.0	12.89	19	167	11.4
W. CLEDDAU	WC	123	2 794	14.0	22.71	18	141	12.8
CLWYD	CL	279	4 521	10.0	16.20	37	316	11.7
CONWY	CN	322	3 747	4.0	11.64	54	376	14.4
COTHI	CO	480	4 234	5.0	8.82	75	555	13.5
DEE	DE	853	11 211	6.0	13.14	102	955	10.7
DULAS	DL	2	16	8.0	8.00	0	2	0.0
DULAS N.	DN	37	277	4.0	7.49	3	40	7.5
DULAS S.	DS	29	267	5.0	9.21	6	35	17.1
DYFI	DF	570	4 470	4.0	7.84	74	644	11.5
DYSYNNI	DY	157	2 071	6.0	13.19	19	176	10.8
DWYFACH	DC	24	180	6.5	7.50	4	28	14.3
DWYFAWR	DR	283	3 430	6.0	12.12	32	315	10.2
DWYRYD	DD	62	1 291	10.0	20.82	12	74	16.2
EDEN	ED	10	146	6.5	14.60	3	13	23.1
ELWY	EL	257	3 102	8.0	12.07	24	281	8.5
ERCH	ER	31	388	10.0	12.52	4	35	11.4
EWENNY	EW	37	810	11.0	21.89	1	38	2.6
GLASLYN	GL	184	2 192	5.0	11.91	39	223	17.5
GWAUN	GW	8	148	13.0	18.50	0	8	0.0
GW. FACH	GC	73	1 233	10.0	16.89	9	82	11.0
GW. FAWR	GR	11	130	8.0	11.82	3	14	21.4

GWILI	GI	135	1656	8.0	12.27	38	173	22.0
GWYRFAI	GF	26	283	7.5	10.88	10	36	27.8
KENFIG	KE	2	9	4.5	4.50	0	2	0.0
LERI	LE	3	22	4.0	7.33	1	4	25.0
LLEDR	LD	103	955	5.0	9.27	13	116	11.2
LLUGWY	LG	18	176	4.0	9.78	5	23	21.7
LOUGHOR	LO	133	2965	12.0	22.29	22	155	14.2
LLYFNI	LF	86	1147	7.5	13.33	12	98	12.2
MAWDDACH	MA	515	5543	5.0	10.74	75	590	12.7
NEATH	NE	94	2392	16.0	25.44	16	110	14.5
NEVERN	NV	108	1713	8.0	15.86	16	124	7.9
OGWEN	OW	78	1889	17.0	24.22	15	93	16.1
OGMORE	OM	137	3257	12.0	23.77	37	174	21.3
RHEIDOL	RH	178	2821	6.0	15.85	26	204	12.7
RHYDHIR	RD	6	36	5.0	6.00	0	6	0.0
RHYMNEY	RM	2	86	43.0	43.00	0	2	0.0
SEIONT	SE	199	3245	8.0	16.30	26	225	20.8
SOLVA	SL	2	27	13.5	13.50	0	2	0.0
SOCH	SC	2	8	4.0	4.00	0	2	0.0
TAF	TA	161	2804	9.0	17.41	44	205	21.5
TAFF	TF	69	1542	10.0	22.35	7	76	9.2
TAWE	TW	182	2616	10.0	14.37	26	208	12.5
TEIFI	TE	1298	18357	7.0	14.14	183	1481	12.4
TYWI	TY	1184	20102	10.0	16.98	220	1404	15.7
USK	US	587	9208	7.0	15.69	68	655	10.4
WNION	WN	166	1672	5.5	10.07	17	183	9.3
WYE	WY	1601	20110	6.0	12.56	224	1825	12.3
WYRE	WR	17	214	6.0	12.59	2	19	10.5
YSTWYTH	YS	96	1025	5.0	10.68	23	119	19.3
OTHERS		23	341	10.0	14.82	4	27	14.8
UNKNOWN		28	281	4.5	10.00	796	824	96.6
TOTAL/OVERALL		11468	159113		13.87	2513	13981	18.0
MEAN PER RIVER		200	2785		14.14	43.3	230	14.4
MEDIAN PER RIVER		86	1507	7.0	13.08	15.0	99	12.5

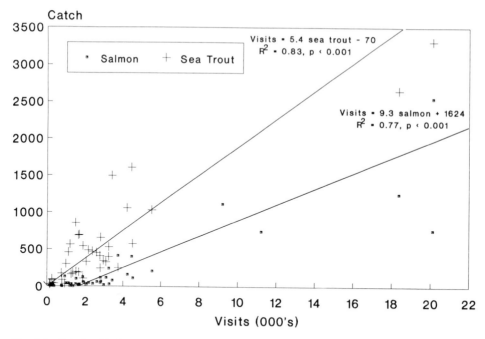

Fig. 2.1   Relationship between catch and effort for each river

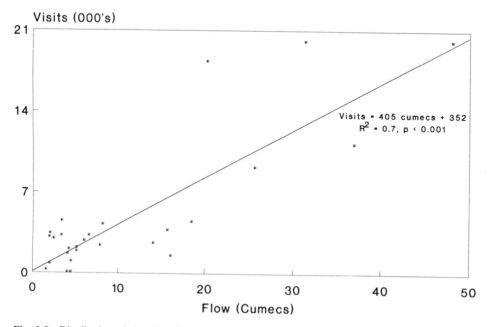

Fig. 2.2   Distribution of river flow (cumecs) and effort for various rivers in 1988

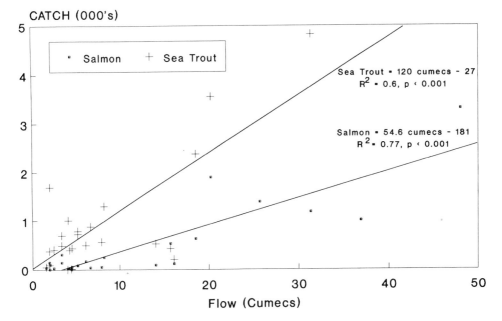

Fig. 2.3 Distribution of river flow (cumecs) and catch for various rivers in 1988

The frequency distribution of CPUE for all rivers is positively skewed, particularly for salmon, which indicates that sea trout are more widely abundant than salmon, and/or easier to catch.

The plot of salmon CPUE against sea trout CPUE (Fig. 2.4) is a convenient graphical representation to compare each river for each species. The mean CPUE for salmon and sea trout are shown and the distance of each river point above or below the 45° line represents its 'performance' as a salmon or sea trout fishery. No relationship is implied, but it should be pointed out that very few rivers are 'above average' for both salmon and sea trout.

Catch per return (per visitor) (C/R) has been used in the past as an index of CPUE when true effort data has not been available. There was a significant correlation between C/R and CPUE for both salmon ($p < 0.001$, $R^2 = 0.78$) and sea trout ($p < 0.001$, $R^2 = 0.76$) (Figs. 2.5 and 2.6).

2.3.5 *Catchability*

River flow is the major stimulus to migration and subsequent capture of migratory salmonids, particularly salmon (Millichamp & Lambert, 1966; Banks, 1969; Alabaster, 1970; Gee, 1980; Clarke & Purvis, 1989). Mean monthly river flow (cumecs) has been calculated for most rivers during the fishing season (February–October for the Wye, Usk and Dee; April–October for all other rivers). Despite being a simple measure, mean flow was found to be significantly correlated with total catch for rivers in 1988 (Fig. 2.3). For each river a flow index

Table 2.3 Summary of catch effort data for all rivers

River	Salmon with effort	Sea trout with effort	CPUE salmon	CPUE sea trout	Total salmon per visitor	Total sea trout per visitor	Mean flow Cumecs	Flow index (%)	Salmon abund. index (Ns)	Sea trout abund. index (Nt)
ABER	0	2	0.0000	0.0307	0.06	0.23	—	—	—	—
AERON	28	857	0.0186	0.5687	0.53	10.42	—	—	—	—
AFAN	1	176	0.0067	0.1283	0.01	2.97	—	—	—	—
ALWEN	12	0	0.1319	0.0000	0.72	0.08	3.98	109	12.10	0.00
ARTH	2	70	0.0089	0.3111	0.20	3.95	—	—	—	—
ARTRO	6	173	0.0076	0.2204	0.09	3.29	—	—	—	—
E. CLEDDAU	55	542	0.0288	0.2841	0.49	4.37	5.10	142	2.03	2.00
W. CLEDDAU	46	410	0.0165	0.1467	0.43	3.50	—	—	—	—
CLWYD	111	584	0.0246	0.1292	0.44	2.20	3.29	109	2.26	11.85
CONWY	412	251	0.1099	0.0670	1.41	1.15	15.70	127	8.65	5.28
COTHI	159	1059	0.0376	0.2501	0.46	2.34	8.15	123	3.06	20.33
DEE	743	100	0.0663	0.0089	1.05	0.15	36.95	109	6.08	0.82
DULAS	0	9	0.0000	0.5625	0.00	6.00	—	—	—	—
DULAS N.	6	84	0.0217	0.3032	0.18	2.10	—	—	—	—
DULAS S.	3	94	0.0112	0.3521	0.11	4.01	—	—	—	—
DYFI	401	1608	0.0897	0.3597	0.97	3.69	18.50	129	6.81	27.88
DYSYNNI	25	332	0.0121	0.1603	0.19	2.26	4.18	115	1.05	13.94
DWYFACH	6	30	0.0333	0.1667	0.21	1.79	—	—	—	—
DWYFAWR	79	1496	0.0230	0.4362	0.30	5.37	1.99	102	2.25	42.76
DWYRYD	27	163	0.0209	0.1263	0.80	2.82	—	—	—	—
EDEN	5	19	0.0343	0.1301	0.92	1.85	—	—	—	—
ELWY	114	325	0.0368	0.1045	0.51	1.35	1.90	109	3.38	9.59
ERCH	1	39	0.0026	0.1005	0.03	1.11	—	—	—	—
EWENNY	0	86	0.0000	0.1062	0.00	2.45	1.96	117	0.00	9.08
GLASLYN	45	484	0.0205	0.2208	0.38	3.52	5.10	105	1.95	21.03
GWAUN	1	40	0.0068	0.2703	0.13	5.10	0.82	148	0.46	18.26
GW. FACH	3	563	0.0024	0.4566	0.05	8.40	—	—	—	—
GW. FAWR	4	80	0.0308	0.6154	0.14	6.21	—	—	—	—
GWILI	16	686	0.0097	0.4143	0.13	5.85	4.02	155	0.63	26.73
GWYRFAI	19	44	0.0671	0.1555	1.00	1.47	1.55	88	7.65	17.63

KENFIG	0	0	0.0000	0.0000	3.50	0.00	–	–	–	–
LERI	0	4	0.0000	0.1818	0.00	2.00	1.50	131	0.00	13.88
LLEDR	130	34	0.1361	0.0366	1.53	0.29	–	–	–	–
LLUGWY	9	6	0.0511	0.0341	1.09	0.61	–	–	–	–
LOUGHOR	17	344	0.0057	0.1160	0.17	2.63	2.41	190	0.30	6.10
LLYFNI	39	458	0.0340	0.3993	0.55	8.19	–	–	–	–
MAWDDACH	206	1033	0.0372	0.1864	0.54	2.16	–	–	–	18.82
NEATH	33	468	0.0138	0.1957	0.47	5.14	7.80	104	1.33	–
NEVERN	26	695	0.0152	0.4057	0.31	7.89	–	–	–	–
OGWEN	123	105	0.0651	0.0556	2.12	1.38	–	–	–	–
MORE	27	529	0.0083	0.1624	0.22	5.06	6.56	150	0.55	10.83
RHEIDOL	63	655	0.0223	0.2322	0.62	4.01	–	–	–	–
RHYDHIR	2	0	0.0556	0.0000	0.50	0.17	–	–	–	–
RHYMNEY	0	14	0.0000	0.1628	0.00	7.00	4.40	143	0.00	11.38
SEIONT	240	398	0.0740	0.1227	1.36	2.17	3.30	95	7.78	12.90
SOLVA	0	6	0.0000	0.2222	0.00	7.00	–	–	–	–
SOCH	0	0	0.0000	0.0000	0.00	0.00	–	–	–	–
TAF	114	243	0.0407	0.0867	0.80	2.42	6.01	160	2.54	5.42
TAFF	101	194	0.0655	0.1258	1.50	2.74	16.10	159	4.12	7.91
TAWE	58	452	0.0222	0.1728	0.44	2.52	14.07	179	1.24	9.65
TEIFI	1251	2650	0.0684	0.1443	1.28	2.40	20.20	141	4.85	10.23
TYWI	765	3310	0.0381	0.1647	0.84	3.44	31.30	159	2.40	10.36
USK	1114	41	0.1210	0.0045	2.11	0.09	25.65	130	9.31	0.35
WNION	26	188	0.0156	0.1124	0.19	1.36	–	–	–	–
WYE	2547	15	0.1267	0.0007	2.15	0.03	48.00	149	8.50	0.05
WYRE	1	2	0.0047	0.0094	0.26	5.32	–	–	–	–
YSTWYTH	40	302	0.0390	0.2946	0.47	3.67	4.45	113	3.45	26.07
OTHERS	0	62	0.0000	0.1818	0.00	3.19	–	–	–	–
UNKNOWN	0	12	0.0000	0.0427	0.02	0.19	–	–	–	–
TOTAL	9262	22626	0.0582	0.1422	1.20	2.74	–	–	4.44	10.85
MEAN/RIVER	–	–	0.0340	0.1896	0.61	3.32	10.52	131	3.61	12.80
MEDIAN/RIVER	–	–	0.0220	0.1603	0.44	2.69	5.10	129	2.40	10.83
MINUS DEE, USK, WYE										
TOTAL/OVERALL	22470		0.1894		3.61		–	–	–	14.46
MEAN/RIVER	–		0.2001		3.15		7.47	131	–	14.23
MEDIAN/RIVER	–		0.1626		2.52		4.43	128	–	11.61

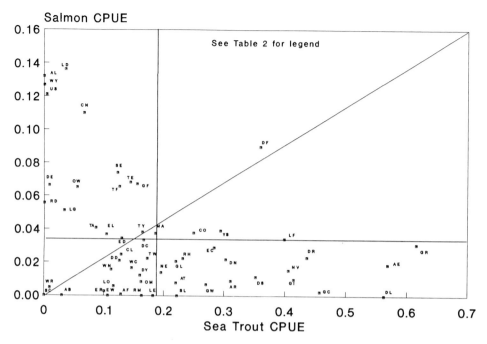

Fig. 2.4   Distribution of salmon and sea trout CPUE for each river examined

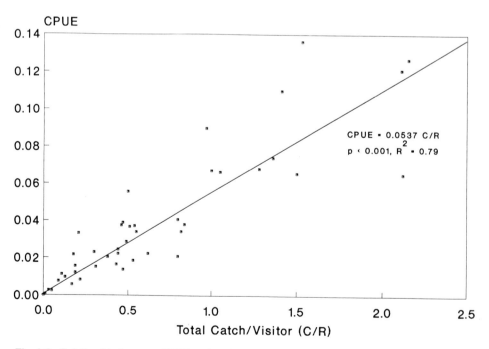

Fig. 2.5   Relationship between CPUE and catch per return (salmon)

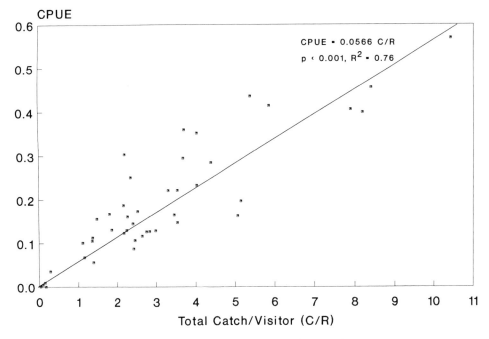

Fig. 2.6 Relationship between CPUE and catch per return (sea trout)

(q) was calculated as the mean monthly flow in 1988 as a percentage of the long term average during the fishing season. For most rivers q was greater than 100% (mean = 131%) indicating that 1988 was a generally 'wet' year in the Welsh Region. Hydrographs indicated that this was due to regular freshets.

To test the suitability of using this flow index for an individual river over a period of time, catch was plotted against q for the River Teifi for the period 1976–88 (Fig. 2.7). Salmon catch was significantly correlated with flow index ($p = 0.005$, $R^2 = 0.53$). The correlation for sea trout was less significant ($p = 0.066$, $R^2 = 0.27$).

2.4   Discussion

2.4.1   *Effort*

Fishing effort alone gives very useful information to the fisheries manager on the intensity of fishing and the number of anglers visiting the river. The mean and median number of visits made by anglers to a river possibly indicate the type of angler utilizing the resource. A low number of median visits (e.g. Conwy, Dyfi) suggests a high proportion of visiting anglers or short-term licence holders whereas a high figure (e.g. Neath, Ogmore) suggests the majority of fishing is by local anglers.

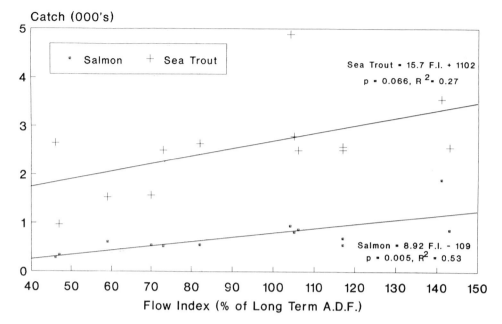

Fig. 2.7 Relationship between flow index and catch in the River Teifi 1976–88

2.4.2   *Catch*

The relationship between catch and effort for all rivers (Fig. 2.1) shows the pattern of use for different rivers in one year. A similar plot over a number of years for each river, when both parameters are likely to vary, will indicate if the rate of exploitation is above or below the point of maximum sustainable yield as the relationship changes from being linear to a plateau (Ricker, 1975). This relationship is complicated by factors which may affect stock availability and catchability. However, Mills *et al.* (1986) found that the highest single determinant of salmon and sea trout catch on Lough Feeagh was fishing effort.

2.4.3   *Catch per unit effort*

Data on CPUE are often obtained from intensive creel surveys of specific, often closely managed, fisheries, whilst more extensive studies covering a range of river types are rarer. CPUEs for the Twyi, Tawe, Wye and Conwy are compared with data obtained in other more intensive surveys (Table 2.4). In these other surveys fishing effort has been measured in hours, so for direct comparison the mean number of hours per fishing visit found on the Tywi and Conwy (4.75 ± 1.22; a figure similar to that of five hours for all game anglers (NOP, 1971)) has been applied.

CPUE for both species on the Conwy are within the range recorded in recent years from intensive studies. However, the 1988 figures are higher in the intensive investigation by Davidson (1989) which may be explained by:

Table 2.4 Comparison of CPUE between rivers in the present study and those studied intensively in recent years.

River	Source	Year	Salmon CPUE	Sea trout CPUE
Tywi	(Clarke, Unpub.)	1985	0.0533	0.3676
Tywi	(Clarke, Unpub.)	1986	0.0565	0.5814
Tywi	(Present study)	1988	0.0381	0.1647
Tawe	(Wightman, 1987)	1986	0.0365	0.0526
Tawe	(Present study)	1988	0.0222	0.1728
Wye	(Gee, 1980)	1977	0.2138	N/A
Wye	(Present study)	1988	0.1267	(0.0007)
Conwy	(Davidson, 1989)	1982	0.3100	0.1470
Conwy	(Davidson, 1989)	1983	0.0930	0.0410
Conwy	(Davidson, 1989)	1984	0.0530	0.0420
Conwy	(Davidson, 1989)	1986	0.1600	0.0860
Conwy	(Davidson, 1989)	1987	0.0970	0.0670
Conwy	(Davidson, 1989)	1988	0.1760	0.0900
Conwy	(Present study)	1988	0.1099	0.0670

(1) The intensive survey was targeted at selected regular season anglers who are possibly more successful than the average visitor (Alabaster, 1986).

(2) Effort was split between the two species.

CPUE for the intensive studies on the Tywi and Tawe were also higher than for this investigation, which is probably due again to targeting of angling club members who have better local fishing knowledge. The differential was much higher for sea trout than for salmon which may be explained by the intensive surveys utilizing logbooks to collect data. There is probably less tendency towards underreporting when details of fishing trips are recorded regularly in a specially designed booklet, especially when there is much information to record.

Results for the Wye were lower than those found by Gee (1980), which may be explained by:

(1) Many Wye anglers make a nil return, stating that they reported the catch to the fishery owner.

(2) The calculated mean length of each visit may be an underestimate for the Wye (anglers fishing the Wye tend to be visitors not locals).

The Dyfi is conspicuously above average in comparison to other rivers for both salmon and sea trout (Fig. 2.4). This may be due to true productivity for both species, and/or having very strict rules and limited membership in the local angling association.

2.4.4   *Catchability*

Gulland (1969) expressed catch (*C*) as:

$$C = f \, q \, N/A$$

where *q* is a constant expressing catchability, *f* the fishing effort, *N* the abundance of stock and *A* the area inhabited by stock.

Catchability of migratory salmonids is affected by many factors including water temperature, skill of anglers, method used (Gee, 1980; Alabaster, 1986), stock composition (Shearer, 1988), weather conditions (Mills *et al.*, 1986) and river flow (Millichamp & Lambert, 1966; Alabaster, 1970; Gee, 1980). Flow is the most important factor (Mills *et al.*, 1986) and is certainly the most easily measured.

Whilst many authors have associated migration and capture with optimum flow regimes it is only recently that increased catchability (of salmon) *per se*, as opposed to mere availability, has been demonstrated to be linked with changes in flow (Clarke & Purvis, 1989). Essentially, salmon are more catchable in the 10–14 days following entry into fresh water, and freshets stimulate entry when fish are available.

Millichamp and Lambert (1966) described optimum flow ranges for migration and capture of salmon on the Usk. In the absence of similar information for all Welsh rivers, a simple measure of flow, the % of the long-term average during the fishing season (flow index) is proposed as an index of catchability. Whilst this has been demonstrated to have had some success, river flow is not adequately described by a mean value (Alabaster, 1970). Flow variability is a more important factor (Banks, 1969; Clarke & Purvis, 1989).

As experienced anglers are receptive to suitable fishing conditions, it is possible that effort alone may be found to be correlated with catchability. In the 1989 drought year, Water Bailiffs reported that anglers were met far less frequently on the river than in 1988 when flows were higher.

2.4.5   *Abundance*

Fishery managers ultimately wish to know stock abundance accurately. In the absence of direct counting methods (counters, traps) for migratory salmonids on most rivers, abundance can only be estimated for those fish that are available during the fishing season.

Working from the assumptions that fishing effort and catchability do not vary between years, rod catch has been used as the adult abundance rating (Dempson, 1980). Such constancies cannot be assumed, as effort over the long term has increased considerably with gaining popularity in angling (Gee & Milner, 1980; Harris, 1988) whilst marked short-term variations are attributable to differences in weather and fishing conditions between each year. There has been, for instance, considerable variation in the mean visits per angler for each season on the Conwy in recent years (Davidson, 1989). However, in several cases a reasonable balance between rod catch and stock abundance has been observed (Elson, 1974; Chadwick, 1982).

Based on the single assumption that vulnerability is constant between years, CPUE has been widely used as an abundance rating of migratory salmonids (Cousens *et al.*, 1982; Prouzet & Dumas, 1988). Unless it can be demonstrated that effort is an accurate index of catchability, some consideration of catchability must be made.

If, in Gulland's equation, N/A is replaced by n, an index of abundance is achieved:

$$n = C/qf$$

which reduces variability in catch that may be attributable to effort and flow. Although flow variability is less important for sea trout migration than for salmon (Banks, 1969; Purvis & Clarke, 1990), and techniques for sea trout angling often differ to those for salmon, n has been applied to sea trout as well as salmon for comparative purposes only (Fig. 2.8).

Observing trends in the abundance index may be more meaningful than those in CPUE or catch. By examining this parameter over a number of years where stock abundance can be measured accurately by traps or counters it may be possible to establish a relationship between n and total stock.

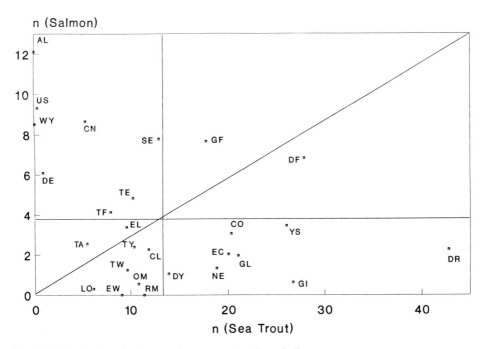

Fig. 2.8 Distribution of salmon and sea trout abundance indices

2.4.6   *Collection, analysis and presentation of data—past and future*

It is recommended that in the Welsh Region summary catch effort data should be presented as demonstrated in Tables 2.2 and 2.3. Ideally it should be possible to identify effort separately for salmon and sea trout fishing. This would enable calculation of actual CPUE for each species. Trends only may be identified when combined effort is utilized, and this assumes that the proportion of effort for each species in mixed stock fisheries does not vary. It is therefore recommended that the method of reporting effort should be expanded to request from anglers:

(1) Number of visits fished for salmon only.
(2) Number of visits fished for sea trout only.
(3) Number of visits fished for both species at the same time.

This method has been used with some success on the Conwy (Davidson, 1989).

There are potential problems with this method. For example, it is possible to catch either species utilizing methods usually applied for the other. However, (1) + (2) + (3) = total visits for salmon and sea trout which allows comparison with data where effort is combined. This has been implemented as from the 1990 season.

Logbooks should be issued to season anglers so that they are able to record this and other information more accurately and therefore fewer returns are based on recollection alone (Wightman, 1987; Welsh Water Authority, 1987). A trial logbook scheme to test if more accurate information is received has been implemented in the 1990 season.

By using visits as the unit of effort it is assumed that the mean hours per visit by an angler is constant. Whilst this may be true for the region between years, there are likely to be differences between rivers. Total hours fished would be more accurate and would improve comparability with intensive studies. The issue of logbooks would be imperative to accurately collect this information.

Flow data should be closely analysed on a regular basis where available for the monitored rivers. A more accurate index of catchability based on flow variability and the number of flow days within the optimum flow range for angling catch should be calculated. Calculation of the abundance index may then be refined in which inter-annual trends should be observed. Ultimately, it should be possible to measure n on a monthly basis, though this would require monthly effort data which is logistically difficult to collect using current reporting methods.

By applying these statistics to models formulated by Beverton and Holt (1957) and Ricker (1975) it may be possible to estimate the size of parental stock to maximize catches in the long term, which is the ultimate aim of the fishery manager (Prouzet & Dumas, 1988). However, such models have been developed for commercial fisheries where CPUE has economic constraints and catchability is determined primarily by efficiency of the gear. In sport fisheries CPUE is affected by anglers' psychology and catchability is determined largely by fishing conditions.

Acknowledgements

The views expressed in this report are those of the author only and do not necessarily reflect those of the National Rivers Authority. I am grateful to the many colleagues who assisted in this investigation. In particular I wish to thank Alan Winstone, Miran Aprahamian, David Clarke and Guy Mawle for their guidance and comments on the manuscript, and Garry Jones and Jeremy Dawe for their computing expertise. A similar report has been submitted in fulfilment of the requirements of the IFM Diploma course.

References

Alabaster, J.S. (1970) River flow and upstream movement and catch of migratory salmonids. *J. Fish Biol.* **2**: 1–13.

Alabaster, J.S. (1986) An analysis of angling returns for trout. *Aq. Fish. Mgmt.* **17**: 313–326.

Banks, J.W. (1969) A review of the literature on the upstream migration of adult salmonids. *J. Fish Biol.* **1**: 85–136.

Beverton, R.J.H. and Holt, S.J. (1957) On the dynamics of exploited fish populations, UK. *Fish. Invest.* (Ser. 2) **19**: 533p.

Chadwick, E.M.P. (1982) Recreational catch as an index of Atlantic salmon spawning escapement. *International Council for the Exploration of the Sea.* C 1982 M, **43**, 4p.

Clarke, D.R.K. and Purvis, W.K. (1989) Migration of Atlantic salmon in the River Tywi system, South Wales. Paper presented to the Atlantic Salmon Trust Symposium on Migration of Salmonids in Relation to Freshwater Flow and Water Quality. Bristol 1989.

Cousens, N.B.F., Thomas, G.A., Swann, G.G. and Healey, M.C. (1982) A review of salmon escapement estimation techniques. *Canadian Technical Report of Fisheries and Aquatic Sciences.* **1108**: 122p.

Davidson, I.C. (1989) Conwy fisheries monitoring programme. Annual Report 1988. W.W.A. Report No. EAN/89/2.

Dempson, J.B. (1980) Application of a stock recruitment model to assess the Labrador Atlantic salmon fishery. *International Council for the Exploration of the Sea.* C 1980 M, **28**, 15p.

Elson, P.F. (1974) Impact of recent economic growth and industrial development on the ecology of Northwest Miramichi Atlantic Salmon (*Salmo salar*). *J. Fish. Res. Bd. Can.* **31**: (5), 521–544.

Gee, A.S. (1980) Angling success for Atlantic salmon (*Salmo salar*) in the River Wye in relation to effort and flows. *Fish. Mgmt.* **11**: (3), 131–138.

Gee, A.S. and Milner, N.J. (1980) Analysis of 70-year catch statistics for Atlantic salmon (*Salmo salar*) in the River Wye and its implications for management of stocks. *J. Appl. Ecol.* **17**: 41–57.

Gulland, J.A. (1969) Manuel des methodes d'evaluation des stocks d'animaux aquations. Premiere partie – Analyse des populations. Manuel FAO de Science halieutique. **4**, 160p.

Harris, G.S. (1988) The status of exploitation of salmon in England and Wales. In *Atlantic Salmon: Planning for the Future* (Ed. by D. Mills & D. Piggins). London: Croom Helm. 169–190.

Millichamp, R.I. and Lambert, A.O. (1966) On investigations into the relationship between salmon catch and flow in the River Usk during the 1965 season. In *Symposium on River Management* (Ed. by P.C.G. Isaac). London: McLaren. 119–123.

Mills, C.P.R., Mahon, G.A.T. and Piggins, D.J. (1986) Influence of stock levels, fishing effort and environmental factors on anglers catches of Atlantic salmon, *Salmo salar* L., and sea trout, *Salmo trutta* L. *Aq. Fish. Mgmt.* **17**: 289–297.

N.O.P. (National Opinion Poll) (1971) National Angling Survey 1969–70. N.O.P. Market Research Ltd. 129p.

N.R.A. Welsh Region (1989) Salmon and Sea Trout Catch Statistics 1988. 109p.

Purvis, W.K., and Clarke, D.R.K. (1990) Migration of sea trout in the River Tywi, South Wales. N.R.A. Welsh Region. In preparation.

Prouzet, P. and Dumas, J. (1988) Measurement of Atlantic salmon spawning escapement. In *Atlantic Salmon: Planning for the Future.* (Ed. by D. Mills and D. Piggins). London: Croom Helm. 325–343.

Ricker, W.E. (1975) Computation and interpretation of biological statistics of fish populations. *Bull. Fish. Res. Bd. Can.* **191**: 382p.

Salmon Advisory Committee (1988) Information on the Status of Salmon Stocks. 30p.

Shearer, W.M. (1988) Relating catch records to stocks. In *Atlantic Salmon: Planning for the Future.* (Ed. by D. Mills and D. Piggins). London: Croom Helm. 256–274.

Talhelm, D.R. (1976) The demand and supply of fishing and boating on inland lakes in Michigan: summary report. Michigan Dept. of Natural Resources, Lansing, MI.

Wightman, R.P. (1987) The results of an angler logbook survey on the River Tawe, 1986 season. W.W.A. Report No. SW/87/14.

W.W.A. (Welsh Water Authority) (1987). Final Report of the Working Group on Rod Catch Statistics. Report No. HQF 87/5.

Chapter 3
Use of telemetric tracking to examine environmental influences on catch effort indices. A case study of Atlantic salmon (*Salmo salar* L.) in the River Tywi, South Wales

D. CLARKE, W.K. PURVIS and D. MEE *Regional Environmental Appraisal Unit, National Rivers Authority (Welsh Region), Penyfai Lane, Llanelli, Dyfed, Wales*

Radio tracking of over 200 Atlantic salmon running into the River Tywi (South Wales) in 1988 and 1989 has demonstrated that flow is an important factor in modifying both run timing and migratory success. Entry of salmon into the river is typically in response to flow events, and periods of low falling flows delay entry and may directly result in reduced runs into the river. Delayed entry may also increase the proportion of the run migrating after the end of both rod and net fishing seasons.

The implications of these results for net and rod catch and catch effort data are discussed, using both statutory reported catch data and data from specific catch effort studies. Flow is demonstrated to be a dominant factor in determining the within-season distribution of rod catch and catch effort during low-flow years. Estuarine seine net catch and catch effort tend to be controlled more by time of return than by flow, although low flows may delay runs. Annual reported rod catch is correlated with flow, which controls in season availability, catchability and consequently the amount of fishing effort. Use of catch or catch effort data should take account of inter-year variations in flow and other environmental factors. Although catch and catch effort are valuable indicators of fishery performance, they are inadequate to represent changing stock levels in the short term.

3.1   Introduction

Data on salmonid fisheries is often restricted to reported catch or catch effort data, primarily because of the cost and practical difficulty involved in direct sampling. While many authors have identified the limitations inherent in salmonid catch data (e.g. Harris, 1988; Shearer, 1988), in the absence of alternative information, catch effort data is often accepted as a reliable indicator of relative stock abundance, both within and between years (Beverton & Holt, 1957; Ricker, 1975; Small & Downham, 1985; Prouzet & Dumas, 1988). Since catch effort is a function of both abundance and catchability (Gulland, 1969; Bunt, 1990), use of catch per effort in this way assumes constant catchability.

The influence of flow on upstream migration and catches of salmonids has long been recognized (Huntsman, 1948; Hayes, 1953; Millichamp & Lambert, 1966, Banks, 1969; Alabaster, 1970; Gee, 1980). These authors have examined catch and catch effort statistics in relation to flow statistics, demonstrating significant effects of flow in most situations. However, data has not been available which allows the mechanism of these relationships to be examined; in particular it has not been possible to distinguish between variations in abundance and catchability.

During the summers of 1988 and 1989, the National Rivers Authority (NRA) undertook a detailed behavioural study of salmonid migration within the estuary and freshwater reaches of the River Tywi in South Wales, involving telemetric tagging and tracking of more than 200 Atlantic salmon (*Salmo salar* L.). As part of this work, detailed catch and effort data have been recorded within both net and rod fisheries, and longer term catch data are also available from 1960. This paper examines the relationships between salmon behaviour and catch, particularly the availability of fish for capture within net and rod fisheries.

3.2   The Tywi system

3.2.1   *Fresh water*

The River Tywi (Fig. 3.1) rises at a height of 425 m in an afforested and moorland region of mid-Wales. It is 111 km in total length, and has a catchment area of 1335 km². Average daily flow for the Tywi at its mouth is 45 m³s⁻¹, although in 1989 minimum flows were as low as 2 m³s⁻¹. A regulating reservoir, Llyn Brianne, exists in the upper catchment and abstractions occur at Manorafon and Nantgaredig, the latter supplying Carmarthen and much of the Swansea area.

Water quality is generally good (NWC class 1A), the only exception being the headwaters of the upper catchment which suffer acidification problems (Stoner, Gee & Wade, 1984).

3.2.2   *The estuary*

The estuary (Fig. 3.2) is shallow, well mixed and tidally influenced for an upstream distance of up to 20 km from the sea. The lower estuary is 2.5 km wide at the mouth, and is shared with the smaller Taf and Gwendraeth catchments. Further upstream the estuary narrows rapidly, and the upper estuary, upstream of Green Castle, is generally less than 50 m wide at high water. Tidal range is large (6.6 m on spring tides) and the tidal regime is asymmetric, with the flood tide period shorter than the ebb.

3.2.3   *Fisheries*

The Tywi is a nationally important salmon and sea trout rod fishery, with an average declared salmon rod catch in excess of 900 per annum since 1983 and a declared annual sea trout (*Salmo trutta* L.) rod catch averaging more than 6000.

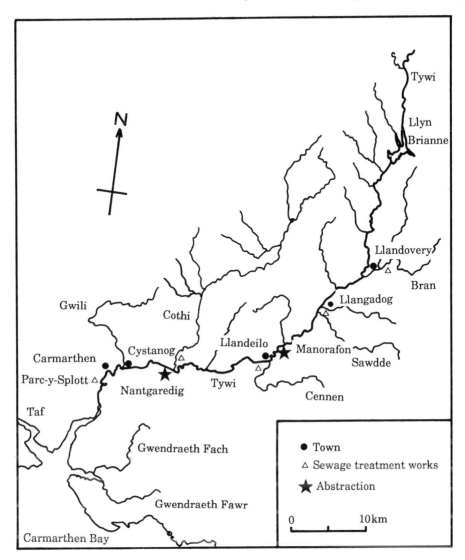

Fig. 3.1 Map of the Tywi catchment

The river also supports commercial salmon and sea trout net fisheries, with eight seine nets licensed in the lower estuary and nine coracle nets in the upper estuary (Fig. 3.2).

3.3 Material and methods

3.3.1 *Catch statistics*

Long-term catch data were obtained for the period 1960–88 from statutory returns. Within seasons, data were obtained from two sources:

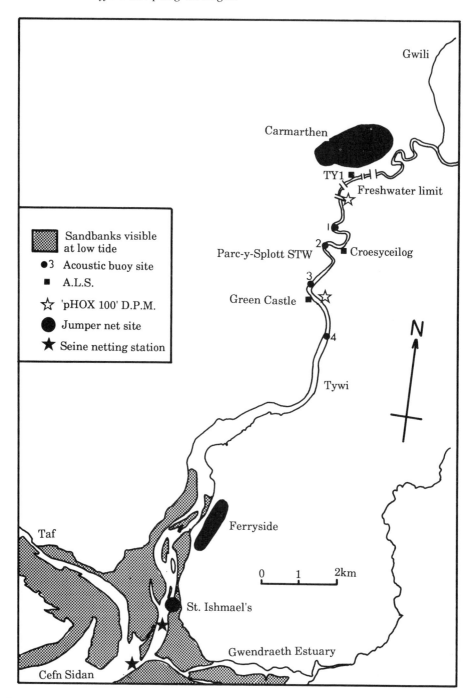

Fig. 3.2 The Tywi estuary showing tagging sites and monitoring points

(1) A specific angler census; the example given in this paper, for 1986, comprises data from 208 anglers who returned detailed logbooks reporting on 8257 fishing hours.

(2) Seine net catch per haul data for 1988 and 1989. This represents data from 239 hauls (60 fishing sessions) in 1988 and 729 hauls (213 fishing sessions) in 1989. These data were recorded by NRA staff as hauls were observed for tagging purposes.

3.3.2 *Tagging and Tracking*

Tagging

Salmon were captured using either jumper nets or seine nets in the estuary downstream of Ferryside (Fig. 3.2). Both these methods were used because they provide free-swimming, undamaged fish for tagging. Following the end of the normal fishing season on 31 August, a dispensation was obtained allowing two of the crews to catch fish for scientific purposes during the period September to December.

Salmon, undamaged by the capture process, were selected for tagging and immediately transferred to a handling bag where they were anaesthetized using 100 ppm 2-phenoxy ethanol. Once anaesthetized, the fishes' sex and fork length were recorded, and a scale sample (two scales) taken. Radio tags or combined acoustic and radio tags (CART) were then inserted into the stomach via the oesophagus, using methods similar to those described by Solomon and Storeton-West (1983). Fish were also externally tagged with orange Floy anchor tags. Following the tagging procedure, each fish was individually held until it was sufficiently recovered to maintain station and swim actively.

Both radio tags and CARTs used were of standard Ministry of Agriculture, Fisheries and Food (MAFF) design, operating on radio frequencies between 173.805 MHz and 173.850 MHz (Solomon & Potter, 1988). These tags are individually identifiable, and have a design life of 6−9 months, depending on pulse rate.

Tracking

The tags were used in conjunction with six acoustic buoys and 36 automatic listening stations (ALS) of MAFF design (Solomon & Potter, 1988). Estuarial acoustic buoys and ALS sites covered the upper estuary, the lowest unit being at Green Castle in 1988 and at Ferryside in 1989 (Fig. 3.2). In fresh water, ALS units were sited on major tributaries and at points roughly equidistant along the length of the main river.

To obtain information on the movement of fish between scanner sites, fish were actively located using a Yaesu FT290 receiver from a small inflatable dinghy. Tracking was conducted between Llandeilo and Carmarthen at least once per week between July and December inclusive. Less frequently, the area between Llandeilo and Llandovery was tracked. In addition, active tracks were undertaken on 12 and 13 November 1988 and on 1 August, 18 and 20 November and 19 December 1989 from a small commercial aircraft. In this way, a large area, including major catchments outside the Tywi, was searched.

Flow Data

Hydrological data used in this paper were obtained from Welsh Water/NRA gauging stations at Ty-Castell (1960–82) and Capel Dewi (post 1982). These stations gauge approximately 90% of discharge to the estuary.

3.4   Results

3.4.1   *Flows in 1988 and 1989*

High rainfall throughout the summer of 1988 maintained abnormally high base flows during the period June–September (mean $46 \, m^3 s^{-1}$). These were followed by a period of unusually dry weather in the autumn and winter. In contrast drought conditions (mean flow $6.1 \, m^3 s^{-1}$) prevailed until October 1989 (Fig. 3.3), with the exception of three small artificial freshets in July–September and a natural freshet in early September.

3.4.2   *Behavioural data*

Sample size and bias

Behavioural data reported in this paper are based on 94 salmon tagged during 1988 and 111 in 1989, a total of 215 (Table 3.1). This is a relatively large number of fish for a behavioural study, allowing quantitative statistics to be calculated and patterns of migration to be identified. We must accept that both the sample as a whole and individual statistics may be biased toward fish whose behaviour makes them more susceptible to net capture. Estuarial time of travel may also be biased toward fish which enter rapidly, thus increasing their individual probability of survival. Nevertheless, consistent behavioural patterns are recognizable, both between months and years, and observed behaviour is consistent with catch patterns in both net and rod fisheries. Even if bias exists, inter-year comparisons are valid, since both years are based on similar experimental techniques and sampling regimes.

Estuarial passage

Entry to fresh water was defined, for analytical purposes, to be movement above Carmarthen, as represented by detection *above* site TY1 (Fig. 3.2). Examination

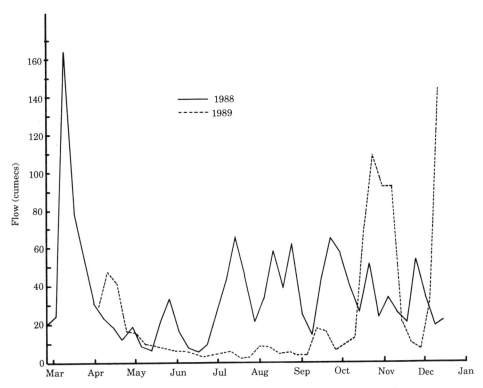

Fig. 3.3 A comparison of the flow hydrographs (weekly mean values) for the years 1988 (————) and 1989 (----------)

Table 3.1 Numbers tagged in each month

Month	1988	1989
May	1	2
June	0	10
July	13	24
August	32	27
September	17	19
October	22	11
November	7	13
December	2	5
Total	94	111

of individual tracks demonstrates that fish tagged during dry weather often remain at sea for long periods, sometimes months, and that subsequent entry is associated with freshet events (see for example Figs. 3.4(a) and (b)).

Median time of travel from tagging to fresh water in June−August 1989 was found to be almost four times that observed in June−August 1988, when flows were high (Table 3.2). In contrast, during the period September−December when

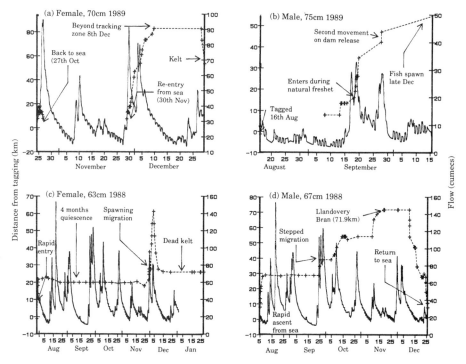

Fig. 3.4   (a) to (d) Examples of salmon tracks

Table 3.2   Median time of travel from tagging to fresh water (hours)

	1988	1989
June – August	50 hrs	195 hrs
September – December	128 hrs	62 hrs

flow levels were generally higher in 1989 than 1988, time of travel in 1989 is less than half that found in 1988.

Only 38% of fish tagged in 1989 entered fresh water in comparison to the 1988 figure of 65%. Detailed examination suggested that this was directly attributable to the effect of drought in 1989; comparison with 1988 showed that the probability of a fish tagged in June – August 1989 subsequently entering fresh water was greatly reduced, but that the figures were similar from September onward (Fig. 3.5).

The reason for this apparent reduction in survival is not totally clear. Tag failure was precluded as a likely cause by comparison with a randomly selected control batch of 25 tags which were highly reliable, 24/25 (96%) still being operational after four months and 22/25 (88%) after five months (Mee & Ellery, 1991). Although occasional regurgitation of tags was evident in reported net and

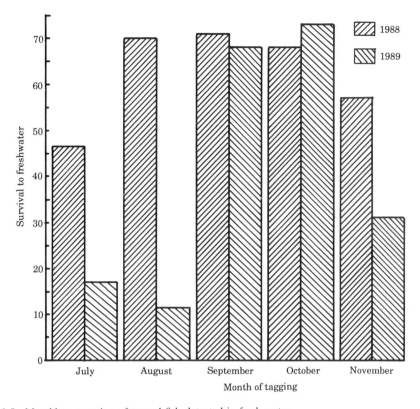

Fig. 3.5 Monthly proportion of tagged fish detected in fresh water

rod recaptures, this only represented 2/24 (8%) of recaptures with one tag regurgitation in each year. Tagging techniques were identical between years, with only one personnel change. We must therefore view the observed reduction in survival as real, rather than an experimental artefact.

Migratory patterns

Most of the salmon entering the Tywi showed a three-stage migratory behaviour (Fig. 3.4(c); Clarke & Purvis, 1989). Following the initial migration in from the estuary, a period of upstream movement occurred, lasting up to 20 days. Most rod recaptures were taken during this active entry phase (8/11), even though it comprised only a small part of within season in-river availability for capture. This was followed by a quiescent stage (during which time the fish were usually located in deep pools, were not stimulated to move by flow events, and were rarely captured by rods). Following this phase a further (secondary) migration often occurred, taking the fish into the spawning areas later in the year. At this time recapture data suggest that fish may become re-available for capture, although we must accept that the number of recaptures during this phase (3) are very small. A variation on this behaviour exhibited by a smaller number of fish was a stepped migration, with two or more quiescent phases, migrating up-river in a series of discrete movements (Fig. 3.4(d)).

3.4.3   *Rod catches, flow and rod catch effort*

The importance of flow in stimulating rod catches is illustrated in Fig. 3.6. Although continuous, albeit varying, effort occurred throughout the season, it is clear that both effort and catch effort increased during and immediately following freshets in late July, August and early September.

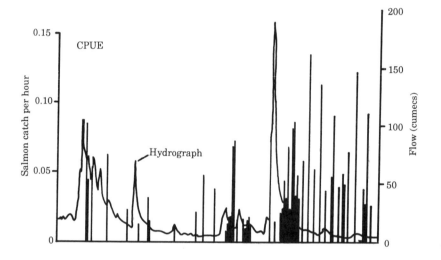

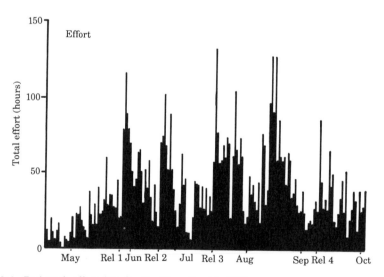

Fig. 3.6   Rod catch effort data for the River Tywi in 1986

The importance of flow is also reflected in longer-term data, with significant correlations existing between long-term catch data and corresponding flows (Fig. 3.7). Both June/July and August/September yielded correlations with flow (Table 3.3), despite the interfering effect of long-term population changes and the relatively crude measure of flow adopted. Removal of a small number of obvious outliers (Table 3.3) increases the annual, June/July and August/September figures to a significance level >99%, and March/May to P > 95%. The lower significance level in March/May is unsurprising; absolute numbers are small and there is evidence within the data of a long-term reduction in catch with time.

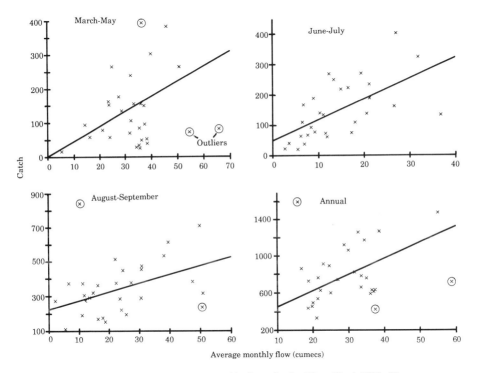

Fig. 3.7 Catch of salmon against average monthly flows in the River Tywi 1960−89

Table 3.3 Significance of flow/catch regressions (exclusions: * 1968, 1979, 1981; ** none as outliers; *** 1968, 1975; **** 1968, 1975, 1981)

Period	Correlation coefficient R	Significance level	Correlation coefficient R	Significance level
March−May	0.207	n/s	0.398*	>95%
June−July	0.614	1%	0.614**	>99%
Aug−Sept	0.306	10%	0.501***	>99%
Annual	0.207	n/s	0.519****	>99%

3.4.4   *Seine net catch per effort including dispensation period*

The net catch per effort data presented here show similar overall trends and absolute values in both 1988 and 1989 (Fig. 3.8). Few salmon were caught prior to June, a peak in catches occurred in August/September, and a significant part of the run occurs outside the normal fishing seasons which end on 31 August. In 1989, the peak of the grilse run appears to have been delayed, maximum catch rates occurring in December, compared to October in 1988.

3.5   Discussion

3.5.1   *Seine net fishing*

Direct effects of river flow on seine net gear efficiency and effort are minimal, the method being dominated by tidal flow rather than freshwater discharge. However, behavioural data demonstrate that fish took almost four times as long to enter fresh water during the early summer period of 1989, when flows were low (Table 3.2).

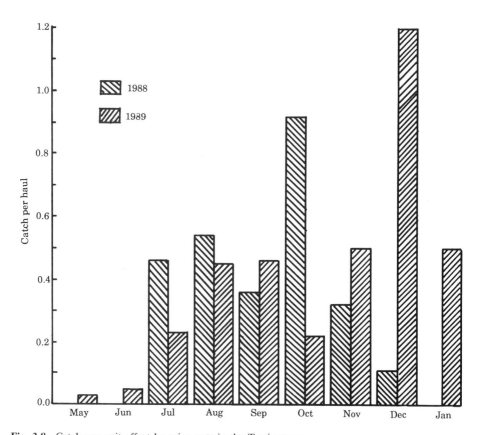

Fig. 3.8 Catch per unit effort by seine nets in the Tywi estuary

Delayed entry in low flows has also been qualitatively observed in other studies (Brawn, 1982; Potter, 1988). Data for the Tywi show that fish in the lower estuary move with the tide during low flow periods, a result compatible with data reported for the Fowey (Potter, 1988) and for the Mirimichi (Stasko, 1975). They may, therefore, pass through the fishery on a number of occasions, and are thus available for capture during a number of fishing sessions. In higher flows, entry is rapid, often effected within one or two tides, and availability of individuals within the netting zone is therefore reduced. Although increased estuarial passage time may not be directly proportionate to net catchability, because of non-random distribution of both fish and fishery within the estuary, it is clear that capture probability will be increased during lower flows. This is consistent with the observation of a higher overall net recapture rate in 1989 than in 1988 (6:1).

These results demonstrate an effective increase in catchability in 1989, and suggest that despite the apparently similar catch effort values in each year, in-season abundance in 1989 may have been substantially lower than in 1988. However, reduced in-season abundance does not automatically imply reduced levels of stocks returning to homewaters, since an increased catch effort of late run 1 sea-winter fish was also observed, presumably as a result of delayed entry in earlier periods of low flows.

This clearly illustrates the difficulty of interpreting catch statistics without detailed understanding of underlying factors; in this case salmon catchability within the seine net fishery is negatively related to flow, because of an increase in individual encounter time with the fishery at low flows.

3.5.2   Rod fisheries

Catchability and availability

The importance of flow in determining rod catch and rod catch effort has been demonstrated by many authors (Alabaster, 1970; Banks, 1969; Gee, 1980; Hayes, 1953; Huntsman, 1948; Millichamp & Lambert, 1966). However, the mechanism of this relationship has not been fully examined, as most datasets do not allow even qualitative estimation of varying catchability.

Consistent with the above authors, behavioural data collected during this study show that during periods of lower flow, entry to the river is stimulated by freshet events. Although we must recognize the limited extent of recapture data, comparison of recaptures with availability in the river suggests that rod catchability varies following entry, being initially high, declining rapidly as the fish enters its quiescent period and then increasing again as the fish undertakes further activity in the autumn. Similar results have been reported by Solomon (1988) working on the Hampshire Avon. Peaks in catch effort associated with freshets (either natural or artificial) therefore principally represent fish attracted into the river by the freshet, and the size and duration of the catch effort peak will be related to the number of fish entering the river as a result of the freshet. This number will probably be defined by the availability of fish in the vicinity of the estuary and the size and duration of the freshet, i.e. its ability to attract fish from farther afield.

Since total exploitation by rod within the Tywi is not high (recapture rates in this project suggest less than 25% of available fish) this implies that much of the in-river stock is effectively unavailable for rod capture (excluding foul hooking). Thus, although a relationship between total in-river stock and catch effort may exist, it will be weak, a result consistent with the observations of Mills *et al.* (1986).

Effect of Drought

In particularly dry summers with few freshets, as occurred in 1989, many fish may not enter during the fishing season, and the total rod catch will be drastically reduced. Unfortunately our data also suggest a significant reduction in successful entry to the river, which more than negates any beneficial effect of reduced rod exploitation on spawning escapement.

Although an increase in legal exploitation within the net fishery was observed, exploitation of salmon within the legal net fisheries is still at a low level, and cannot account for the >75% reduction in entry success observed from June– August 1989 as compared with 1988. No evidence of increased straying was found in aerial searches of other catchments; indeed, the proportion of missing fish relocated out of the catchment in 1989 was less than half that found in 1988, even though three times as much search time was input in 1989. We must therefore conclude that the result represents a high inshore mortality rate, either as a result of water quality related problems, predation or illegal take in drift and set nets which are fished off the estuary mouth.

Longer-term data

The effect of flow is also reflected in longer-term (inter-year) results (Fig. 3.7). A high proportion of inter-year variation in catch may be explained in this way, probably as a result of delayed entry, reduced run size, and increased effort when flows are high. The observed increase in effort during and immediately after freshets is worthy of particular note; fishing effort on many rivers is related to anglers expectation of catch, and increased availability of fish during high flows therefore results in increased effort and further catch increases.

3.5.3   Interpretation of catch effort data

These results have important implications for the interpretation of both catch and catch effort data. At least on the Tywi, substantial variations in both rod and net catchability are related to flow, and an important and varying proportion of the salmon run occurs outside the fishing season. These observations are behaviourally based, and are likely to be applicable to other systems. Variation in catchability, both between and within seasons, could result in misleading data and conclusions if either catch or catch effort are assumed to represent abundance.

Variation in the proportion of fish running within the season may also result in misrepresentation of stock performance. While it can be argued that analysis of long-term data using smoothing techniques will even out the influence of such factors, this will not be true if factors such as long-term climatic change alter catchability or the proportion of the stock available within the fishing season. Long term changes in salmon : grilse ratio in Scotland have been associated with temperature change at sea (Martin & Mitchell, 1985).

Analysis of catch or catch effort data for stock performance purposes should therefore include an assessment and correction for flow and changes in out-of-season runs. While catch and catch effort data are obviously valuable indicators of fishery performance, they are inadequate measures of stock performance.

Acknowledgements

We would like to thank the following people for their assistance with this work. Tony Morse, Tim Davies and Sally Crudginton for help with fieldwork. Guy Mawle, who instituted the Tywi anglers logbook scheme. The Tywi bailiffs, particularly Mike Todd and the Hydrology section, in particular Jean Frost and Rhian Phillips for providing data concerning river flows throughout the year. Finally we would like to thank Dr Alun Gee for his advice and support during the course of this work and Alan Winstone for valuable comments on the draft paper. The views expressed in this paper are those of the authors, and do not necessarily reflect those of the National Rivers Authority.

References

Alabaster, J.S. (1970) River flow and upstream movement and catch of migratory salmonids. *Journal of Fish Biology.* **2**: 1–19.

Banks, J.W. (1969) A review of the literature on the upstream migration of adult salmonids. *Journal of Fish Biology.* **1**: 85–136.

Beverton, R.J.H. and Holt, S.J. (1957) On the dynamics of exploited fish populations. *U.K. Min. Agric. Fish., Fish Invest.* (Ser. 2) 19, p 533.

Brawn, V.M. (1982) Behaviour of Atlantic salmon (*Salmo salar*) during suspended migration in an estuary, Sheet Harbour, Nova Scotia, observed visually and by ultrasonic tracking. *Can. J. Fish. Aquat. Sci.* **39**: 248–256.

Bunt, D. (1991) Use of rod catch effort data to monitor migratory salmonids in Wales. In *Catch Effort Sampling Strategies: Their Application in Freshwater Fisheries Management* (Ed. by I.G. Cowx). Oxford: Fishing News Books, Blackwell Scientific Publications Ltd.

Clarke, D.R.K. and Purvis, W.K. (1989) Migrations of Atlantic salmon in the River Tywi system, South Wales. Paper presented at the Atlantic Salmon Trust Conference, Bristol, April 1989.

Gee, A.S. (1980) Angling success for Atlantic salmon (*Salmo salar* L.) in the River Wye in relation to effort and flows. *Fisheries Management.* **11**: 131–138.

Gulland, J.A. (1969) Manuel des methodes d'evaluation des stocks d'animaux aquations, Premiere partie. Analyse des populations, Manuel FAO de Science la Letique 4, p 160.

Harris, G.S. (1988) The status of exploitation of salmon in England and Wales. In *Atlantic Salmon, Planning for the Future* (Ed. by D. Mills and D. Piggins), London: Croom Helm, pp 169–190.

Hayes, F.R. (1953) Artificial freshets and other factors controlling the ascent and population of Atlantic salmon in the La Have River, Nova Scotia, Burton. *Fisheries Research Board of Canada.* **99**: 1–47.

Huntsman, A.G. (1948) Freshets and Fish. *Transactions of the American Fish Society.* **75**: 257–66.

Martin, J.M.A. and Mitchell, K.A. (1985) Influence of sea temperature upon the numbers of grilse and multi-sea-winter Atlantic salmon (*Salmo salar* L.), caught in the vicinity of the River Dee (Aberdeenshire). *Journal of Fish. Aquatic Sci.* **42**: 1513–1521.

Mee, D.M. and Ellery, D.S. (1991 in preparation) The tag failure experiment. Regional Environmental Appraisal Unit Report. National Rivers Authority, Wales.

Millichamp, R.I. and Lambert, A.O. (1966) On investigations into the relationship between salmon catch and flow in the River Usk during the 1965 season. In *Symposium on River Management.* (Ed. by P.C.G. Isaac), London: McLaren, 199–123.

Mills, C.P.R., Mahon, G.A.T. and Piggins, D.J. (1986) Influence of stock levels, fishing effort and environmental factors on anglers' catches of Atlantic salmon (*Salmo salar* L.) and sea trout (*Salmo trutta* L.). *Aquaculture and Fisheries Management.* **17**: 289–297.

Potter, E.C.E. (1988) Movements of Atlantic salmon (*Salmo salar* L.) in an estuary in South-west England. *J. Fish. Biol.* **33** (Suppl. A): 153–159.

Prouzet, P. and Dumas, J. (1988) Measurement of Atlantic salmon spawning escapement. In *Atlantic Salmon: Planning for the Future* (Ed. by D. Mills and D. Piggins). London: Croom Helm, 325–343.

Ricker, W.E. (1975) Computation and interpretation of biological statistics of fish populations. *Bull. Fish. Res. Bd. Can.* **191**: 382 pp.

Shearer, W.M. (1988) Relating catch records to stocks. In *Atlantic Salmon: Planning for the Future* (Ed. by D. Mills and D. Piggins). London: Croom Helm, pp 256–274.

Solomon, D.J. (1988) Tuning in to Avon Salmon. *Trout and Salmon Magazine*, November.

Solomon, D.J. and Potter, E.C.E. (1988) First results with the new estuarine tracking system. *J. Fish. Biol.* **33** (Suppl. A): 127–132.

Solomon, D.J. and Storeton-West, T.J. (1983) Radio tracking of migratory salmonids in rivers, developments of an effective system. *Fish. Res. Tech. Rep., MAFF Direct. Fish. Res., Lowestoft. No. 75*, 11 pp.

Stasko, A.B. (1975) Progress of migrating Atlantic salmon (*Salmo salar* L.) along an estuary, observed by ultrasonic tracking. *J. Fish. Biol.* **7**: 329–338.

Stoner, J.H., Gee, A.S. and Wade, K.R. (1984) The Effect of Acidification on the Ecology of Streams in the Upper Tywi Catchment in West Wales. *Env. Pollution Ser. A.* **35**: 125–127.

Chapter 4

Comparison of rod catch data with known numbers of Atlantic salmon (*Salmo salar*) recorded by a resistivity fish counter in a southern chalk stream

W.R.C. BEAUMONT, J.S. WELTON and M. LADLE *Institute of Freshwater Ecology, East Stoke, Wareham, Dorset, BH20 6BB, UK*

Yearly and monthly rod catch records from a southern chalk stream taken over a 16-year period were compared with numbers of salmon recorded by a resistivity fish counter. The ability of an individual angler's catch per unit effort (CPUE), beat CPUE, beat rod catch and river rod catch to predict stock numbers are assessed. The effect of fishing effort upon rod catch and the exploitation rate at different stock levels are examined.

4.1 Introduction

The widespread assumption that catch data can be used as an indicator of stock abundance has never been critically examined for any river in England and Wales (Harris, 1986). Despite this, the use of catch information to assess the status of fish populations in rivers is becoming increasingly common (Fahy, 1978; Gee & Milner, 1980; Gee, 1980; Alabaster, 1986; Mills, Mahon & Piggins, 1986). Various forms of data have been used to quantify the state of a fishery, e.g. rod catch, rod days. However, catch per unit effort (CPUE) is the parameter most widely used to evaluate the state of fish populations and to forecast changes (Nikolski, 1969).

The Institute of Freshwater Ecology (IFE) owns 2.4 km of mainly single bank on the East Stoke beat of the River Frome in Dorset, southern England. The beat is situated approximately 12 km above the head of tide. Since 1970, data on numbers of salmon (*Salmo salar* L.) ascending the River Frome have been collected using a resistivity fish counter located at East Stoke. Full details of the site are given in Hellawell, Leatham and Williams (1974), Mann, Hellawell, Beaumont and Williams (1983) and Beaumont, Mills and Williams (1986).

Since 1973 salmon anglers fishing at East Stoke have been required to log all details of time spent fishing as well as fish caught.

This chapter examines the use of these data in conjunction with differing beat catch statistics and Water Authority rod catch records to predict the numbers of salmon ascending the river. The effect of fishing effort upon rod catch and both seasonal and annual exploitation rates at differing stock levels are also examined.

4.2   Methods

Rod catch data were collected from the East Stoke beat between 1973 and 1988. By reciprocal agreement with the owners of the North Bank fishery the beat was fished by IFE lessees on Monday to Thursday inclusive (both banks), Friday to Sunday being controlled by the other owner. Although times and catch were usually recorded by anglers from the North Bank fishery the data were not considered sufficiently complete for analysis and the data referred to in this paper were collected from the IFE fishery, Monday to Thursday only. The IFE rod catch between 1973 and 1988 averaged 9.5% of the total river catch and as such is comparable to other studies of rod statistics, e.g. 10.7% for the River Wye analysed by Gee (1980). Lengths of time fished were recorded to an accuracy of 15 minutes. Details of all fish caught (even if returned to river) were also recorded.

The fishing season is between 1 March and 30 September and two anglers per day are allowed to fish. There is a bag limit of two fish per rod per day, although this limit has been reached on less than 1% of the rod days since 1973. Three anglers who had fished for most of the study period were chosen for analysis of individual catches. Alabaster (1986) found that individual angler efficiency increased with the number of seasons spent fishing. Choosing these anglers may therefore bias the results, but any analysis of long-term angler records are likely to suffer this same bias.

Rod catch information for the whole river was obtained from Wessex Water Authority. This was based on returns of forms issued with salmon licences. Unfortunately no indication of the number of licences issued nor the number returned could be obtained. Wessex Water Authority, however, estimate that 95% of issued licences were accounted for (Anon, 1986), and this probably accounted for 98% of the fish caught (Small, 1991).

Estimates of counter accuracy were carried out at intervals by analysis of photographic records, voltage wave form analysis and video recordings (Beaumont *et al.*, 1986). Although the accuracy of the raw data collected could vary greatly from day to day, overall accuracy of data after correction of errors from obvious sources was in the region of 90% (Welton, Beaumont & Johnson, 1988). There were some occasions when the counter malfunctioned, and a period of renovation of the site in 1986 when records of salmon migration were unavailable.

For compatibility of data total numbers of upstream salmon counts were used. Comprehensive recording of downstream movement of fish did not start until 1983 and only four complete years of data were available. Data from these four years suggested that the total number probably overestimated the run by about 15%. Both total year (January–December) and part year (January–September) counter data were used to allow for the possibility of fish migrating after the end of the fishing season and thus distorting the relationships. In fact most of these later migrating fish were probably available to anglers on the lower river beats.

Previous studies have shown that very few fresh run fish enter the river after

September and salmon movement over the counter between September and December consisted of maturing fish ascending the river to the higher spawning gravels (Welton, Beaumont & Clarke, 1989). No allowance was possible for the numbers of salmon residing in the stretch of the river below the counter. However, only a small percentage of the spawning on the River Frome occurs in the lower reaches of the river (Anon, 1986) and most fish were considered to pass the counter site prior to spawning. Counter data therefore represent a good estimate of the total fish entering the river system. Similarly it was not possible to estimate the proportion of salmon caught by anglers below the counter.

Catch per unit effort was calculated for two measures of effort, number of outings per fish caught (CPUE outings), and hours fished per fish caught (CPUE hours).

4.3   Results

Six parameters based on rod catch were assessed for their correlation with counter numbers. Results are summarized in Table 4.1. The parameters used (arranged in order of ease of obtaining the information) were: total catch of salmon for the river, IFE rod catch, IFE CPUE outings, IFE CPUE hours, individual angler CPUE outings and individual angler CPUE hours. Each parameter required a further degree of precision in terms of detail in rod catch data. This range of parameters was chosen to give an indication of the level of data needed to accurately assess salmon numbers in the river.

There was a significant correlation ($p < 0.05$) between total river rod catch and both seasonal and yearly counter number (Fig. 4.1). The value of r for both

Table 4.1 Correlation values for differing measures of catch and effort (NS − non-significant; * − significant at the 5% level; ** − significant at the 1% level)

		N	r	r²	P
Annual IFE Rod Catch/Counter	Total	14	0.30	0.09	NS
Annual IFE Rod Catch/Counter	Jan−Sep	14	0.41	0.17	NS
Total River Rod Catch/Counter	Total	14	0.60	0.36	*
Total River Rod Catch/Counter	Jan−Sep	14	0.60	0.36	*
IFE CPUE − Hours/Counter	Total	14	0.56	0.32	*
IFE CPUE − Hours/Counter	Jan−Sep	14	0.74	0.56	**
IFE CPUE − Outings/Counter	Total	12	0.37	0.14	NS
IFE CPUE − Outings/Counter	Jan−Sep	12	0.49	0.24	NS
Angler H CPUE − A Hours/Counter	Jan−Sep	14	0.16	0.03	NS
Angler H CPUE − A Outings/Counter	Jan−Sep	12	0.005	0.00002	NS
Angler I CPUE − A Hours/Counter	Jan−Sep	14	0.84	0.70	**
Angler I CPUE − A Outings/Counter	Jan−Sep	12	0.71	0.51	**
Angler M CPUE − A Hours/Counter	Jan−Sep	13	0.53	0.28	NS
Angler M CPUE − A Outings/Counter	Jan−Sep	11	0.17	0.03	NS
Angler I CPUE − Hours/Counter	Total	14	0.76	0.57	**
Angler I CPUE − Outings/Counter	Total	12	0.67	0.44	*

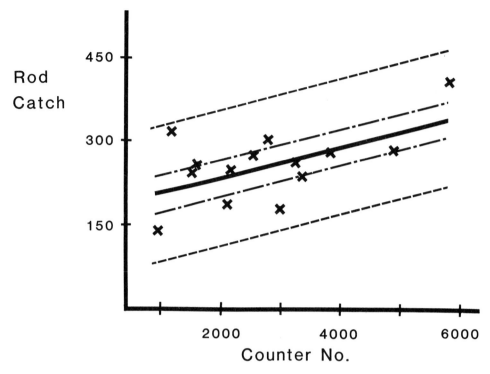

Fig. 4.1 Relationship between total rod catch and annual counter number for the River Frome. Lines either side of the regression line represent 95% confidence limits for the line and new individuals, respectively

seasonal and yearly counter number was the same. No significant correlations, however, ($p > 0.05$), were obtained between IFE rod catch and either seasonal or yearly counter number.

The IFE beat CPUE measured in terms of catch per outing was not significantly correlated with either season or yearly counter number ($p > 0.05$). However, IFE beat CPUE measured in catch per hour fished was highly correlated with counter number in the corresponding season ($p < 0.01$), and significantly correlated with yearly counter number ($p < 0.05$). Correlation of three individual anglers' CPUE, measured both in terms of hours and outings, with season counter number, showed considerable variation. Only one angler's CPUE, both measured in terms of hours and outings, was significantly correlated with counter numbers ($p < 0.01$).

The effect of number of anglers, number of outings and hours fished upon the number of salmon caught on the IFE beat was assessed (Table 4.2). Only the total number of hours fished had any significant influence ($p < 0.01$) on the number of salmon caught.

4.3.1   *Ability of catch data to assess salmon numbers*

Although significant correlations were found for various rod catch parameters, the ability of these relationships to predict the number of fish was still poor. To assess

Table 4.2 Correlation values for catch and exploitation statistics (NS − non-significant; ** − significant at the 1% level)

	N	r	r^2	P
E. Stoke Catch/No. Anglers	16	0.40	0.16	NS
E. Stoke Catch/Outings	14	0.12	0.02	NS
E. Stoke Catch/Hours Fished	16	0.65	0.43	**
Annual Exploitation/Counter Total	14	−0.72	0.51	**

the accuracy, predicted annual run values were calculated from the relationship between total rod catch and total salmon number as, with the exception of Angler I's catch statistics, this relationship had the highest r^2 value ($r^2 = 0.36$). The regression equation derived from Fig. 4.1:

$$N = 12.87 \times R - 512.94 \tag{1}$$

where N is the number of salmon and R the total rod catch, was used.

Although half the values predicted in this way were within 25% of the counter figure for the annual run, the error for other years varied considerably (range −41% to +207%).

Monthly exploitation rates of the salmon were calculated from the equation:

$$E_m = R_m/C_m \tag{2}$$

where E_m is the monthly exploitation, R_m the monthly rod catch, and C_m the monthly counter number.

Calculation of the cumulative monthly exploitation was also based on the cumulative number of salmon caught in the season and the cumulative counter number over the year. Although monthly exploitation gives a better indication of exploitation of fresh run salmon, it was not considered because it does not allow for fish migrating in one month to be caught in subsequent months, nor does it allow for the low numbers of fish entering the river at the end of the season (i.e. September), whilst still having a substantial number of salmon present in the river from earlier immigration. Allowance was made in the cumulative calculations for the number of fish removed from the river by angling in the previous months. Equation 2 therefore became:

$$E_{cm} = R_{cm}/(C_{cm}R_p) \tag{3}$$

where E_{cm} is cumulative monthly exploitation rate, R_{cm} is cumulative monthly rod catch, C_{cm} is cumulative monthly counter number and R_p is the cumulative rod catch for the previous months.

Average exploitation rates for each month over the whole period of study were calculated as a weighted mean of exploitation rates for the particular month in all years. Rod catch for each year was weighted by the total number of fish recorded by the counter in that year. This gave less weight to years when fewer salmon were available for capture in the particular month. There were significant differences in both sets of monthly exploitation rates with the highest levels of exploitation taking place early in the season and the lowest in the latter part of the season (Figs. 4.2 and 4.3).

Annual exploitation rate (E_a) was also calculated for each year over the study period and correlated with annual counter number according to:

$$E_a = R_a / C_a \qquad (4)$$

where R_a is total annual rod catch and C_a is yearly counter number.

Errors and natural variation in the estimates of both annual rod catch and counter number render it invalid to determine the relationship between exploitation (E) and catch (C) by standard regression analysis. The relationship was derived via the functional geometric mean regression (Ricker, 1973) of $\mathrm{Log}_e R$ on $\mathrm{Log}_e C$ which allows for errors in both variables and Equation 1 becomes:

$$\mathrm{Log}_e R = a + b \, \mathrm{Log}_e C$$

$$R = A \times C^b \qquad (5)$$

where $A = \exp(a)$.

The linear relationship between yearly exploitation and stock is shown in Fig. 4.4. If the exploitation rate is not related to catch, the value of b will not be significantly different from 1, giving on average:

$$E = R / C = A \qquad (6)$$

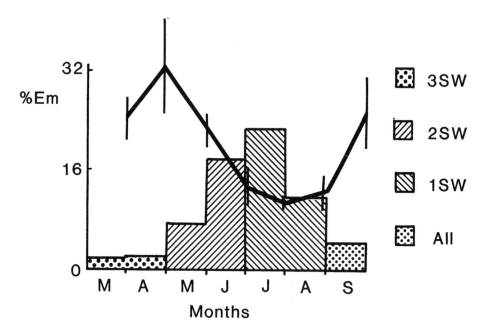

Fig. 4.2 Monthly exploitation rate (E_m) with standard error. Histogram shows the mean number of salmon ascending each month and their predominant sea age (SW − sea winter)

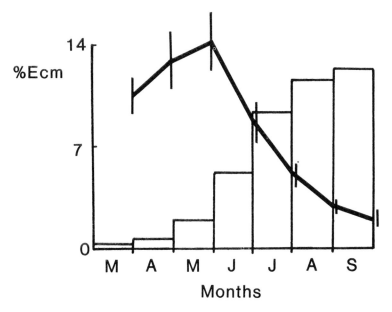

Fig. 4.3   Mean cumulative monthly exploitation rate (E_{cm}) with standard error. Histogram shows the cumulative mean number of salmon ascending

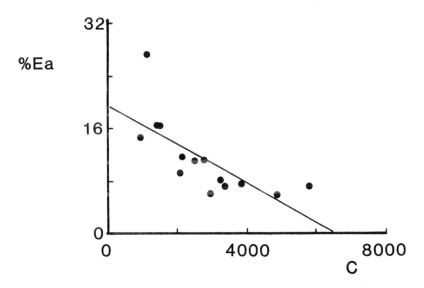

Fig. 4.4   Relationship between annual exploitation rate (E_a) and counter number (c)

For the 14 years of data available, this gave $A = 3.44$, $b = 0.54$, and the Standard Error $(b) = 0.13$. The 95% confidence limits for b were 0.26 to 0.83 indicating that b was significantly less than 1 and hence that exploitation declined with increasing counter number. Equation 5 therefore becomes:

$$E_a = 3.44 \ C^{-0.46} \tag{7}$$

Figure 4.5 shows the predicted values of exploitation from the geometric mean regression (Equation 7).

4.4   Discussion

Various authors (op. cit.) have used differing rod catch parameters to assess salmon numbers in rivers. Of those parameters considered in this study only four showed significant correlation with the total numbers of salmon ascending the River Frome, as assessed by the fish counter. The highest r^2 value was obtained for angler I's CPUE hours ($r^2 = 0.57$) followed by angler I CPUE outings ($r^2 = 0.44$), total river rod catch ($r^2 = 0.36$) and IFE CPUE hours ($r^2 = 0.32$).

Despite the high correlation in both CPUE hours and CPUE outings with counter numbers for angler I the variability between individual anglers CPUE makes it extremely speculative to use any one angler's CPUE as an indicator of salmon numbers. Although IFE CPUE hours has a highly significant correlation with January to September counter numbers ($r^2 = 0.56$), the correlation with total counter numbers is less good. Therefore this information still does not give a consistently good estimate of the annual run. Although an estimate of CPUE, either for the whole river or several beats, is likely to give a better correlation with counter numbers, attempts to obtain an accurate measure of the effort (hours fished) for other beats of the River Frome have proved impractical. Whole river rod catch, however, is comparatively easy to obtain (from the records of the National River Authority or their predecessor the Water Authority) and likely to be consistently available. The major drawback in using such returns lies in estimating the error caused by non-return of catch information. Small and Downham

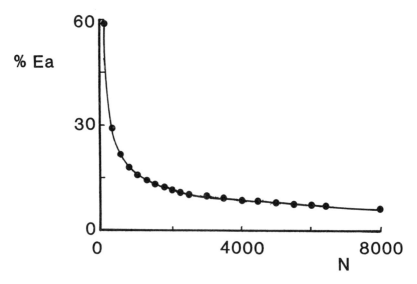

Fig. 4.5 Predicted annual exploitation (E_a) and stock number (N)

(1985) and Small (1987) discuss this problem and present a formula for estimating the errors so caused. Catch returns in their studies ranged from 5−95% of the estimated true value, with 40% considered good. Information on numbers of rod catch forms returned to the Wessex Water Authority indicate that with the aid of postal reminders 95% of issued licences were accounted for (Anon, 1986). However, no correction factors were applied in this study.

Although superficially the pattern in numbers of fish caught and IFE CPUE hours mirror the pattern in counter numbers (Fig. 4.6), there is no statistical

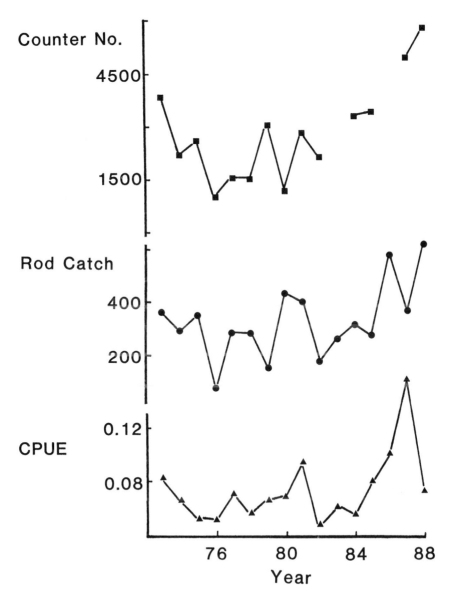

Fig. 4.6   Annual counter numbers (■), total river rod catch (●) and IFE CPUE hours (▲)

validity in the relationship. Of the 16 years studied, 10 showed trends of catch and CPUE in the same direction as trends of numbers. However, eight (50%) would be expected to agree purely on the basis of chance. Lakhani (1986) examining statutory catch statistics, also concluded that catch statistics were inherently poor indicators of available stock for Scottish fisheries.

Of the catch statistics assessed in the present study only one pair had any significant correlation (catch *v.* hours fished, $p < 0.01$). This agrees with the findings of Mills *et al.* (1986) who also found a significant correlation in numbers of salmon and sea trout caught with effort in a rod fishery in western Ireland.

4.4.1   *Exploitation*

Monthly exploitation (E_m) varied over the season. Welton, Beaumont and Clarke (1989) showed that different age groups of salmon migrate into the river at different times of year. Various age groups are therefore subject to differing levels of exploitation. In this study the highest E_m rates were associated with the time when the larger, 3-sea-winter salmon (average weight *c.* 9 kg) enter the river in the early part of the year. The lowest E_m rate was associated with the time of immigration of the 1-sea-winter grilse (average weight *c.* 3 kg) in July and August (Fig. 4.1). Gee and Milner (1980) also found a high exploitation of spring fish and a low exploitation of grilse in the River Wye. The apparent increase of E_m in September is partly a function of the low counter numbers in that month. Therefore the E_{cm} (monthly exploitation of the cumulative run) rates (Fig. 4.3) show a more realistic value for the exploitation rate in that month.

Catchability of salmon is governed by many factors and almost certainly closely follows activity patterns of the fish. Clarke and Purvis (1989) found that after salmon enter the River Tywi they are active for 10 to 20 days, they then remain dormant for a period, becoming active again prior to the spawning period. This change often takes place at or after the end of the rod fishing season (September) prior to the fish spawning in late December. Catchability appears to follow similar phases, although the higher exploitation rate of fish in the spring compared to the summer may be partially related to fish remaining catchable for longer periods in the cooler spring water conditions than in the warmer summer conditions.

Other factors also apply to rates of exploitation. Peterman and Steer (1981) showed that there is increased exploitation of salmon at low stock levels and Mills *et al.* (1986) found that the catchability of sea trout had an inverse relationship with stock. In this study it was demonstrated that increased exploitation did occur at low stock levels (Figs. 4.4 and 4.5). Anecdotal evidence from the River Frome (and other rivers) suggests that good salmon 'lies' are occupied before poorer lies (a 'lie' being an area of river bed which is attractive to a resting salmon) and anglers are likely to fish these good lies more intensively. At low stock levels, therefore, a higher proportion of the fish are likely to be in the good (better known, more accessible) lies. Any new (catchable) fish migrating into the river will be forced to occupy poorer lies and hence will be less likely to be caught. This state of affairs is likely to continue until the fish occupying the good lies are caught or otherwise displaced and possibly replaced by a fresh-run, more catchable, fish. Over the year the mean size of fish entering the river becomes smaller. These later-running smaller fish may be less able to compete for the best lies (already occupied by larger fish), and hence will be more likely to reside in poorer areas, thus having a lower catchability. Peterman and Steer (1981) also suggested that increased exploitation at low stock levels might be due to non-random clumping of the fish in areas which were under high angling pressure.

Different river types are likely to have differing characteristics in their run of salmon. The River Frome (Welton *et al.*, 1988), however, appears to be typical of southern chalk rivers. Exploitation in the River Frome covers the entire period when fish are entering the river. The River Frome therefore does not suffer from the drawbacks experienced by Shearer (1986) with regard to a high and variable proportion of the stock entering the river *after* the period of exploitation.

The annual exploitation by rod in the River Frome is fairly low (averaging 11% of the upstream count) and is unlikely to be having an adverse effect on the population structure. Even the early running fish, which suffer the highest rate of exploitation, are probably unaffected, the long-term counter records showing no evidence of systematic change in the proportion of differing age groups. However, if the same differential levels of exploitation occur in other rivers where the overall annual exploitation is higher, e.g. 72% for the Hampshire Avon (Anon, 1986) there may be a serious risk of affecting the population structure of the stock. Anecdotal evidence (Anon, 1982; George, 1977; Shearer, 1985) of the decline of spring fish in many rivers may be attributable to this factor.

Acknowledgements

The authors would like to thank all the anglers who helped in this study by recording details of their angling efforts, especially angler 'I' for at least showing some correlation with the number of fish. We would also like to thank Ralph Clarke for helping us through the minefield called statistics. A large proportion of this work was funded by the Natural Environmental Research Council.

References

Anon (1982) *The Times* 24.8.82.

Anon (1986) Investigation into the alleged decline of migrating salmonids. Final Report. Wessex Water Authority.

Alabaster, J.S. (1986) An analysis of angling returns for trout *Salmo trutta* L., in a Scottish river. *Aquaculture and Fisheries Management* **17**: 313–316.

Beaumont, W.R.C., Mills, C.A. and Williams, G.I. (1986) Use of a microcomputer as an aid to identifying objects passing through a resistivity fish counter. *Aquaculture and Fisheries Management* **17**: 213–226.

Clarke, D. and Purvis, W.K. (1989) Migration of Atlantic Salmon in the River Tywi System, South Wales. Atlantic Salmon Trust/Wessex Water Workshop, Bristol, April 1989, 22 pp.

Fahy, E. (1978) Performance of a group of Sea Trout Rod Fisheries, Connemara, Ireland. *Fisheries Management* **9**: 22–31.

George, A. (1977) Salmon in a year of drought. *Trout and Salmon*, January 1977. pp. 42−45.

Gee, A.S. and Milner, N.J. (1980) Analysis of 70−year catch statistics for Atlantic salmon (*Salmo salar*) in the River Wye and implications for management of stocks. *Journal of Applied Ecology* **17**: 41−57.

Gee, A.S. (1980) Angling success for Atlantic salmon (*Salmo salar* L.) in the River Wye in relation to effort and river flows. *Fisheries Management* **11**: 131−138.

Hellawell, J.M., Leatham, H. and Williams, G.I. (1974) The upstream migratory behaviour of salmonids in the River Frome, Dorset. *Journal of Fish Biology* **6**: 729−744.

Harris, G.S. (1986) The status of exploitation of salmon in England and Wales. In *Atlantic Salmon: Planning for the Future*, (Ed. by D. Mills and D. Piggins). London: Croom Helm. pp. 69−90.

Lakani, K.H. (1985) Salmon population studies based upon Scottish catch statistics: statistical considerations. In *The status of the Atlantic Salmon in Scotland*. (Ed. by D. Jenkins and W.M. Shearer). ITE Symposium 15, Banchory Research Station, pp. 116−121.

Mann, R.H.K., Hellawell, J.M., Beaumont, W.R.C. and Williams, G.I. (1983) Records from the automatic fish counter on the River Frome, Dorset 1970−1981. *Freshwater Biological Association Occasional Publication* No. 19.

Mills, C.P.R., Mahon, G.A.T. and Piggins, D.J. (1986) Influence of stock levels, fishing effort and environment factors on anglers' catches of Atlantic salmon (*Salmo salar* L.) and sea trout (*Salmo trutta* L.) *Aquaculture and Fisheries Management* **17**: 289−297.

Nikolski, G.V. (1969) *Theory of fish population dynamics*. Edinburgh: Oliver and Boyd.

Peterman, R.M. and Steer, G.J. (1981) Relation between sport-fishing catchability coefficients and salmon abundance. *Transactions of the American Fisheries Society* **110**: 585−593.

Ricker, W.E. (1973) Linear regressions in fishery research. *J. Fish. Res. Bd Can.* **30**: 409−434.

Small, I. and Downham, D.Y. (1985) The interpretation of anglers records (trout and sea trout, *Salmo trutta* L., and salmon, *Salmo salar* L.). *Aquaculture and Fisheries Management* **16**: 151−171.

Small, I. (1987) Thoughts on anglers game fishing returns and grading of fisheries for quality. *Salmon and Trout*, Autumn 1987.

Small, I. (1991) Exploring data provided by angling for salmonids in the British Isles. In *Catch Effort Sampling Strategies: Their Application in Freshwater Fisheries Management* (Ed. by I.G. Cowx). Oxford: Fishing News Books, Blackwell Scientific Publications Ltd.

Shearer, W.M. (1985) An evaluation of the data available to assess Scottish salmon stocks. In *The status of the Atlantic salmon in Scotland* (Ed. by D. Jenkins and W.M. Shearer). ITE Symposium No. 15 Banchory Research Station, pp. 91−112.

Shearer, W.M. (1986) Relating catch records to stocks. In *Atlantic Salmon : Planning for the future* (Ed. D. Mills and D. Piggins). London: Croom Helm. pp. 256−274.

Welton J.S., Beaumont, W.R.C. and Johnson, I.K. (1988) Salmon counting in chalk streams. *Institute of Fisheries Management Annual Study Course 1988*. Southampton.

Welton, J.S., Beaumont, W.R.C. and Clarke, R.T. (1989) Factors affecting the upstream migration of salmon in the River Frome, Dorset. Atlantic Salmon Trust/Wessex Water Workshop, Bristol. April 1989, 29 pp.

Chapter 5
Catch for effort in a New Zealand recreational trout fishery – a model and implications for survey design

M. CRYER and G.D. MACLEAN *Department of Conservation, Turangi, New Zealand*

The distributions of catch per unit effort (CPUE) for individual anglers in Lake Taupo and its major tributary the Tongariro River are always highly non-normal, but of a characteristic form. Both binomial and Poisson models yield distributions of CPUE qualitatively similar to that encountered in the field, although the assumptions and predictions of the Poisson model are more realistic. Field samples of CPUE contain a higher proportion of extreme values than predicted by either model, but these can be generated theoretically by refining the Poisson model to include a measure of angler skill and some degree of environmental patchiness. Patchiness appears to be more important for samples of CPUE taken over several days fishing (implying some temporal heterogeneity), whereas angler skill becomes more important within a single day's sampling. The implications of non-normality for monitoring programmes are discussed, and analytical options for the particular field survey design employed in Taupo are explored.

5.1 Introduction

Catch rates reported by anglers or determined by creel census are a vital source of information for managers of recreational fisheries yet relatively little work has been conducted on their statistical properties. Previous studies of angler catch rates have suggested that their distributions are rarely normal (Small & Downham, 1985; Parkinson *et al.*, 1988; Stephens, 1989). In the past it has been usual either to ignore this fact and apply parametric analyses based on the normal distribution regardless (e.g. Von Geldern & Tomlinson, 1973), or to use a larger measure of effort (e.g. total angler days on a given fishery or sector) which yields a CPUE distribution which is close to, or can be transformed to, normality (e.g. English *et al.*, 1986; Small, 1987; Parkinson *et al.*, 1988). Markedly skewed or multi-modal distributions violate the important assumptions of normality made in the first approach, whereas the latter method may mask important aspects of the data by aggregating effort units that are not equivalent.

This chapter concerns angler catch rates (CPUE, measured as fish caught per hour) in a New Zealand recreational fishery for rainbow trout (*Oncorhynchus mykiss*). The distribution of CPUE in this fishery is always slightly non-normal, difficult to transform to normality, and of very characteristic form (e.g. D. Maclean, unpublished data). Attempts have been made to determine the origins of the distribution, and explore several alternative analyses for comparing catch rates among seasons using the survey design already in use in the fishery.

5.2   Methods

Field data were collected for the years 1985 to 1989 using face-to-face interviews with anglers fly fishing the Tongariro River in August or trolling on Lake Taupo during December. The river was divided into six sections (upper, middle and lower, right and left banks), and the lake into four sections (north, south, east and west). On each of six randomly chosen days in the respective months, interviewers patrolled their allocated sections of the fishery from 8.00 a.m. to 12.00 noon, and again for three hours shortly before dusk. Surveys were undertaken on foot for the river fishery, and by boat for the lake. All anglers encountered were asked how many hours they had been fishing (to the nearest $\frac{1}{4}$ hour), how many fish they had kept, and how often they used the fishery each year. Catch Per Unit Effort (CPUE) for each angler was calculated as the number of legal fish (>35 cm) caught divided by hours fished on that day.

Other information relating to angling method, total angling experience (years fishing), licence type, and town of residence was also collected but not presented here. This survey was carried out annually to provide a regular monitoring programme for the Taupo fishery.

5.3   Results

5.3.1   *Field data*

Anglers were probably not precise in reporting their effort as fishing times of integral numbers of hours were greatly overrepresented (Fig. 5.1, χ-squared $= 662.1$, $p < 0.001$). However, as a more accurate measure of effort was not available, all CPUEs were calculated using angler reports of effort.

The frequency distribution of angler CPUEs (Fig. 5.2) was always highly non-normal (Kolmogorov-Smirnow $D_{735} = 0.305$, $p < 0.001$), but of a characteristic form. About two thirds of anglers interviewed had no fish, and catch rates for the remainder ranged up to over four fish per hour. The mean catch rate for most surveys was between 0.2 and 0.3, and this figure was used to monitor the fishery (as an indicator of overall angling success), or for calculations of yield. However, it was a very poor indicator of what most anglers experience on a given day: over three quarters of those interviewed had either no fish (the modal value) or had caught more than 1.0 fish per hour (about four times the mean CPUE).

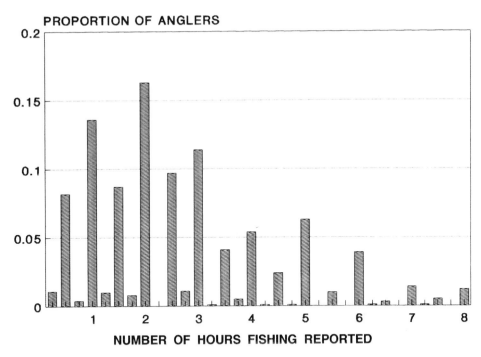

Fig. 5.1 Distribution of angler reports of time spent fishing prior to being interviewed on the Tongariro River in 1987

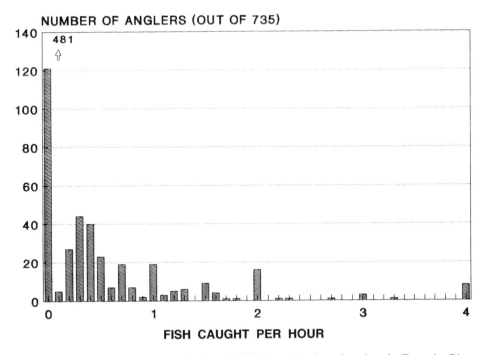

Fig. 5.2 Distribution of catch per unit effort (CPUE) for anglers interviewed on the Tongariro River in 1987

5.3.2   *Modelling the CPUE distribution*

The shape of the CPUE distribution was so consistent among surveys that it was decided to explore the underlying characteristics of the fishery and to develop appropriate analytical methods. Survey results from year to year tended to be rather similar, therefore the 1987 river survey was chosen (arbitrarily) as the base for binomial and Poisson models. Angler reports of effort and the overall CPUE were the only inputs into these primary models.

The binomial model is defined by:

$$P_R = N!/(R! \times (N-R)!) \times \pi^R \times (1-\pi)^{N-R}$$

where P_R is the probability of exactly R successes (fish caught) in N trials ($\frac{1}{4}$ hour fishing periods) assuming an overall probability of success of π (overall mean catch per $\frac{1}{4}$ hour). This model assumes a 'handling time' of 15 minutes for each fish, and sets an arbitrary limit of four fish per hour as the maximum possible CPUE.

The Poisson model takes the form:

$$P_R = e^{-\mu} \times \mu^R/R!$$

where P_R is the probability that exactly R fish will be caught within a given time period if μ is the average number of captures within that given interval. The Poisson model assumes that captures are independent (catching a fish does not preclude or enhance the chances of catching a second within any given time period, i.e. there is no 'handling time') and simulate a random process.

For both models, probability distributions for catching given numbers of trout were generated for successive $\frac{1}{4}$ hour increments in effort, leading to a two-dimensional matrix of probabilities with 0 to 8 fish on one dimension and 1 to $32\frac{1}{4}$ hour sessions on the second (reports of over eight hours fishing were very rare in this fishery). The matrix was then weighted according to the field proportion of each $\frac{1}{4}$ hour effort band. Thus a value of CPUE could be calculated for each matrix cell (e.g. 3 fish in $7\frac{1}{4}$ hours leads to a CPUE of 0.414), and a CPUE distribution extracted from the weighted matrix by summing the contents of all matrix cells whose fish/effort ratios fell within given bands of CPUE.

Although the assumptions of the two models are quite different, at the low catch rates prevalent in the fishery (probability of catching a fish in $\frac{1}{4}$ hour = 0.07), the outcomes were similar (Fig. 5.3). However, a handling time of 15 minutes as required by the binomial model was not typical of the time taken to land a fish, and the maximum CPUE of 4 fish per hour restricted the model in an unrealistic manner at the top end of the distribution. Typically, about 1% of anglers reported a CPUE of greater than 4 fish per hour, and while this was a small proportion, their influence on the mean CPUE was quite large. Effort was recorded to the nearest $\frac{1}{4}$ hour, so it was not possible to refine the binomial model for shorter handling times and greater maximum CPUEs without smoothing the effort distribution curve and interpolating. For these reasons, and because the Poisson

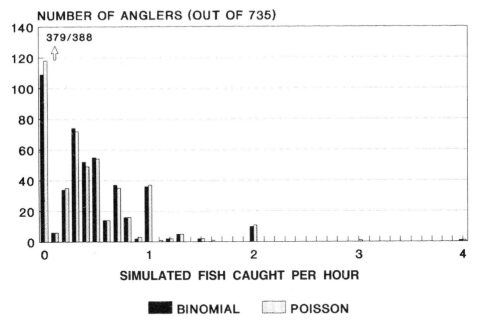

Fig. 5.3 Distribution of CPUE generated by simple binomial and Poisson models

model generated a CPUE distribution slightly closer to reality, the Poisson was selected for refinement.

The simple Poisson model produced a CPUE distribution qualitatively similar to the field data, but generated too few extreme values (both high and zero). The overall fit of the model was poor (Table 5.1). This suggested that, although the random process indicated by the Poisson model was probably responsible for the general form of the CPUE distribution, some variability had also been introduced by factors not included in the model.

In broad terms, there was usually a strong relationship between angler familiarity with the fishery and overall catch rates (Fig. 5.4). Stephens (1989) used a multiple

Table 5.1 Models used to generate angler CPUE frequency distributions for five-day and single-day sampling sessions. SS = sum of squared deviations from expectation, R% = reduction in SS compared with simple Poisson model, *** = <0.001, * = <0.05, NS = >0.05.

	Full survey				One day			
	SS	R%	χ^2_{15}	P	SS	R%	χ^2_8	P
Binomial	13 499	–	78.1	***	--- Not tested ---			
Poisson	11 407	–	68.0	***	305	–	13.9	*
Poisson + Skill	7 752	32	45.9	***	116	62	9.9	NS
Poisson + Patch	1 037	91	30.0	*	105	66	11.2	NS
Poisson/Skill/Patch	706	94	15.1	NS	111	64	8.7	NS

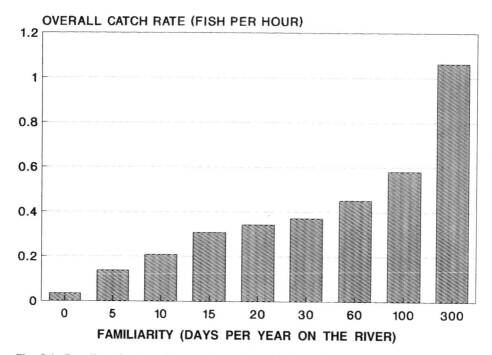

Fig. 5.4 Overall catch rates of groups of anglers with increasing levels of familiarity with the Tongariro River (zero familiarity indicates a given angler's first session on the river)

regression model to examine the importance of various angler attributes on the Tongariro fishery, and concluded that angler familiarity (number of days spent on the fishery per year) was the most important. To include this measure of skill in the Poisson model, nine catch-effort matrices were built using the field estimates of overall catch rate for the nine levels of familiarity (shown in Fig. 5.4 for the 1987 survey). A CPUE distribution was extracted from each (as described for the simple model), and these were combined according to the proportion of anglers in each group. This refinement improved the fit of the model, reducing the total *SS* (sum of squared deviations from expectation) by 32%, and increasing the proportion of extreme values. Statistically, however, the fit was still poor.

Familiarity was the most influential variable in determining angler CPUE yet, after its inclusion, the model still failed to reproduce important aspects of the field distribution of CPUE. Clearly, another factor must have been involved in generating the extreme values.

Fish are rarely encountered at random in the environment, and a measure of patchiness was the next model refinement. An arbitrary assumption that certain parts of the river held concentrations of fish ('shoals' or 'runs') split the angler population into those that were fishing in 'good' areas, and those that were outside these areas. Model catch rates for the 'good' and 'bad' areas were calculated by assuming that effort was randomly distributed throughout the total area, but fish were distributed according to the level of patchiness applied. Two catch effort matrices were produced corresponding to the two catch rates, and the two resulting CPUE distributions were combined according to the relative sizes of high and low catch rate areas.

Several levels of patchiness were tested in the model, the best fit being obtained when one third of the area was assumed to sustain catch rates 10 times greater than the other two thirds. This reduced total *SS* by 91% compared with the simple model and, while the CPUE distribution generated was significantly different to the field distribution, the two looked very similar indeed. The low number of high CPUE values (>3.0) predicted by this model (1 compared with 9 in the field data and 2 in the familiarity model) was its greatest weakness.

A final model combined both angler skill and arbitrary patchiness to produce 18 catch effort matrices (nine levels of angler familiarity inside and outside 'patches' of fish: Table 5.2) and 18 CPUE distributions which were summed according to the relative proportions of anglers within each.

This model generated a distribution which was not significantly different from the field distribution of CPUE (Fig. 5.5), and total *SS* was reduced by 94% compared with the simple model. The combination of skilful anglers and dense patches of fish in the model generated 0.7% CPUE values > 3.0 compared with 1.2% in the field, and slightly over 65% of anglers were fishless in both distributions.

It is highly likely that the model could be further refined to reflect the field situation more closely, but this simple simulation shows that the characteristic (and highly non-normal) CPUE distribution illustrated in Fig. 5.2 can be generated using clearly identifiable factors – a random fish catching process within groups of more or less skilful anglers who may or may not be lucky enough to have a shoal of fish in front of them. The imprecise reporting of effort by anglers may not be important – the distribution of CPUE produced by the simple Poisson model using smoothed effort data (5-point moving average) was only slightly different from that produced using field data (χ-squared $= 27.35$, $p = 0.03$).

Table 5.2 The proportions and predicted catch rates of anglers with successively more familiarity with the Tongariro fishing inside and outside modelled concentrations of fish. Mean CPUE was $0.277\,\mathrm{hr}^{-1}$.

Familiarity (Days/year)	Outside Patches		Inside Patches	
	Proportion	CPUE (%)	Proportion	CPUE (%)
0	3.17	0.008	1.59	0.085
1−5	16.24	0.034	8.12	0.343
6−10	12.52	0.052	6.26	0.521
11−15	8.44	0.076	4.22	0.768
16−20	7.17	0.085	3.58	0.856
21−30	6.80	0.092	3.40	0.926
31−60	6.53	0.113	3.27	1.131
61−100	2.99	0.145	1.50	1.459
101−365	2.81	0.265	1.41	2.661

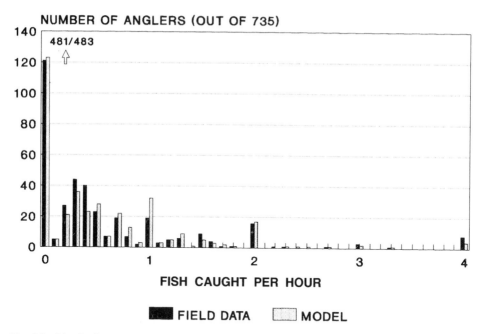

Fig. 5.5 Distribution of angler CPUE from the Tongariro River in 1987 and generated by a Poisson model incorporating angler 'skill' and environmental patchiness

The composite Poisson model was constructed using data from the entire 1987 river survey. It was next applied to data collected during a single day to determine whether the large proportions of high and low CPUEs were merely a result of differences among survey days. The simple Poisson model produced a moderately good fit for these data, predicting 55% fishless anglers and 3% with CPUE > 1.0 (63% and 9% in the field).

Including angler skill in the model (using catch rates specific to groups of anglers of varying experience encountered on the day) reduced total SS by 62%, and led to a CPUE distribution not significantly different from field data. Patchiness at the level used in the first model resulted in too many extreme values, but a lower level of patchiness (one third of the area sustaining a CPUE only five times greater than the rest) led to a much better fit. Using this level of patchiness and the field recorded groups of angler skill in the composite model predicted a distribution of CPUE very close to reality (65% fishless anglers, and 6% with CPUE > 1.0).

The model is thus capable of generating expected distributions of CPUE similar to field survey data for both short and extended time periods, but the relative importance of angler skill compared with patchiness is much greater for the shorter time period. This suggests that patchiness has temporal and spatial components, with the temporal aspect becoming much less important for data collected within a single day.

5.4   Discussion

5.4.1   *Analysis of field data*

As field data of CPUE (per hour) were so demonstrably non-normal, it would be clearly inappropriate to use parametric analytical methods, based on assumptions of normality, to test for differences among and within angling seasons. One alternative would be to use simple non-parametric tests such as the Mann-Whitney 'U', but such tests are usually of lower statistical power than parametric tests (e.g. Zar, 1984), and the results can be distorted by the large number of zero values.

Bannerot and Austin (1983) tested the utility of a range of other descriptors of the CPUE distribution, and found that, for their data, the proportion of zero catches was the best index of fish abundance. Their method was aimed at estimating fish density, however, and the present chapter considers a monitoring programme for fishery performance in which the relative proportions of high and low non-zero catch rates (implicitly ignored by Bannerot and Austin's approach) could be important.

As the two survey designs for the Taupo monitoring programme consist of discrete sections, information on the relative performance of each section within and among seasons should be available to managers. This information could be extracted by repeat testing (and consequent modification of α levels) if simple tests are employed, but this approach is cumbersome, and analysis of variance methods followed by *a posteriori* comparison of means (e.g. Studentized Newmann-Keuls testing) would be more appropriate given suitably distributed data.

Summing catch and effort for anglers interviewed within given sections on each of the six random days leads to distributions of CPUE that are approximately log-normal (Kolmogorov-Smirnov tests for raw and log transformed data: $D_{24} = 0.354$, $p < 0.05$ and $D_{24} = 0.097$, $p > 0.20$). Grouping anglers in this way reduces the degrees of freedom for the analysis and may mask some interesting features of the dataset, but the transformed data violate none of the assumptions of parametric analysis.

One-way ANOVA could now be used to test for differences among years using each of the section-specific CPUEs on each of the six random days as replicates, although the sections were not true random replicates as they were identical for all surveys. Two-way ANOVA using season and river section as the two factors gives six truly random replicates of reach-specific catch-for-effort within each cell. Such a design allows comparisons of section performance within and among years, as well as providing the manager with the vital information on overall fishery performance on an annual basis.

5.4.2   *Conclusions and implications for survey design*

This study confirms previous suggestions that angler catch rates per hour are not normally distributed. Furthermore, other distributions such as log-normal, simple Poisson and negative binomial do not provide adequate fits.

A model has been developed that can accurately reproduce the form of the CPUE distribution using a Poisson process and components for angler familiarity with the fishery and environmental patchiness. This model suggests that catching a fish is a random process, but one affected by the skill of the angler, and whether or not a 'run' of fish is encountered during a given session.

It would seem from this analysis that distributions of angler catch per hour in fisheries with low mean catch rates (<1.0 fish per hour) and angling sessions of up to eight hours will always be non-normal and probably of the general form shown in Fig. 5.2. Surveys, therefore, should be designed such that the unit of effort chosen for analysis produces a distribution of CPUE that can be transformed to normality. This will almost certainly entail a pilot survey to assess the likely distribution of data collected by given methods. In the Taupo fishery, the survey areas were divided into sections on geographical or operational lines and all catch and effort data were summed within these areas to produce a CPUE curve which was approximately log-normal. An increase in the degrees of freedom for testing could be obtained by dividing into smaller areas or sampling on a larger number of days within the respective months. However, both of these alternatives have associated costs in manpower and equipment requirements.

An alternative approach to the parametric one described might be to employ a multi-dimensional contingency table analysis (Bishop *et al.*, 1975; SAS, 1988, Ch. 14), although lack of space precluded a full examination of the possibilities here. This approach would not depend to such a great extent on assumptions about the distribution of data, although sample size requirements for some of the methods might present problems for surveys undertaken within the Taupo/Tongariro fishery.

Acknowledgements

Our grateful thanks go out to those hardy souls who helped in the collection of data from the Tongariro and Lake Taupo fisheries. Theo Stephens, Tom Caithness, Brian Lloyd and Richard Sadleir reviewed early versions of the manuscript and made many perceptive comments.

References

Bannerot, S.P. and Austin, C.B. (1983) Using frequency distributions of catch per unit effort to measure fish-stock abundance. *Trans. Am. Fish. Soc.* **112**: 608–617.

Bishop, Y.M.M., Fienberg, S.E. and Holland, P.W. (1975) *Discrete Multivariate Analysis: Theory and Practice*. Cambridge MA, MIT Press.

English, K.K., Shardlow, T.F. and Webb, T.M. (1986) Assessment of Georgia sport fishing statistics, sport fishing regulations and trends in chinook catch using creel survey data. *Can. Tech. Rep. Fish. Aquat. Sci.* No. 1375.

Parkinson, E.A., Berkowitz, J. and Bull, C.J. (1988) Sample size requirements for detecting changes in some fishery statistics from small trout lakes. *North. Am. J. Fish. Mgmt.* **8**: 181–190.

SAS (1988) *SAS/STAT User's Guide, Release 6.03 Edition*. Cary NC, SAS Institute Inc.

Small, I. (1987) The performance of a small catch-and-release fishery stocked with rainbow trout, *Salmo gairdneri* L. Richardson, and thoughts on stocking densities. *Aquacult. Fish. Mgmt.* **18**: 291–308.

Small, I. and Downham, D.Y. (1985) The interpretation of anglers' records (trout and sea trout, *Salmo trutta* L., and salmon, *Salmo salar* L.). *Aquacult. Fish. Mgmt.* **16**: 151–170.

Stephens, R.T.T. (1989) Flow management in the Tongariro River. *Dept. Conservation Sci. Rsch. Ser.* No. 16. 115 p.

Von Geldern, C.E. and Tomlinson, P.K. (1973) On the analysis of angler catch rate data from warmwater reservoirs. *Calif. Fish & Game* **59**: 281–292.

Zar, J.H. (1984) *Biostatistical Analysis*. Englewood Cliffs: Prentice-Hall International. 718 p.

Chapter 6

The relationship between catch, effort and stock size in put-and-take trout fisheries, its variability and application to management

M.G. PAWSON *Ministry of Agriculture, Fisheries and Food, Fisheries Laboratory, Lowestoft, NR33 OHT, UK*

The most important features of a put-and-take fishery are the daily catch rates and the relationship between them and stock levels. The latter must be maintained to provide the catches to satisfy customers, and this has to be balanced against the costs of stock fish and management. The manager's aim is to provide satisfactory sport with good quality fish at the right price.

This paper reviews catch and effort statistics in small trout fisheries for which comprehensive records are available and where accurate estimates of stock size have been possible. The relationship between catch per unit effort and stock size − the trouts' catchability − and the influence of fish naivety, trout strain, species and pre-stocking husbandry and angler behaviour are examined. It is concluded that the characteristic catchability of trout in a put-and-take fishery can be modified by the choice of stock fish, genetic strain and species, and that stock number is the principal determinant of overall catch rates. Because there is a tendency for the distribution of catches to be highly skewed towards the most skilful anglers, it is suggested that regulations controlling the conduct of fishing, particularly those aimed at spreading the catch across the majority of participating anglers, are the most effective means of stock management.

6.1 Introduction

To the angler, the number of fish caught is the test of a fishery's worth; to the owner, caught fish cost money and have to be replaced. To the scientist, however, catch statistics provide an insight to the behaviour of fish and of anglers, and their analysis and interpretation should enable better advice to be given for fishery management. Artificial though they may be, put-and-take trout fisheries represent an experimental device with which to investigate the relationship between catch, fishing effort and stock size, and this paper presents the main findings of a 10-year study in this field and their implications for management.

6.2　Materials and methods

The investigations were conducted in three small put-and-take trout fisheries in eastern England, chosen partly because good catch and effort data were available from daily record sheets, but also in view of the managements' willingness to subject their fisheries to scrutiny and experimentation. Padley Water (approximately 1 ha) near Woodbridge, Suffolk, was used to develop and test a model of the relationship between catch, effort and stock levels in 1979 (Pawson, 1982); Loompit (12.5 ha) near Trimley, Suffolk, was the scene of a long-term investigation of this relationship for rainbow trout during the 1980–83 fishing seasons (Pawson, 1986), and on the influence of rainbow trout strain (in 1985, Pawson & Purdom, 1987) and pre-stocking husbandry (in 1987, Pawson & Purdom, 1991). Whittle Lake (4 ha) near Sawston, Cambridgeshire, was used to compare the performance of brown and rainbow trout through the period 1981–89 (Pawson, 1991). All these waters were near sea level, had a productive, alkaline environment, and were run as fly-only fisheries for a private membership.

In each case the fishery was operated on a strict put-and-take basis, with all fish caught being killed and reported in the fishery log. A daily limit of four or six fish per rod was imposed. Records of hours fished during individual angler's visits were also available for Padley Water and Loompit. Trout were stocked three or four times during each fishing season, at intervals of between three and seven weeks until mid-August, which in some cases enabled stock numbers to be estimated by the technique of successive removal (Lesley & Davis, 1939). Some control and experimental batches of trout were given distinctive marks on the belly with Alcian blue dye using a Panjet inoculator (Hart & Pitcher, 1969), from which stock estimates could be made by mark and recapture analyses. Two assumptions implicit in both these techniques – that trout mortality in addition to that due to anglers' catches is either negligible or can be determined, and that catchability remains consistent and is similar between marked and unmarked fish – were objects of the studies, and various iterative techniques were used to elucidate their levels and variations.

In these fisheries, unreported mortalities appeared to be a minor source of error in the catch statistics, and could be accommodated by using a weekly instantaneous 'natural' mortality coefficient (M), which was rated to produce stock levels consistent with other criteria. Catchability (q) did vary, but in a predictable manner, and it could be assumed that the relationship between catch per unit effort and stock level of a particular species was reasonably constant under a given set of circumstances. These involve the choice of stocked fish – size, age, strain, source and pre-stocking husbandry – and the quality of fishing effort directed at them, both in terms of the skill of the anglers and their attitude towards catching trout, factors that will be explored later in this paper.

Once stock levels in the fishery (N_t) had been estimated with some confidence throughout the period of the respective study, weekly catch (C_t) and effort (f_t) data were used to determine the corresponding coefficient of instantaneous fishing mortality (F_t) and to investigate the relationship between catch, effort and stock. The principal equations are:

$$N_{t+1} = N_t \times e^{-(F+M)} t$$

or

$$\ln N_t \times (N_{t+1})^{-1} = F_t + M_t \tag{1}$$

which describe an exponential decline in stock numbers with fishing effort and time:

$$F_t = q_t \times f_t \tag{2}$$

that is, catchability is the relationship between fishing effort and the resultant mortality and:

$$C_t \times f_t^{-1} \sim q_t \times N_t \tag{3}$$

which shows catch per unit effort as an index of stock sizes, given that catchability is constant.

It follows that the catchability of a stock of fish is the proportion of that stock caught by one unit of fishing effort.

6.3   Results

6.3.1   *Is recorded catch (C_t) the same as loss from the stock due to fishing mortality?*

There is probably little reason to doubt the veracity of anglers' catch data in the fisheries studied. Fishing time during the season was not limited and anglers' subscriptions were perceived to be more a payment for an exclusive right to fish in pleasant surroundings rather than to take a limit catch of trout. Given that recapture levels of rainbow trout in particular could exceed 90% of the fish stocked, in waters in which some additional removal of trout by poaching anglers and predatory birds (heron, cormorant, osprey) was known to occur, it is suggested that these catch records do accurately represent the impact of anglers' effort on fish stocks.

The use of an additional mortality coefficient (M) to account for unreported losses from the stock enables fish numbers to be estimated with some confidence, though assumptions have to be made about level and consistency of the value of M during and between fishing seasons. Where trout stock levels can be determined directly, as when Loompit was drained before and at the end of the 1980−83 study period, this approach has been seen to be valid. In situations where this is not possible, the mark and recapture technique can be used to estimate stock size, though the prime requisite, that marked fish are no more nor less catchable than the unmarked population, requires testing. Alternatively, a more empirical approach, based on the assumption that with a regular stocking policy and steady angling levels and management regime, the stock levels of trout and their catch-ability will display a marked consistency, has also been used. Figure 6.1 provides evidence of the validity of this approach, though it should be noted that the unseasonal increase in catchability during July and August 1983 may have been due to an intensification of effort by anglers in order to fish the stock down prior to draining the fishery in mid-August that year.

6.3.2   *The measurement of anglers' fishing effort*

It soon becomes apparent that angling effort, quantified simply as rod hours or angler days, is not always a good measure of fish catching capacity, partly because it can be difficult to determine just how many anglers fished and for what time on a given day (in Whittle Lake unsuccessful anglers were not required to record the fact!), but chiefly because their individual skill level can vary enormously (Fig. 6.2). As a consequence, it has been found that where daily catch limits are imposed the most successful anglers spent less time fishing than did the lesser mortals. However, by aggregating effort over periods of a week or multiples thereof, it can be assumed that similar qualitative arrays of effort are being sampled. It is a feature of angling effort in private put-and-take fisheries that the duration of an individual's visits appears to be determined more by a desire to be occupied in fishing, rather than being governed by the need to catch a given number of fish.

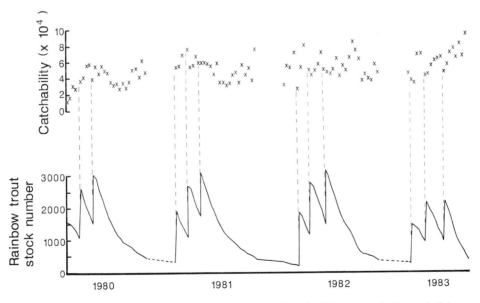

Fig. 6.1 Estimates of stock number and weekly catchability of rainbow trout in Loompit fishery, 1980–83 (after Pawson, 1986)

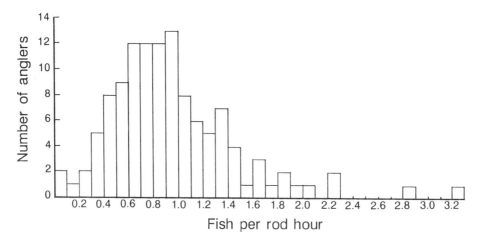

Fig. 6.2   The distribution of mean catch rates recorded for individual anglers who made five or more visits to Loompit fishery during May and June in 1982 and 1983

Figure 6.3 suggests that anglers at Loompit preferred a fishing session to last for around three hours, and as a consequence, when trout were easy to catch, may have fished less assiduously in order to avoid an early homeward departure. Similarly, the choice of fishing time during the day is likely to be influenced by factors other than just the best time for catching fish. These considerations might apply otherwise in public fisheries and those operating on a day-ticket basis, where value for money has come to be represented by fish in the bag at the end of the day.

6.3.3   *Catch per effort as an index of stock levels*

The traditional use of catch per unit of effort (CPUE) as a credible index of stock level is based on the assumption that their relationship, i.e. catchability, remains reasonably constant. The main aim of the studies covered by this paper was to investigate the relationship between anglers' catch rates and trout stock density, to evaluate it and to elucidate the causes of its variability. In summary, it has been found that in fisheries where caught fish are not returned to the water, the catchability of rainbow trout remains relatively constant from year to year, though with a seasonal pattern. Under these circumstances, CPUE trends throughout the fishing season (averaged over weekly or longer periods) can provide a useful indication of stock density (Fig. 6.4).

The main factors influencing the level of catchability are the skill (angling ability) and attitude (to taking trout) of the participating anglers, the strain of rainbow trout stocked and to a lesser extent the pre-stocking husbandry of these fish. The consequence is that stocked rainbow trout will be caught in proportion to angling effort, and that mean catch rates (per unit of effort) are chiefly a function of stock levels and can be manipulated by stock adjustments, at least

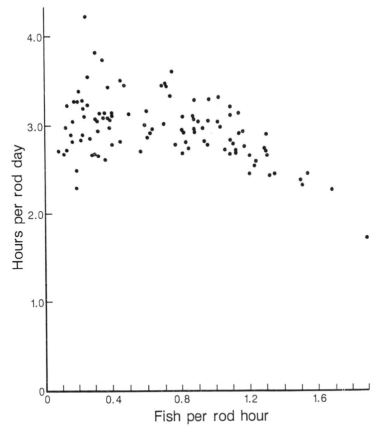

Fig. 6.3 The relationship between catch per unit of effort and the weekly mean duration of individual angler's fishing sessions at Loompit fishery, 1980−83 (from Pawson, 1986)

within the limits required for satisfactory fishery management. This principle has been found to hold true by many of those who run put-and-take fisheries.

Brown trout, on the other hand, do not respond so readily to anglers' efforts to catch them. Figure 6.5 indicates that in Whittle Lake they were consistently less susceptible to fly fishing than rainbow trout − the relationship between brown and rainbow trout fishing mortalities for the same levels of effort was approximately 1:2. Though both species exhibited similar seasonal fluctuations in their catchability, as shown by the characteristic seasonal groupings of fishing mortality values, these were more distinct for brown trout than for rainbows. This implies that to achieve equivalent catch rates of brown and rainbow trout, higher stock levels of the former are required and that catches of brown trout cannot be expected to be sustained to the same degree as rainbows throughout the fishing season. A corollary is that a large proportion of stocked brown trout survive through successive stockings, and that these fish may be much less catchable than newly introduced brown trout − or indeed any rainbows − further reducing the catchability of the brown trout population as a whole. A result is that between season

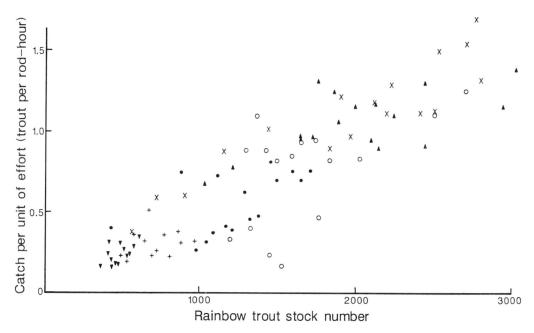

Fig. 6.4 The relationship between rainbow trout stock numbers and the weekly mean catch per unit of effort at Loompit fishery, 1980–83, for each month of the fishing season; ○ = May, ▲ = June, x = July, ● = August, + = September and ▼ = October (after Pawson, 1986)

(over-wintering) stocks of brown trout commonly represent a much higher proportion of the number of fish stocked each year than is found with rainbow trout. In Whittle Lake, these proportions were 52.7% and 17.6% respectively over the years 1984–89.

6.4   Discussion

It is apparent that there is a strong and persistent relationship between the stocking regime in put-and-take fisheries and the catch rates of both brown and rainbow trout realized by a given angling effort. Unfortunately, the high level of catch accountability reported here is seldom possible in commercial put-and-take fisheries that rely chiefly on day-ticket clients, and where a substantial proportion do not make catch returns. In all cases, however, it is probable that the magnitude and frequency of stocking largely determines the level and consistency of the fishery's overall catch, though it is obvious that this is not distributed uniformly among the participating anglers (Cane, 1980; Pawson, 1982).

The main cause for concern, therefore, is not the cost of stocking, which should be reflected in the price of fishing, but the influence that variation in the quality of individual angler's fishing effort has on the way in which the fishery's catch is taken. In short, modestly skilful anglers require much higher stock levels to achieve the average catch rate (two to four fish per rod day, say) than do the

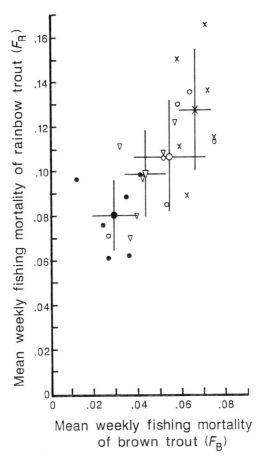

Fig. 6.5 The relationship between mean weekly fishing mortalities of brown and rainbow trout at Whittle Lake, 1984–89, for each inter-stocking period; x = April to mid-May, ○ = mid-May to mid-June, ● = mid-June to early August and ▽ = mid-August to October. Means and two standard deviations are given for each of these periods

expert minority who, if not restricted by catch limit regulations, will be expected to catch many more fish for the same amount of angling effort. This might not be a problem in syndicate or club waters, where weekly or monthly quotas of up to four times the daily bag limit can be used to even out individual angler's share of the annual catch, but is likely to be much more important where day-tickets are the norm.

It is suggested, therefore, that stocking policy (numbers, frequency, species and strain of trout introduced) can be used to regulate average catch rates which, however, must not be so low that the less skilful majority catch fish so infrequently that they cease to patronise the fishery. Whilst technical regulations controlling the anglers' conduct (fly only, no night fishing, etc.) might also be used for the same purpose, the most important restriction on anglers is the bag limit. Daily catch limits should be high enough to enable those anglers that are successful on the day to take home a satisfactory bag of fish, but must effectively avoid excessive plunder of the fishery's stocks. It follows that the practice of selling repeat day-tickets when trout are particularly easy to catch, or to the most successful anglers, will result in an increase in the mean catch per angler-day, which, through good public relations, militates against sound economic management.

On a more aesthetic note, the choice of stock fish, species and strain, is largely a matter of supply and economics, though it should be noted that different strains of rainbow, and presumably brown, trout might prove to be more or less catchable in a particular fishery, and stocking will have to be adjusted accordingly to maintain catches at the required level. Although brown trout are not usually recaptured as rapidly as rainbows, in the absence of non-human predators they persist from season to season and their eventual recapture level can be just as high. All the evidence in these studies suggests that over-wintering mortality of brown and rainbow trout is low, and certainly no higher on a weekly basis than the losses to the stock that go unrecorded during the fishing season.

References

Cane, A. (1980) The use of anglers' returns in the estimation of fishing success. *Fisheries Management* **11**: 145–156.

Hart, P.J.B. and Pitcher, T.J. (1969) Field trials of fish marking using a jet inoculater. *Journal of Fish Biology* **1**: 383–385.

Leslie, P.H. and Davis, D.H.S. (1939) An attempt to determine the absolute number of rats in a given area. *Journal of Animal Ecology* **8**: 94–113.

Pawson, M.G. (1982) Recapture rates of trout in a 'put-and-take' fishery: analysis and management implications. *Fisheries Management* **13**: 19–31.

Pawson, M.G. (1986) Performance of rainbow trout, *Salmo gairdneri* L. Richardson, in a put-and-take fishery, and influence of anglers' behaviour on catchability. *Aquaculture and Fisheries Management* **17**: 59–73.

Pawson, M.G. and Purdom, C.E. (1987) Relative catchability and performance of three strains of rainbow trout, *Salmo gairdneri* L. Richardson, in a small fishery. *Aquaculture and Fisheries Management* **18**: 173–186.

Pawson, M.G. and Purdom, C.E. (1991) The influence of pre-stocking feeding levels on catchability of rainbow trout (*Oncorhynchus myskiss* Walbaum) in a small fishery. *Aquaculture and Fisheries Management* **22**: 105–11.

Pawson, M.G. (1991) A comparison of the performance of brown trout (*Salmo trutta* L.) and rainbow trout (*Oncorhynchus myskiss* Walbaum) in a put-and-take fishery. Submitted to *Aquaculture and Fisheries Management* **22**: 241–57.

Chapter 7
Exploring data provided by angling for salmonids in the British Isles

I. SMALL *Dept. of Environmental and Evolutionary Biology, P.O. Box 147, University of Liverpool, L69 3BX, UK*

Some analytical methods which have evolved during a study of the data recorded by angling for migratory fish, and trout in still-waters, in Ireland and Great Britain are explored.

They include the estimation of total rod catch (C) from the records on unprompted returns (Cu) and the number of returns as a proportion (Pu) of licence or permit sales is estimated by:

$$C/Cu = Y/Pu + (1 - Y)$$

Negative binomial distributions were fitted to frequency counts of individual angler's catch per unit effort (CPUE) and their use in estimating total rod effort, or the effects of limiting the catch were examined.

Trends of annual effort in areas where data are incomplete or not available were estimated from records of rod catches and counts of fish, by a non-linear iterative solution to:

$$Catch = K(K_1 + K_2\ Year - K_3\ Year^2)\ Abundance^{\,(b + 1)}$$

7.1 Introduction

Fisheries managers in the British Isles are often hindered in their work by the reluctance of licensed anglers in England, Ireland and Wales, and those issued with permits for Scottish Association waters, to make a complete return on the results of their activities. Table 7.1 indicates, for migratory fish, typical proportions of licence or permit sales, unprompted returns made and total fish recorded on the returns and it can be seen from this that the problem can be acute in some areas. On still-waters the proportion of returns made can vary from 0.05 on a remote upland loch to more than 0.95 on a well controlled small lake.

In Scotland there is the additional problem of estimating the total angling effort on individual rivers containing migratory fish.

7.2 Estimating the total catch

To improve the estimates of the total rod catch, several authorities have taken action and reminders were issued to defaulting anglers by the South West Water

Table 7.1 Migratory Fish. British Isles − proportions of licence or permit sales, returns made, and total catch made on returns for migratory trout (MT) and salmon (Sal) in the period 1980−85

Licence type		SW Eng	Wales	NW Eng	Ireland	Scotland	
Licence or permit sales by type and region							
Season	Sal	0.48	0.63	0.22	0.46	Scots	0.30
	MT			0.20			
Part Seas.				0.31	0.16		
14d or 7d	Sal	0.14	0.13	0.17	0.38	Visitors	0.70
	MT			0.10			
Day	Sal	0.38	0.24				
Unprompted returns made/licence or permit sales							
Season	Sal	0.59	0.34	0.36	<0.1	Scots	0.1−0.5
	MT			0.08			
Part Seas.				0.21	<0.1		
14d or 7d	Sal	0.59	0.33	0.14	<0.2	Visitors	0.1−0.5
	MT			0.10			
Day	Sal	0.45	0.25				
Catch recorded on returns/total catch							
Season	Sal	0.92	0.94	0.53		Not available	
	MT			0.10	0.85		
Part Seas.				0.33			
14d or 7d	Sal	0.03	0.04	0.01	0.15	Not available	
Day	Sal	0.05	0.02				

Authority (SWWA) (1969−81), Foyle Fisheries Commissioners (FFC) (1972−78) and Welsh Water Authority (WWA) (from 1976). Other bodies take similar action but their records do not contain sufficient detail for analysis. The data obtained and derivations are laid out in Table 7.2.

Table 7.2 Catch data estimates

	Total	Unprompted	Prompted	Missing
Number of anglers reporting	A	Au	Ap	Am
Catch recorded or estimated	C	Cu	Cp	Cm
Proportion of effort	1.0	Pu	Pp	Pm
Calculated mean CPUE	M	Mu	Mp	Mm

As it is convenient to estimate the total catch (C) from the unprompted returns (Cu) it can be shown (Small & Downham, 1985) that:

$$C/Cu = 1/Pu \times M/Mu \tag{1}$$

Analyses of the numerical results by WWA (1981) and Small and Downham indicate that Mu is usually greater than Mp and Mm is likely to be a relatively small value. Second reminders issued by SWWA (1977) and WWA (1983) confirmed that the missing catches could be estimated as relatively small numbers for these regions; about 5% of the total unprompted and prompted catch for salmon and less than 15% for sea trout.

This difference in catch rates was confirmed on still-waters (Cane, 1980; Small, 1982, 1983) where the number of returns volunteered by anglers increased directly in proportion to their success. A guide for consideration is:

$$Pu = Constant + 0.3 \; Relative \; mean \; catch \; rate. \; r = 0.8 \qquad (2)$$

Small and Downham (1985, Fig. 7.1) found that:

$$M/Mu = a + b \; Pu \pm c \; Pu^2 \qquad (3)$$

Small (1988a) simplified this to:

$$M/Mu = Y + (1 - Y) \times Pu \qquad (4)$$

where Y is the intercept on the M/Mu axis. Hence from Equation 1 (see Fig. 7.2):

$$C/Cu = Y/Pu + (1 - Y) \qquad (5)$$

The values of Y, derived from the linear regressions for all the migratory fish data and those from the ranked results of multi-team still-water trout competitions, available to date, are given in Table 7.3. They suggest that the following values of Y can be recommended for use in assessing trends of catches over a number of years:

All trout: 0.5 ± 0.18

Salmon: 0.3 ± 0.08

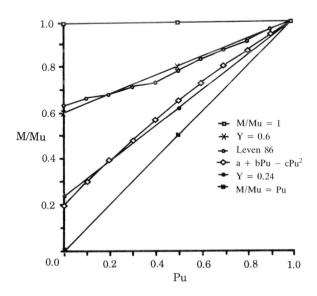

Fig. 7.1 Range of regression lines for various solutions of M/Mu against Pu

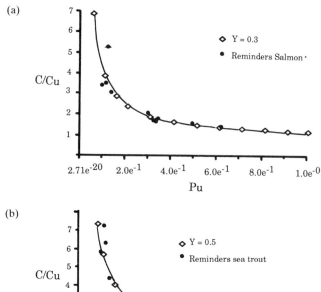

Fig. 7.2 Plots of *C/Cu* against *Pu* for salmon (a) and trout (b)

If more accurate estimates are required for particular waters effort should be made to establish a value of *Y* appropriate to local conditions, by issuing reminders or arranging a competition between at least 20 teams each made up of four or more anglers.

Season licence or permit holders usually catch 70–90% of the total migratory fish and the number of short-period visitors should either be converted to equivalent season permits or their catch estimated separately. Note that if M/Mu approaches 1.0 all anglers or teams tend to catch the same number of fish, or have a very high mean CPUE. When all successful anglers only make perfect returns, $M/Mu = Pu$ and relates to a frequency distribution truncated at zero.

7.3   Estimating the total effort from frequency counts

Three hundred sets of frequency counts of individual angler's catch per licence, permit or rod day etc. have been collated. Investigation shows that they are usually not significantly different from the negative binomial distribution (Small & Downhan, 1985). This enables a solution to the problem for fishery managers which occurs when they have a good record of the successful anglers' catches but

Table 7.3 Observed values of *Y* obtained from linear regressions

	Multi-team still-water competitions Effort: 10–20 teams, 60 to >200 rod days				Rivers–migratory fish 1000 to 30000 rod days			
Year	Rainbow	Trout	Brown	Trout	Period	Data Set	Sea Trout	Salmon
1983	Rutland B&H	0.53			1969–81	SWWA Seas	0.47	0.33
1984	Rutland B&H	0.61			1977	SWWA 2nd Rem	0.45	0.33
1985	Bewl Br	0.40			1976–89	Welsh W Seas	0.40	0.26
1986	Grafham	0.51	Leven B&H	0.60	1983	Welsh 2nd Rem	0.41	0.35
			Leven B&H	0.40	1976–90	Welsh all licences	0.33	0.22
			Kielder McE	0.47	1972–78	Foyle FC Seas	0.51	0.30
1987	Draycote Skp	0.53	Kielder McE	0.59				
	World	0.34	Leven B&H	0.45		Overall Seas + Conc Lic	0.45	0.35
1988	Grafham B&H	0.56	Leven B&H	0.48				
	Rutland B&H	0.65	Tasmania	0.35				
	Grafham S're	0.45						
1989	Langford	0.58	Leven S Fin	0.52				
	Rutland Yth	0.14	Leven SANA	0.55				
	Chew Eng Fin	0.35	Leven Int	0.52				
	Grafham B&H	0.41	Fitty Yth	0.45				
Mean		0.48		0.49			0.44	0.31
Sd		0.13		0.07			0.05	0.05
t-test			0.29			0.92	4.2	

uncertain or no figures for the number of blank periods, i.e. a truncated distribution. The total fishing effort can be estimated by arranging the known data in the form of a frequency count, i.e. the number of anglers catching 1, 2, 3, 4, etc. fish in unit time.

An example is shown in Table 7.4 for a brown trout loch in Galloway, Scotland.

Table 7.4 Estimating the total effort from frequency counts

Number of fish caught per day	0	1	2	3	4	5	Total caught 57
Number of anglers	?	19	8	3	2	1	Total successful anglers 33

A computer program based on an iterative maximum likelihood method (available on request) estimates the number of anglers who had blank days, and when the data above are processed, this is shown to be about 67, i.e. the total effort was about 100 angler days. This brings the catch rate based on the returns only, 1.73 (σ 1.14), down to the figure of 0.57 (σ 1.09) per rod-day. The program can also estimate the consequences of imposing a bag limit, i.e. a censored distribution (Crisp & Mann, 1977).

A feature of the negative binomial distribution is that the variance is greater than, and related to, the mean. Table 7.5 indicates possible relationships between the two parameters for various types of fishing. As published frequency counts are rare it is commonly found in fishery reports that only the mean value of CPUE has been recorded and not the variance. If it is desired to estimate the appropriate distribution the method of calculation described in Small and Downham (1985) is recommended.

7.4   Trends and angling effort

The Salmon Fishery Act 1865 (Harris, 1986) introduced licensing to England and Wales and, as a consequence, a measure of the effort put into angling for

Table 7.5 Frequency distributions of anglers' CPUE according to relationship *Variance* = $A * Mean^b$

Species	Data set	No.	Range of Mean	A	B	r
Salmon	Season	60	0.5−9.0	4	1.25	0.92
	Week + Day	36	0.2−2.0	2	1.0	0.97
Seatrout	Season	60	1.0−15	8	1.5	0.90
	Week	11	0.2−1.5	5	1.0	0.70
	Day	25	0.2−2.5	2.5	1.0	0.80
Trout	Still-waters					
Rainbow	Day	90	0.3−8.0	2	1.2	0.92
Brown	Day	22	0.5−3.5	1.5	1.0	0.85
Grayling	Day	5	1−15	4−6	1.0	NA

migratory fish. By combining the records in Grimble (1913), county archives and recent annual reports by Water Authorities, it is possible to plot trends of effort since about 1867, e.g. for the Wye and Welsh Dee (Fig. 7.3); similar plots from 1950 may be found for other major salmon rivers. Two main trends are evident − the slow increase to about 1950 and the rapid, almost linear, increase to about 1980, since when there has been some uncertainty in the trend.

The associated decline of CP season licence since about 1920, when catch records commenced, is indicated in Fig. 7.4. This inverse relationship has also been found on the Foyle since 1952 and may be worth further investigation.

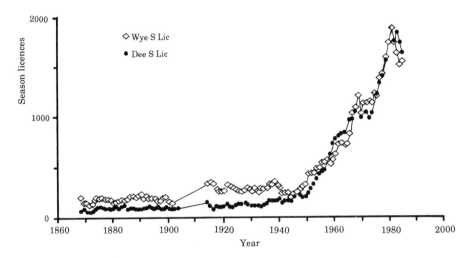

Fig. 7.3 Trends in rod effort for the Rivers Wye and Dee

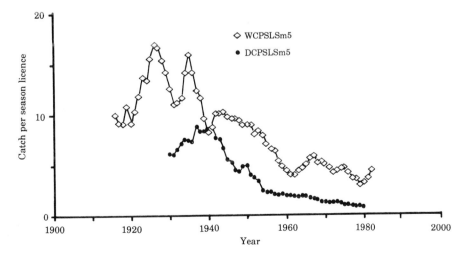

Fig. 7.4 Rod catch per season licence for the Rivers Wye and Dee

In areas such as Scotland when published records of effort are rare, fragmented or only available for recent years, it has been demonstrated (Gardiner, 1991) that it may be possible to estimate the trend of rod effort on migratory fish rivers, by applying non-linear iterative least squares regression methods to records of rod catch and counts for fish, where these are available for about 15 years or more, to solve the fisheries equation (Peterman & Steer, 1981):

$$Rod \ catch = K \times Effort \times Abundance^{(b + 1)} \tag{6}$$

where

$$Effort = K_1 + (K_2 \ Year) + (K_3 \ Year^2) \tag{7}$$

and K_i are constants. Any model, linear, parabolic or cubic, can be tried in equation 7, but a parabolic plot seems the most useful. If a non-linear regression programme is not available, an indication of the trend may be obtained by repetitive use of a linear polynomial regression, with $(b + 1)$ ranging in steps of 0.1 from -0.2 to 1.2, to solve equation 8 for the optimum correlation:

$$Rod \ catch / Abundance^{(b + 1)} = k \times Effort \tag{8}$$

where $K = 1$. By assembling SAS (Statistical Analysis System) procedures the method has been tested on known 30-year records of licence sales or lettings for the Foyle and Conon (Small, 1988b, 1989) and shorter periods for the Coquet, Welsh Dee, Tummel, Bush and L. Feeagh. Its predictions of the trends of day permits sales seem close but it also highlights the uncertainty of determining actual effort when a season licence is used as a nominal unit of effort. The computed trend may be markedly different from that indicated by licence sales. The present exercise in Wales, where anglers have been asked to keep logbooks recording actual days and hours of fishing throughout the season, may resolve this problem.

7.5   Relation of CPUE and catchability coefficients to abundance

CPUE and catchability coefficients (q) have been related to abundance (N) by Peterman and Steer (1981) according to:

$$CPUE = k \times N^{(b + 1)} \tag{9}$$

and

$$q = k \times N^b \tag{10}$$

Processing such data as are available by the non-linear routine has enabled exploration of these relationships for a number of waters. Preliminary estimates of the parameters are shown in Table 7.6 together with some previously published values for still-waters (Small, 1987). Note that in theory $(b + 1)$ and (b) for each water, should differ by 1.0; for the table the exponents were determined by separate non-linear procedures.

Table 7.6 Exploratory estimates of the parameters relating angling CPUE and catchability coefficients to abundance, (values derived from annual totals and means by non-linear least squares), where $CPUE = K \times abundance^{(b+1)}$ and the catchability coefficient is calculated from catch/effort/abundance $= K \times abundance^b$. (S indicates significance at the 5% level)

Water	Years	Nominal unit of effort	k^4	$(b+1)$	b	SE of b	r (CPUE)
ATLANTIC SALMON							
R Foyle[2]	1952–87	Seas Lic	0.025	0.52	−0.49	0.15	0.51 S
		Variable	0.010	0.47	−0.29	0.13	0.60 S
R Faughan[123]	1963–87	Seas Lic	0.004	0.60	−0.06	0.16	0.64 S
		Variable	0.003	0.65	0.08	0.19	0.64 S
R Coquet[123]	1957–88	Seas Lic	0.004	0.63	−0.29	0.13	0.70 S
		Variable	0.008	0.75	−0.38	0.08	0.84 S
Welsh Dee[2]	1975–88	Seas Lic	0.021	0.46	−0.70	0.20	0.53 S
		Variable	0.035	0.61	−0.81	0.17	0.65 S
R Bush	1972–88	Rod day	0.005	0.52	−0.45	0.25	0.53 S
R Conon	1955–87	Rod day	0.030	0.33	−0.69	0.10	0.56 S
R Mourne[12]	1963–87	Rod day	0.003	0.64	−0.26	0.20	0.87 S
R Tay[1]	1952–85	Rod day	0.002	0.52	−0.56	0.09	0.70 S
R Tummel	1975–88	Rod day	0.012	0.38	−0.52	0.18	0.50 S
SEATROUT							
R Coquet[123]	1957–88	Seas Lic	0.004	0.52	−0.43	0.12	0.59 S
		Sal + ST	0.003	0.65	−0.35	0.13	0.55 S
L Feeagh[3]	1970–84	Boat day	0.029	0.48	−0.25	0.15	0.64 S
RAINBOW TROUT Still-waters (Linear regression)							
6 Lakes in UK Weekly Means	1982–86	Full day	0.860	0.34	−0.66	0.10	0.96 S
Toft Newton	1978	Half day	0.002	0.80	−0.20	na	
Loompit	1980–83	Half day	0.0014	1.00	0.00	−	0.84 S
Hurleston	1985	Half day	6.00	−0.15	−1.15	0.26	0.31 S
Hall	1986	Half day	0.05	0.59	−0.45	0.17	0.23 NS
	1987	Half day	0.35	0.29	−0.70	0.13	0.32 S

Notes:
(1) Based on trends of effort tied to benchmark values.
(2) Rod catches adjusted to allow for variation in the annual proportion of returns or validated by bailiff's counts.
(3) Count at tidehead.
(4) Std Error of k is high and approx. = mean value.

Variable = variable proportion of regional season licence holders fished this water or variable number of days fished per season licence.

7.6 Catch per unit effort

It will be noted that for annual data the relation with abundance is a positive and significant power law, with the exponent $(b+1)$ lying in the range 0.3 to 0.8. On most of the migratory fish rivers $(b+1)$ tends towards the square root, while one river and the records for rainbow trout point to the cube root. Two sets of weekly data for still-waters are close to unity. In future it may be possible to devise a classification for waters based on either the component $(b+1)$ for CPUE or (b) for catchability.

7.7   Catchability (q)

For catchability coefficients, calculated as the proportion of the count or population caught per unit effort, the relation with abundance is negative, the range of (b) being -0.8 to zero. This inverse relationship agrees with reports on several marine and anadromous fish (Crecco & Overholtz, 1990).

It is interesting that in only one salmon water, the River Faughan, and for weekly data from two still-water fisheries, Loompit and Toft Newton, where (b) is not significantly different from zero, q appears to have a near constant value indicating that CPUE has a linear relation to abundance.

Peterman and Steer (1981) and Mills *et al.* (1986) have discussed the management implications of the high values of catchability at low abundance of migratory fish.

7.8   Discussion

The method for adjustment of returns was developed at the request of North West Water because their proportion of unprompted anglers, including short period licences, had dropped to less than 20%, although for season licences it turned out to be about 40%. It has also been found useful when analysing data from waters such as the Foyle where *Pu* has varied over the years from 30% through 70% down to 5%. The method puts all years on a comparable basis, eliminating one source of variance. However, it does entail keeping records for each class of licence and processing them separately, an extra administrative burden. While the recommended values of *Y* will produce comparability between years, further work on validation for particular waters is necessary if there is an urge to obtain estimates of absolute values of rod catches.

Sorting records to obtain frequency counts is tedious if done manually but simple if done via computerized databanks. As the counts provide much useful information, publication should be encouraged. Programs for fitting negative binomial distributions in addition to estimating total effort, will indicate the possible effects of imposing limit bags on angling for migratory fish, a subject of topical interest.

The power law relation of annual CPUE to abundance of migratory fish (also rainbow trout in still-waters) which seems to be indicated by the results in Table 7.6, will lead to a better understanding of the effects of the cessation of netting on recreational high value angling. Average rod catch rates would seem to be proportioned to somewhere between the cube and square roots of the number of fish entering a river. To double the rod catch rate therefore requires the run to increase by a factor lying between 8 and 4.

Acknowledgements

The author is indebted to the many organizations which provided their records for analyses, to many fishery officers and colleagues at the University, for guidance and assistance in interpreting the results, and to the referee for helpful criticism of the manuscript.

References

Annual reports of the Foyle Fisheries Commission, DANI R. Bush Project, Northumbrian Water, Salmon Res. Trust of Ireland, South West Water, N. of Scotland Hydro Electric Board, Welsh Water, Pitlochry Angling Club and River Tay District Salmon Fisheries Board.

Anon (1983 to date) Reports on competition results. Trout Fisherman. Sept. to Dec.

Cane, A. (1980) The use of anglers' returns in the estimation of fishing success. *Fisheries Management* **4**: 145.

Crecco, V. and Overholtz, W.J. (1990) Causes of density-dependent catchability for Georges Bank Haddock, *Melanogrammus aeglefinus*. *Can. J. Aquat. Sci.* **47**: 385–394.

Crisp, D.T. and Mann, R.H.K. (1977) Analysis of fishery records from Cow Green reservoir, Upper Teesdale, 1971–75. *Fisheries Management* **8**: 23–34.

Gardiner, R. (1991) Modelling rod effort trends from catch and abundance data. In *Catch Effort Sampling Strategies: their Application in Freshwater Fisheries Management* (Ed. by I.G. Cowx). Oxford: Fishing News Books, Blackwell Scientific Publications Ltd.

Gee, A.S. and Milner, N.J. (1980) Analysis of 70-year catch statistics for Atlantic salmon (*Salmo salar*) in the River Wye and implications for management of stocks. *J. Applied Ecology* **17**: 41–58.

Grimble, A (1913) *The salmon rivers of Scotland, Ireland, England and Wales*. London: Kegan Paul, 3 vols.

Harris, G.S. (1986) The status of exploitation of salmon in England and Wales. In *Atlantic Salmon: Planning for the Future* (Ed. by D.H. Mills and D.J. Piggins). London: Croom Helm.

Mills, C.P.R., Mahon, G.A.T. and Piggins, D.J. (1986) Influence of stock levels, fishing effort and environmental factors on anglers' catches of Atlantic salmon, *Salmo salar* L. and sea trout, *Salmo trutta* L. *Aquaculture and Fisheries Management* **17**: 284–297.

Pawson, M.G. (1986) The performance of rainbow trout, *Salmo gairdneri* Rich., in a put-and-take fishery and the influence of anglers' behaviour on catchability. *Aquaculture and Fisheries Management* **17**: 59–73.

Peterman, R.M. and Steer, G.J. (1981) Relation between sport-fishing catchability coefficients and salmon abundance. *Transactions of the American Fisheries Society* **110**: 585–593.

Small, I. (1982) Loch Dee – Galloway. Collation and analyses of fishery records. Report to the Forestry Commission.

Small, I. (1983) Clatworthy reservoir. Analysis of angling records. Note to Wessex Water.

Small, I. (1984) Analysis of anglers' catches of migratory fish 1977, 1980 & 1981. Report to South West Water.

Small, I. and Downham, D.Y. (1985) The interpretation of anglers' records (trout, sea trout and salmon). *Aquaculture and Fisheries Management* **16**: 151.

Small, I. (1987) The performance of a small catch-and-release fishery stocked with rainbow trout, *Salmo gairdneri* R., and thoughts on stocking densities. *Aquaculture and Fisheries Management* **18**: 291.

Small, I. (1988a) Thoughts on anglers' game fishing returns and grading of fisheries for quality. *The Salmon and Trout Magazine* 235.

Small, I. (1988b) River Foyle basin – Analyses of the rod catches of migratory fish from 1952 to 1987 and their relation to abundance. 1987 Annual Report of the Foyle Fisheries Commission.

Small, I. (1988c) Migratory Fish: Some patterns of angling effort since 1867, etc. Internal Note, Freshwater Fisheries Group, University of Liverpool.

Small, I. (1989) R. Conon Scotland. Exploratory analyses of the angling records for Atlantic salmon in period 1953–87, (in preparation).

Welsh Water (1981) Review of commercial fishing – for salmon and sea trout. Welsh Water.

Welsh Water (1985) Welsh salmon and sea trout fisheries. Welsh Water.

Wessex Water (1987) Investigation into the alleged decline of migratory salmonids. Wessex Water.

Chapter 8
Modelling rod effort trends from catch and abundance data

ROSS GARDINER *The Scottish Office, Agriculture and Fisheries Department, Freshwater Fisheries Laboratory, Pitlochry, Perthshire, PH16 5LB, UK*

For many rivers where information on annual rod catches and counts of salmon (*Salmo salar* L.) are available, there are no data on the effort expended in catching the fish. Provided certain conditions are met, however, it may be possible to estimate the likely trend in rod effort over the period from the available data and to establish the likely form of the relationship between catch per unit effort and abundance as indicated by the count of fish. The inclusion of an effort term often results in a marked improvement in the fit of catch to count data. It is hoped that use of the methods developed will supplement information from the few rivers where satisfactory data are available on all three variables − abundance, catch and rod days.

8.1   Introduction

A model by Paloheimo and Dickie (1964) predicted that, for some exploited stocks, catch, fishing effort and population abundance would be related by:

$$catch/effort = k. \times abundance^{(b + 1)}$$

with $(b + 1)$ significantly less than 1. Such a relationship is expected where there is non-random searching by fishermen in response to changes in stock distribution and abundance, as is the case in sport fisheries for migratory salmonids (Peterman & Steer, 1981).

In Scotland, estimates of numbers of migratory fish are available from electronic counters associated with hydro-electric dams on a number of rivers. However, there is frequently no readily available information on the effort expended in the sport fisheries on the rivers to allow a conventional analysis of the relationships between catch, effort and count.

In this paper it is suggested that, where there is data on catch and abundance but effort trends are unknown, it may be possible to search amongst hypothetical effort trends, say of the form:

$$effort = (k_0 + k_1 \times t + \ldots + k_n \times t^n)$$

where t is time in years and $k_0 \ldots k_n$ are constants ($n = 1$ and $n = 2$ which correspond to linear and parabolic models) for good fits of the observed annual catch and abundance data to Paloheimo and Dickie's (1964) relationship. Generally, their constant of proportionality, k, is incorporated into the effort term, which is in arbitrary units.

The main conditions/assumptions are:

(1) That the count of fish represents, or is proportional to, the fish available to the angler. The most favourable situation is where the counter is just downstream of the fisheries. In some situations, where the counter is just upstream of the fisheries, it may be appropriate to adjust the count by adding the number of fish taken by anglers.

(2) That there is no trend in the proportion of anglers reporting their catch or in the relationship between catch, effort and abundance.

(3) That a smooth linear or parabolic trend of effort with time is an appropriate model.

8.2   Methods

8.2.1   *Multiple regression (MR)*

This is used repetitively with trial values of $(b + 1)$ to search the equations for the linear effort model:

$$catch/abundance^{(b + 1)} = k_0 + k_1 \times t$$

and the parabolic effort model

$$catch/abundance^{(b + 1)} = k_0 + k_1 \times t + k_2 \times t^2$$

systematically for a maximized adjusted R^2 value.

This is the preferred method for initial exploration of the data.

8.2.2   *Non-linear regression (NL)*

This model is used to solve the linear effort model

$$catch = (k_0 + k_1 \times t) \times abundance^{(b + 1)}$$

or the parabolic effort model

$$catch = (k_0 + k_1 \times t + k_2 \times t^2) \times abundance^{(b + 1)}$$

for best-fit values of k_0, k_1 (and, if appropriate, k_2) and $(b + 1)$. This is an iterative procedure requiring preliminary estimates for the values of these constants and it is recommended that the multiple regression method is used first to provide these.

This is the preferred method for obtaining the final fit, and it provides a direct estimate of the Standard Error of $(b + 1)$.

8.3   Results

Both methods were found to be straightforward to use with readily available computer packages, such as Statgraphics and SAS. To illustrate the types of results obtained two examples are considered here.

8.3.1   *Example 1: River Tay*

Data on the number of salmon running up the River Tay in the years 1952–88 were obtained from an electronic counter at Pitlochry Dam (source of data: North of Scotland Hydro-Electric Board Annual Reports). Unfortunately, the counts only represent some of the available fish on the River Tay catchment because the counter is located on a main tributary. No attempt was made to convert them into total numbers of available salmon. The annual rod catches for the Tay District 1952–86 were obtained from records held by the Scottish Office, Agriculture and Fisheries Department.

As can be seen in Fig. 8.1., these two independent data sets show a closely similar pattern of spikes and there is a modest correlation between catch and count ($catch = 6.950 \times count^{0.815}$, $R^2 = 0.42$, non-linear regression). The strong spiking in abundance creates favourable conditions for well-defined best-fit

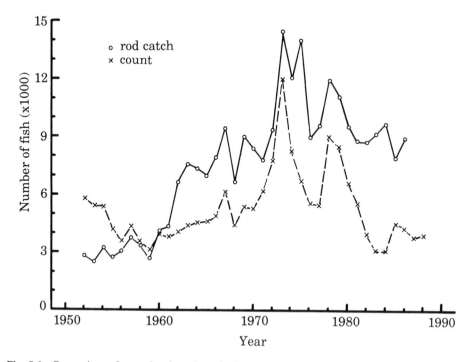

Fig. 8.1 Comparison of annual rod catches of salmon on the River Tay with the annual counts at Pitlochry Dam

solutions for the effort trend and $(b + 1)$ (Table 8.1). The trends in relative effort predicted by the various best-fit models are plotted in Fig. 8.2 and the actual catches are compared with those predicted by the various models in Fig. 8.3. The modelled trends of effort seem generally consistent with local observation of steady increases in available rods and occupancy in the 1950s, 1960s and 1970s with a possible levelling out or drop in the early 1980s, although there is no detailed quantitative information. One-off estimates of 52 500 rod days were made in 1982 (Tourism and Recreation Research Unit, 1984) and 44 000 in 1988 (Mackay Consultants, 1989) but while these two figures may provide information useful in relating the arbitrary units of effort predicted by the models to actual rod days fished they are not directly comparable.

Table 8.1 Best-fit relationships by the various models for the Tay data $(y = year - 1970)$

Model	Best-fit relationship	R^2
MR linear	$catch/count^{0.48} = 126.6 + 3.888y$	0.76
MR parabolic	$catch/count^{0.34} = 474.9 + 11.935y - 0.5421y^2$	0.84
NL linear	$catch = (85.57 + 2.5304y) \times count^{0.526}$	0.82
NL parabolic	$catch = (262.67 + 6.2752y - 0.2961y^2) \times count^{0.409}$	0.88

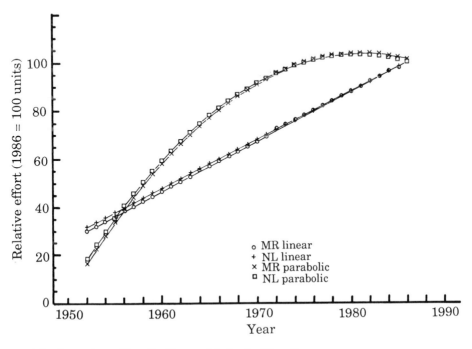

Fig. 8.2 Comparison of best-fit effort models for the River Tay

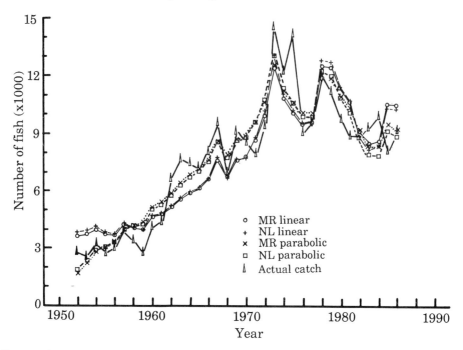

Fig. 8.3 Comparison of the annual rod catches of salmon on the River Tay with those predicted by the various models

8.3.2   *Example 2: River Bush 1972–88*

Data on the number of salmon entering the river each year, the annual rod catch on river and the annual rod days fished were collected from the River Bush Salmon Project Annual Reports. (Source data collated by I. Small, University of Liverpool.)

This is a less favourable set of data for applying the methods but it is possible to get a reasonable fit to the catch-abundance data (Fig. 8.4) over a fairly wide range of combinations of values of $(b+1)$ and of the effort model. The trends in relative effort predicted by the various best-fit models (Table 8.2) are plotted in Fig. 8.5, whilst the actual catches are compared with those predicted by the various models in Fig. 8.6. The availability of data on the actual rod days fished allows the trends in effort predicted by the various models to be tested against this measure of 'true' effort. The results of this are also included in Table 8.2. The parabolic models are particularly good (Fig. 8.7).

8.4   Discussion

The methods have been tested on a number of other data sets where information on effort is available, and generally, as with the Bush data, the predictions of effort for the best-fit model correlate reasonably well with the 'true' effort as recorded by day permit sales (Small, 1991).

Fig. 8.4 Comparison of the annual rod catches of salmon on the River Bush with the numbers of salmon entering the river each year (abundance)

Table 8.2 Best-fit relationships by the various models for the Bush data (y = year−1980; abund = annual count of salmon; r_e = correlation coefficient of relative effort predicted by models against the actual rod days)

Model	Best-fit relationship	R^2	r_e
MR linear	catch/abund$^{0.94}$ = 0.1275 + 0.005595y	0.33	0.53
MR parabolic	catch/abund$^{0.325}$ = 10.88 + 0.7055y + 0.1742y^2	0.5	0.80
NL linear	catch = (0.1277 + 0.005359y) × abund$^{0.940}$	0.68	0.53
NL parabolic	catch = (0.7789 + 0.0378y + 0.00557y^2) × abund$^{0.688}$	0.71	0.82

The linear and parabolic effort models are not the only ones possible. Cubic models of effort ($n = 3$), for example, have also been investigated, but, as they were found to have a tendency to 'overfit' the data, are not considered further in this paper. In some cases partial or anecdotal information on effort may be useful in helping to select an appropriate model.

Use of the methods developed will hopefully supplement information from the few rivers where satisfactory data are available on abundance, catch and rod days to better define the relationships between these variables and so, for example, allow long term trends of abundance to be assessed from information on annual catch and numbers of rod days fished. The methods described have allowed

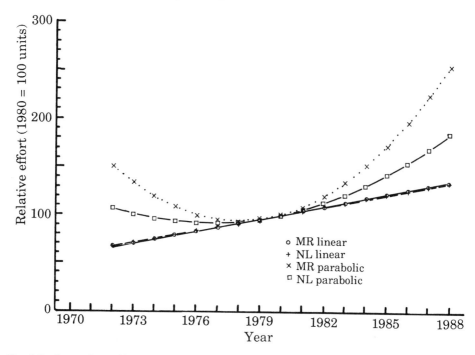

Fig. 8.5 Comparison of best-fit effort models for the River Bush

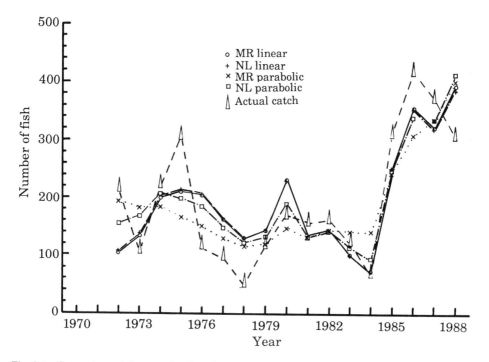

Fig. 8.6 Comparison of the annual rod catches of salmon on the River Bush with those predicted by the various models

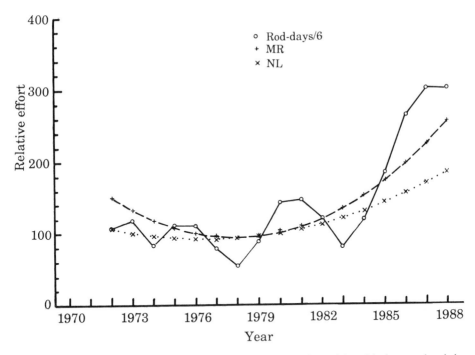

Fig. 8.7 Comparison of predictions of effort, based on parabolic models, with the actual rod days fished each year in the River Bush

progress at an apparent impasse and it is hoped that this type of approach may be useful in tackling other similar problems.

Acknowledgements

The author wishes to thank Ian Small of the University of Liverpool for encouragement and stimulating comment.

References

Mackay Consultants (1989) Economic Importance of salmon fishing and netting in Scotland. Inverness: Mackay Consultants.

Paloheimo, J.E. and Dickie, L.M. (1964) Abundance and fishing success. *Rapp. R-V. Reun. Cons. Int. Explor. Mer.* **155**: 152–163.

Peterman, R.M. and Steer, G.J. (1981) Relation between sport-fishing catchability coefficients and salmon abundance. *Trans. Am. Fish. Soc.* **110**: 585–593.

Small, I. (1991) Exploring data provided by angling for salmonids in the British Isles. In *Catch Effort Sampling Strategies: their Application in Freshwater Fisheries Management* (Ed. by I.G. Cowx). Oxford: Fishing News Books, Blackwell Scientific Publications Ltd.

Tourism and Recreation Research Unit (1984) A study of the economic value of sporting salmon fishing in three areas of Scotland. Edinburgh: University of Edinburgh.

Chapter 9
Carp (*Cyprinus carpio* L.) 'put-and-take' fisheries in the management of angling waters in Czechoslovakia

J. VOSTRADOVSKÝ *Research Institute for Fishery and Hydrobiology, Reservoir and River Research Station in Dol' Libčice nad Vltavou, Czechoslovakia*

Common carp (*Cyprinus carpio* L.) is the main species fished for in anglers' waters in the Czech Republic but little natural breeding occurs. Therefore all stock regeneration and replenishment is by addition of C_2 carp (age 1+). The relationship between the number of fish stocked and their recovery (by angling) was examined in the Lipno reservoir and the Vltava river. The numbers of fish tagged and released into the reservoir and river were 16 600 and 9357 respectively, and the recovery of the tags was 16.57% and 8.3%. The rate of exploitation depended on the number of the fish used for stocking, individual weight, on the numbers of anglers fishing and the time that had elapsed since stocking. The results of the tagging allow a better assessment of the effectiveness of the 'put-and-take' system which has been practised more or less intensively in the Czech Republic for the last 50 years.

9.1   Introduction

Common carp is traditionally the main fish species in Czechoslovakia. As angling developed towards the end of the 19th century, and into the 20th century, carp were taken from ponds to stock rivers and reservoirs. Between 1985–89 the mean annual catch of carp by anglers was 1647 thousand fish (2965 tonnes) representing 64% of the fish caught in Czechoslovakia. Average carp catches were 29 fish (53 kg) per ha in the whole of Czechoslovakia. The value of the carp stocked into the rivers and reservoirs every year is approximately 40 million Kčs, i.e. about 3 million US $ (one third in Slovakia, two thirds in Czech part).

Considerable differences exist in the numbers of fish stocked and caught per unit area. For example, 270 fish ha^{-1} (135 kg ha^{-1}) are stocked into the small but considerably overpopulated (by roach, bream) and overexploited (predators, carp) Hostivař reservoir in Prague (42 ha), whilst the respective figure for the Lipno reservoir (in 1988) was 12 fish ha^{-1} (7.43 kg). The respective catches for Hostivař and Lipno reservoirs are 160 and between 4.5–17.4 kg ha^{-1}. This suggests that there is a direct dependence on the quantity of fish stocked and their exploitation. Similar data are reported from Czech rivers, as far as they are stocked with reared carp.

This high interest in angling for carp, but the low rate of recovery, have given rise to concern about the economic status of 'put-and-take' systems. A study was therefore conducted to improve our understanding of the recapture rates of carp stocked at different sizes and at different places in reservoirs and rivers.

9.2 Materials and methods

The Lipno reservoir on the Vltava river was chosen for extensive stocking with individually tagged carp. This reservoir is located in South Bohemia near the Austrian border (between Linz and Ceské Budejovice). It is 70 m above sea level with a surface area of 4650 ha, maximum depth 20.5 m, average depth 6.6 m, shore length of 118 km, coefficient of shore development R = 4.8, and a water volume of 306×10^6 m^3.

During the study period, the catches of all the fish species were between 20 and 30 kg ha^{-1}. The recapture of the stocked and tagged fish was recorded with respect to the time that had elapsed between stocking, and the dispersion of the fish since introduction. The survival *s*, mean annual mortality *a* and the rate of exploitation *û* were calculated after Ricker (1956). The proportions of tagged fish caught by different methods were also determined.

A total of 16 639 stock carp (mainly one-and-a-half years old) were tagged; most had been reared in ponds (76%). The rest were netted along the shore of the reservoir, particularly from its middle regions. The tags were fixed to the hard ray of the dorsal fin. Stock fish between 161 and 280 mm in size (Table 9.1), i.e. the II and III age groups (Table 9.2) constituted the largest proportion (98%). The reservoir was divided into three zones. The lower part was stocked with 27.8%, the middle with 63.5%, and upper part with 8.7% of the tagged fish (Table 9.3), mostly in the second and third age group (Table 9.4). The experiment lasted five years.

A similar experiment was also conducted on the Vltava river (tributary of the Elbe river). The mean flow in the river in the study site in the Prague region was

Table 9.1 Percentage of stocked and recaptured tagged carp and coefficient of efficiency (Lipno Reservoir)

Size Group (mm)	% of Stocked	% of Recaptured	Coefficient
121–140	0.01	–	–
141–160	0.89	0.61	0.10
161–180	10.07	5.79	0.08
181–200	25.22	17.12	0.10
201–220	24.47	23.85	0.14
221–240	19.95	26.58	0.20
241–260	12.06	17.45	0.21
261–280	6.08	7.05	0.17
281–300	0.91	1.02	0.16
>301	0.31	0.53	0.25
Total number	16 638	2453	0.15

Table 9.2 Age distribution of 16 639 tagged carp

Age Group	%
II	59.07
III	39.52
IV	1.35
V	0.06

Table 9.3 Tag Recovery (%) of the carp stocked in different months of the year

Month	Apr	May	Jun	Jul	Aug	Sept	Oct	Nov
Number	2487	251	451	2442	1116	705	8342	832
Recovery %	16	7	22	18	21	17	17	6

Table 9.4 Number of stocked and recaptured tagged carp according to localities (Lipno Reservoir)

	Lower	Part of Reservoir Middle	Upper
Stocked %	27.85	63.47	8.67
Recaptured %	19.92	63.13	16.94
Recapture from stocked in lower, middle and upper part %	11.95	16.61	32.62

$111-137 \, m^3 s^{-1}$. 8199 carp (mostly one-and-a-half year old), taken from ponds, were tagged and released into the river. In the given part of the river the long-term average anglers' catch of carp was 16% of the number of carp stocked. Most of the tagged carp were within the length range of 241–360 mm (74%) and 141–240 mm (18%).

The majority of the fish (92%) were stocked into the river and reservoir in the autumn (Fig. 9.1) and the rest in spring at 15 different sites. Great publicity was given to the tag return campaign (posters, press, radio, TV, texts on angling licence forms) and premiums were offered for returned tags.

9.3   Results

In the Lipno reservoir, 16.57% of all tagged fish were recovered. A direct relationship was found between the time of stocking and recapture (Fig. 9.1, Table 9.3). The lowest recovery was recorded for fish stocked in November (6%)

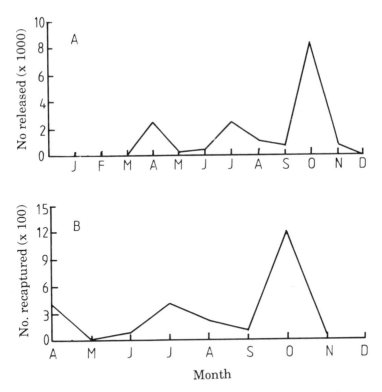

Fig. 9.1 (a) Number of tagged carp stocked in different months and (b) number recovered with respect to month of release

and the highest amongst those stocked in summer (Table 9.3). Professional fishermen and anglers contributed almost equally (55% and 45%, respectively) to the recapture of the tagged fish. The length group 221–240 mm, which comprised 19.9% of the fish at the time of stocking, was predominant amongst the recaptured fish (Table 9.1). The proportion of the length groups 181–200 mm and 201–219 mm were also prominent both in stocked and recaptured fish.

Differences were found between the site of stocking and the site of recapture (Fig. 9.2). The greatest number of recoveries were from fish originally stocked in the upper shallow part (recovery 33%) of the reservoir (Table 9.4). By contrast, the highest number of tagged fish stocked and recovered in the same location was recorded in the lower part of the reservoir (up to a distance of 1 km from the stocking site, Fig. 9.2). The fish released in the upper and middle parts of the reservoir were caught farther from the site of stocking, compared with those released in the lower part (Fig. 9.2). After stocking, 46% of the recaptured fish moved upstream, 38% downstream, and 17% stayed where they were originally stocked. Most of the fish (75%) that had moved upstream were recaptured from within 5 km of the place of stocking and only two were recaptured from a distance

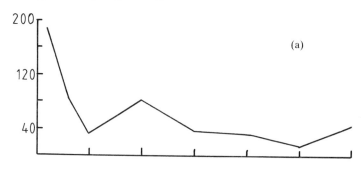

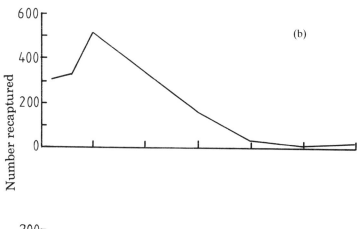

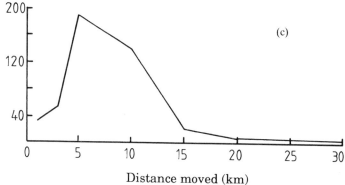

Fig. 9.2 Number of tags recovered from carp according to distance from stocking site in the (a) lower, (b) middle and (c) upper part of the reservoir

greater than 20 km. One carp moved 68.5 km in 92 days. Seven per cent of the tagged fish were recaptured from a distance greater than 15 km (Fig. 9.2).

The time taken between stocking and recovery varied considerably. Recaptures were recorded almost immediately after stocking; 2% of the tags were returned within 24 hours of stocking from very close to the place of stocking. Most tags were returned within one year and only 6% were returned later. (See Fig. 9.3 and Table 9.5)

In the Vltava river a recovery rate of 15% was recorded for fish stocked in spring. This dropped to 3–4% for autumn stocked fish. However, if the fish stocked in the autumn were large, the recovery was higher, 10%.

The exploitation rate, measured by the coefficient $\hat{u}$, was highest in tagged fish that had been stocked in the reservoir in spring and caught in the same year ($\hat{u} = 0.131$). The second year after stocking $\hat{u}$ declined to 0.075 and this trend continued in the subsequent years ($\hat{u} = 0.001 - 0.007$). The annual survival s and mortality rate a ranged between 13–21% and 79–87%, respectively.

9.4   Discussion

Common carp has always been the most popular fish from the anglers' point of view. However, if angling for carp is to be maintained, the waters in the Czech Republic must be regularly stocked, because there is no natural reproduction in running waters or reservoirs. The stocking rates tend to be high but the efficiency of stocking (ratio of number caught to stocked, Table 9.1) is low. In relation to this, the rate of exploitation is highest shortly after stocking when the introduced fish were stocked at the legal size of 35 cm.

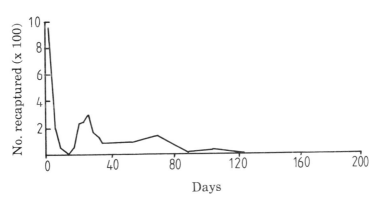

Fig. 9.3 Relationship between the number of tagged carp and their time of recovery after release

Table 9.5 Tag recovery (%) sequence in 30-day intervals (Lipno reservoir)

Interval in days	30	60	90	120	150	180	210	240	270	300	330
Recovery %	34	41	43	44	44	46	55	64	75	81	86

In the Hostivař reservoir the mean catch rate was 168 kg ha^{-1} equivalent to a mean exploitation coefficient $\hat{u}$ of 0.124 (0.37 maximum) (Pivnička & Čihař, 1988). By contrast, in the Lipno reservoir, which is 100 times larger and the fishing effort is assumed to be a few times lower, the exploitation coefficient is about the same at 0.131 (for fish stocked and caught in the same year). Somewhat higher $\hat{u}$ values ($\hat{u} = 0.32$) were reported by Linfield (1988) from England, where the measurements were taken three weeks after stocking.

In both the Lipno (Table 9.1) and Hostivař reservoirs, carp below the legal size (35 cm) prevailed in the stocking material. This was reflected in seine net catches where 96% of the fish were below 35 cm. However, 30% of the 4066 carp caught by anglers were above the legal size. By contrast, 301 fish (64%) of the 470 carp caught by anglers in the Hostivař reservoir were below the legal size (Pivnička & Čihař, 1988). The selective effect of angling appears to be involved here (the size of the bait, etc.). However, in the Lipno reservoir, anglers only record the number of legal-sized fish caught. If they reported the actual number of carp caught this would also show a high proportion of fish below the legal size limit were caught. Thus the low recovery of the stocked fish is partially influenced by the fact that the carp used for stocking are usually smaller than the legal size (Table 9.1).

There is evidence to show that mortality of autumn stocked carp is high because they overwinter under worse conditions than in the pond from where the fish come (Vostradovská, 1975). Hence, it is better to use the spring stocking rather than autumn.

The attempts to improve the results of the put-and-take system of fishing for carp by delaying the fishing after stocking (e.g. by a one-month ban on fishing after stocking) fail to increase the volume of the catch (with the presence of larger fish), indeed the converse was observed (Vostradovský, 1973).

Carp stocking in running waters is liable to the same effects as in the case of reservoirs. Carp are taken from ponds in autumn and are released into the rivers where the number of adverse factors is even higher than in the reservoirs. Again the differences in tagged carp stocked into the Vltava river in autumn and spring were large; 3% of autumn-stocked fish and 15% of spring fish were recovered. Although the method of tagging was the same, the total recovery in the river was only half that in the reservoir. Most fish were found not to migrate great distances, probably because of the presence of weirs and spillways (up to 1.5 m high) which the carp appear to avoid.

References

Linfield, R.S.J. (1988) Catchability and stock density of common carp *Cyprinus carpio* L. in a lake fishery. *Fish. Mgmt.* **11**: 11–22.

Pivnička, K. and Cihař, M. (1988) Analysis of the sport-fishing use of the Hostivař Reservoir in Prague. *Animal Husb.* **31** (10): 938–960.

Ricker, W.E. (1956) Handbook of computation for biological statistic of fish populations. *Bull. Fisheries Research Board of Canada* **119**: 111–127.

Vostradovská, M. (1975) The Use of Fish Tagging for the Evaluation of the Effectiveness of Stock Carp (C_2) Releasing in Dam Lake. *Bull. of VURH Vodnany* **3**: 10–31.

Vostradovský, J. (1973) Put-and-take coarse fish fisheries. Fifth Two Lakes, Fishery Management Training Course. London: Jansson, 48–54.

Vostradovský, J. (1980) Fish Tagging in the Vltava River in Prague. *Animal Husb.* **25** (11): 863–870.

Chapter 10
The recreational fishery in Lake Constance (Bodensee)

H. LÖFFLER *Institute for Lake Research and Fisheries, Environmental Protection Agency, Baden-Württemberg, Untere Seestraße 81, D 7994 Langenargen, Germany*

Since 1983 the border countries of Lake Constance (G: Baden-Württemberg, Bavaria; A: Vorarlberg; CH: St Gallen, Thurgau) have recorded the number of licences issued to recreational fishermen and their annual catches. Trends in the number of licences sold, the importance of angler catches to the total yield (including that of commercial fishermen) of the fishery, the total average catch per angler as well as for the most attractive species (trout (*Salmo trutta f. lacustris* and *Oncorhynchus mykiss*), perch (*Perca fluviatilis*), pike (*Esox lucius*)) in the different countries were examined. Although catch returns are obligatory, only a small proportion of the fishermen responded. A method of improving the catch return information is described.

10.1 Introduction

The collection of catch statistics is a common activity on many waters to regulate management measures. Nevertheless, systematic evaluation is often rather scarce (Staub, 1988). Increasing pressure on inland waters by anglers (Wetzlar, 1988) and a changing awareness for environmental management has demanded the need for well substantiated arguments to support management decisions. This is particularly true for fisheries such as that on Lake Constance which are under multi-national control.

In Lake Constance, the yield of professional fishermen has been recorded systematically since 1914, but catch statistics for the recreational fishermen from all the border countries have only been obligatory for the last seven years (IBK, 1982). This paper compares the angling catch statistics for the different countries around the lake and describes, using an example from one region, Baden-Württemberg, the approach adopted to improve the quality of the data.

10.2 Material and methods

Lake Constance is situated at 47°N, 09°E, north of the Alps. The lake is shared by Austria (country Vorarlberg), Germany (the states of Bavaria and Baden-Württemberg) and Switzerland (the Cantons of St Gallen and Thurgau). It consists of two basins, the Upper Lake (47 600 ha) and the Lower Lake (6300 ha), connected by the River Rhine. The present study refers to the Upper Lake. The lake, and in particular the management situation, has been described in detail by Löffler (1988). In Lake Constance a variety of gears are allowed. Anglers are permitted to use a maximum of two rods at any one time and only two hooks, which must be baited, are allowed per rod. Additional permissible gear includes a small liftnet, bait bottles and a dipnet. A special drop-line with five hooks ('Hegene') is used for catching perch (*Perca fluviatilis*). Trolling is allowed with up to a maximum of eight hooks; double and triple hooks have to be barbless.

According to the laws of one of the border countries, anyone wishing to fish needs a permit. In Baden-Württemberg, for example, the anglers have to pass a public examination to obtain their permit which then allows them to buy an angling licence. In the case of sport fishing the licences are categorized according to the type or place of fishing, i.e. different permits are issued to allow fishing from the shore, from boats and in all or specific areas of the lake.

The number of licences issued and the yield from the recreational fishing are reported annually by the member countries. These figures (Table 10.1) are the basis of this paper.

In the federal state of Baden-Württemberg angling data from Lake Constance has been recorded since 1978 (Löffler, 1983). The purchase of a new licence requires the angler to make a return of the catch record. Unfortunately, effective control is difficult and a number do not respond (Table 10.1). Between 1978−84

Table 10.1 Evaluation of the yield of recreational fishermen in the different border countries of Lake Constance, Bodensee

	Germany		Switzerland		Austria
	Baden-Württemberg	Bavaria	St Gallen	Thurgau	Vorarlberg
Legal base					
civil law	x	x	−	−	−
public law	−	−	x	x	x
Declaration					
catch fishing-day^{-1}	x	−	x	−	−
catch month^{-1}	−	x	−	x	−
catch year^{-1}	−	−	−	−	x
number per species	x	x	x	x	−
weight	x	x	x	x	x
Surveillance					
by fishery warden	(x)	x	x	x	−
Return					
annual licence %	50−70	55−70	98−100	60−70	70−80
monthly licence %	>10	−	−	−	−
Extrapolation	linear since 1985	linear	−	linear	no

these data were evaluated *per se* by an angling association. However, since 1985 the evaluation has been carried out by the Institute for Lake Research and Fisheries at Langenargen. To improve the statistics and estimate the catch of all licence holders, the reported catches are extrapolated to account for non-returns. Within the extrapolation the different types of licences and fish species are treated separately. To achieve this there is a need to compare the returns for the different types of licence and each one was transformed into comparable units, such that three monthly licences or 30 one-day licences corresponded to one annual licence (Wetzlar, 1988). This is particularly important in Austria where a high number of daily licences are issued.

10.3   Results

10.3.1   *Angling yield in the border countries*

The number of angling licences (units) issued in the border countries over the period 1983−88 was relatively stable (8229−8251 (Fig. 10.1)). However, when broken down into the constituent countries, the number issued in Switzerland increased by 55%, whilst in Bavaria and Baden-Württemberg they have decreased by 17.4% and 14%, respectively.

Despite closure of the off-shore fishery (trolling) for three years between 1985−87 to re-establish the population of migrating brown trout (*Salmo trutta fario lacustris* L.) (Ruhlé, 1987) no marked decline in the number of fishermen was evident. Within the period 1983−88 the ratio of recreational anglers to professional fishermen was 20:1 in Switzerland, 30:1 in Baden-Württemberg, 34:1 in Bavaria and 238:1 in Vorarlberg.

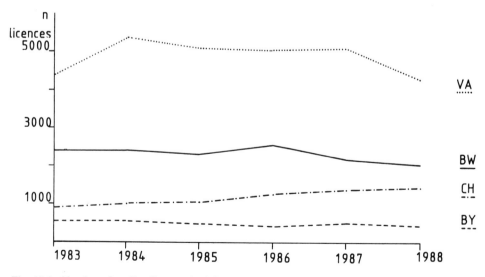

Fig. 10.1 Number of angling licences (units) issued in the border countries of Lake Constance

The catch rate, expressed as catch per unit of effort CPUE (Fig. 10.2a), exhibited more or less the same trend in all countries. It was, however, higher in Switzerland and Bavaria, the two countries with the lowest numbers of licences. Total angling yield was primarily dependent on the perch catches (Fig. 10.2b). This was especially true in Switzerland (82% of the total angling catch) where

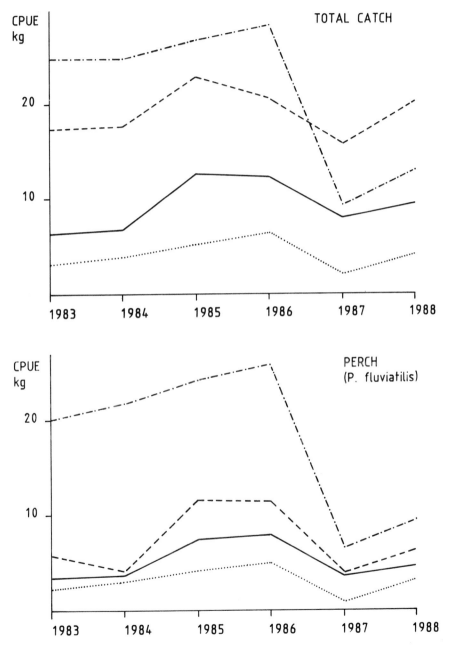

Fig. 10.2 Total CPUE (a) in the recreational fishery of Lake Constance and the contributions by perch (b), pike (c) and trout (d) (for key see Fig. 10.1)

perch is highly desired. CPUE for pike (*Esox lucius*) (Fig. 10.2c) and trout (Fig. 10.2d) rarely exceeded 1 kg per angler day. Since 1984−85 trout catches have been stable despite changes in the closed season and an increase in the minimum size. This was because migrating brown trout and rainbow trout (*Oncorhynchus mykiss*) have not been analysed separately and the majority of trout caught were probably rainbow trout which were excluded from the above restrictions. Other species caught were bream (*Abramis brama*), cyprinids like roach (*Rutilus rutilus*), burbot (*Lota lota*) and eel (*Anguilla anguilla*).

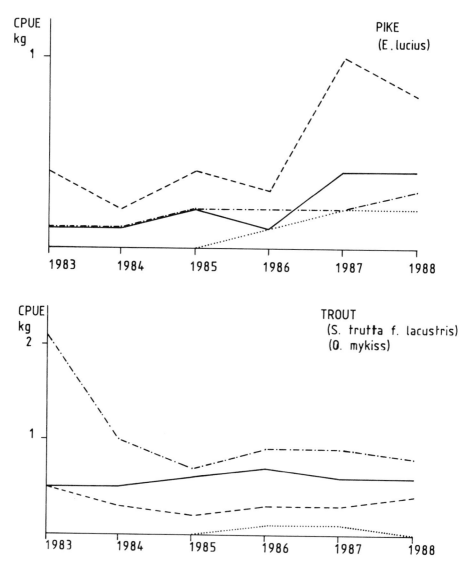

Fig. 10.2 Continued

The mean overall angling yield generally constitutes less than 10% of the total catch (Fig. 10.3), although the percentage is much higher for the more attractive species (e.g. in trout it ranges from 38% in Austria to 50% in Baden-Württemberg and in pike from 25% in Switzerland to 60% in Bavaria).

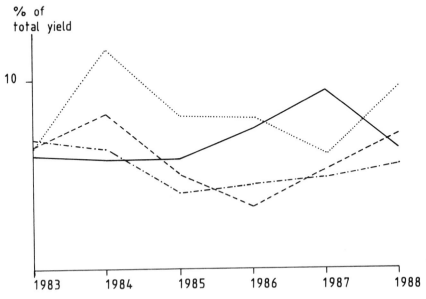

Fig. 10.3 Relation between the recreational fish yield and the commercial yield (for key see Fig. 10.1)

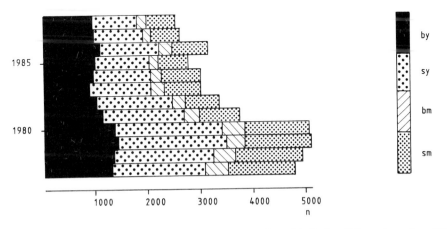

Fig. 10.4 Different types and numbers of angling licences issued in Baden-Württemberg since 1978 (by = yearly boat, sy = yearly shore, bm = monthly boat, sm = monthly shore licences)

10.3.2   *Extrapolation of catch data − example of Baden-Württemberg*

The number of angling licences issued in Baden-Württemberg in 1978 was 4791. By the early 1980s this had declined by 39%. Particularly prominent was the reduction in the number of monthly licences issued (Fig. 10.4). This was related to the new Fishing Act of Baden-Württemberg which became effective in 1981. Since that time all licence holders must pass a public examination to obtain an angling permit.

No obvious relationship was found between the mean angling catch and the number of licences issued, whereas a significant correlation ($r = 0.92$) was found between the total angling yield and the perch yield of the professional fishery.

Differences were found between the species composition of the boat and shore fisheries; more perch, rainbow trout and pike were caught in the boat fishery whereas cyprinids (others) and eel were more important in the shore catches. In contrast to their high value, migrating brown trout, pike-perch (*Stizostedion lucioperca*) and arctic charr (*Salvelinus alpinus* and *Salvelinus profundus*) play a minor role quantitatively in both types of fishery.

Catches by anglers with yearly boat licences represented about 70% of the total boat catch whilst the annual shore licences accounted for approximately half (50%) of the shore catch; 16% of the annual boat licensees and 24% of the shore anglers failed to make a catch declaration. About 15% of the successful anglers had a yield which was twice the mean catch and less than 3% realised catches five-fold the mean (Fig. 10.5).

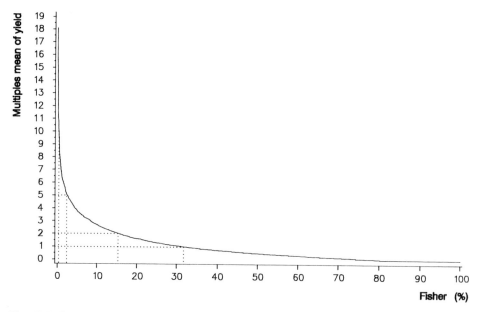

Fig. 10.5 Frequency distribution achieving catches above and below the mean annual yield

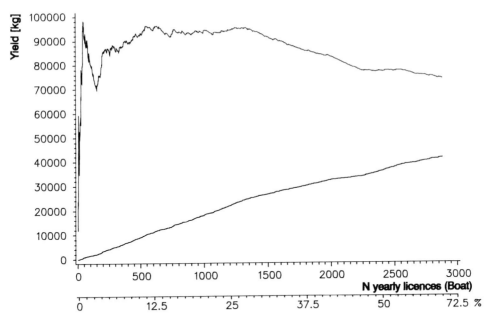

Fig. 10.6 Nominal (lower line) and extrapolated (upper line) catch curves for the annual boat licences

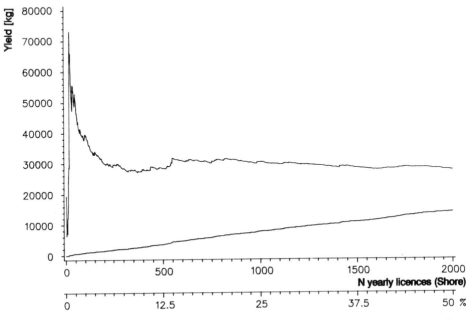

Fig. 10.7 Nominal (lower line) and extrapolated (upper line) catch curve within the annual shore licences

Figures 10.6 and 10.7 present the nominal and extrapolated catch curves for the annual boat and shore licences over the study period. The nominal and extrapolated curves approach each other and a stabilization of the extrapolated curve becomes obvious at c. 35% (shore) and 50% (boat) of the issued licences.

10.4   Discussion

In Lake Constance, perch contribute more than 33% and whitefish over 40% of the mean annual yield (Löffler, 1988) and reflect the importance of perch within the recreational fishery. An exception to this can be observed in Bavaria, where bream and other cyprinid catches constitute a higher proportion of the catch. The high CPUE for pike in Bavaria is an artifact of intensive fishing activities in the littoral zone in this district. The slight increase in pike catch since 1986 is probably a result of improved management measurements in the fishery, e.g. spawning catches and biotope improvements in all countries.

Hartmann (1984) describes a regional distribution of the littoral fishes of Lake Constance with decreasing densities from east to west. The perch catch of the recreational fishery does not reflect this observation, probably because the angling yield is biased by a catch limit of 50 perch per day.

Despite different evaluation methods of the angling yield within the border countries, similar trends in CPUE were observed. Extrapolation of the data introduced in Baden-Württemberg in 1985, however, increased the yield estimate by 48%−65%. Nevertheless, the main trend remained unchanged. This suggests that not all records indicating zero catch are in fact unsuccessful and non-returns do not necessarily indicate unsuccessful fishing. On the other hand successful anglers have more insight into the need for catch statistics and make accurate returns (Staub, 1988).

The Swiss catch data support these suppositions, especially in the canton of St Gallen where anglers who delay or refuse to make returns are punished (Trunz, *pers. comm.*) and the highest return rates and consequently CPUE of total fish, perch and trout are recorded.

Acknowledgements

The author appreciates the assistance provided by colleagues of all border countries in gathering unpublished information. I also acknowledge the invaluable help of Siegfried Blank in data processing.

References

IBK, (1982) Decisions of the International Deputies Conference for Lake Constance Fishery. Unpubl. Documents.

Hartmann, J. (1984) Zur gebietsweisen Verteilung der Fische im Bodensee. *Österr. Fisch.* **37**: 231−233.

Löffler, H. (1983) Bericht über die Baden-Württembergische Fischerei im Bodensee-Obersee im Jahr 1982. *Fischwirt* **33**: 72–73.

Löffler, H. (1990) Fisheries Management of Lake Constance: An Example of International Co-operation. In *The Management of Freshwater Fisheries* (Ed. by W.L.T. van Densen, B. Steinmetz and R. Hughes). Wageningen, PUDOC.

Ruhlé, Ch. (1987) Schutzprobleme bei der Bodensee-Seeforelle. *Terra Plana*, 28–30.

Staub, E. (1988) Fangstatistik aus Fließgewässern – weshalb so wenig Auswertungen? *Schweiz. Fischereiwiss.* **5**: 5.

Wetzlar, H. (1988) Entwicklung in der Bewirtschaftung angelfischereilich genutzter Gewässer. MS.

Chapter 11
Study of the daily catch statistics for the professional and recreational fisheries on lakes Geneva and Annecy

D. GERDEAUX I.N.R.A., *Institut de Limnologie, B.P. 511, F 74203 Thonon Cedex, France*

Since 1986 for Lake Geneva and 1987 for Lake Annecy, professional and recreational fishermen have been obliged to report fish catches (trout, char and coregonid) on a daily basis. More than 20 000 records are available for each year's fishing on Lake Annecy. These records show that the seasonal changes in CPUE for the two lakes in both the sport and professional fisheries show similar trends. Few fishermen catch the maximum number of fish allowed; a factor which has implications on the future management of these fisheries since it is proposed to reduce fishing pressure.

11.1 Introduction

Until recently, catch per unit effort (CPUE) data have not been used in France for the management of recreational and professional fisheries in the inland waters. However, the increasing costs of restocking have resulted in the need for an efficient method of evaluation of the status of the fisheries, such as CPUE data offers. These methods are particularly applicable in the French sub-alpine lakes, where, until 1986, only monthly catch statistics were recorded. This paper presents the results of a daily monitoring scheme introduced on the lakes and how it has been used to aid management decision making associated with exploitation of the fisheries.

11.2 Materials and methods

11.2.1 *The fisheries of Lake Geneva and Lake Annecy*

Lake Geneva and Lake Annecy are stocked with three species of salmonids, which are much in demand: trout *Salmo trutta lacustris*, char *Salvelinus alpinus* and whitefish *Coregonus schinzi palea*. These fish are caught by the professional fishermen with gillnets or with long-lines (hand-line) by the recreational fishermen. Fishing intensity is controlled through the issuing of licences, which is strictly regulated. This allows restriction in the number of fishermen operating and the ability to impose certain conditions on the licence holder. In the French part of Lake Geneva, 55 professional licences and about 380 licences for long line/hand line fishing are issued annually. On Lake Annecy, six licences for professional fishermen and up to 800 licences of recreational fishing are issued. Thus the smaller Lake Annecy (2700 ha) experiences a much higher fishing pressure than Lake Geneva (58 000 ha).

11.2.2 *Statistics*

In Lake Geneva until 1986, professional fishermen were required to submit to the local authorities monthly statistics, whereas the recreational fishermen only had to submit returns at the end of the fishing year. A similar situation was observed in Lake Annecy except that monthly declarations were required from both professional and recreational fishermen. Unfortunately, these declarations only requested information on the catch and no indication of the corresponding effort was requested.

In 1982, a research programme to assess the efficiency of restocking was started. This required the collection of better statistical information on catches. As a result the fishermen were educated in the value of catch statistics and a transition to daily declarations was brought about. The fishermen were requested to declare the number and weight of every species caught daily. Professional fishermen each have authorization to set a limited number of nets and for the purposes of these statistics they set all these nets each night they go fishing. Consequently, the catch and effort is known and the CPUEs are measureable. The recreational fishermen maintain a special notebook in which they have to record the date of their fishing game and the species of fish subsequently caught. At the end of the fishing, they record the total weight of the catch per species. Failure to submit the report on time is a violation of the law and can be reprimanded. In these recreational fisheries, effort is only considered in terms of each fishing experience, which can be of a varying number of hours. It was not possible to ask the fishermen to report the total number of effective fishing hours and the species upon which the effort focused. However, an experiment involving voluntary declarations on Lake Annecy showed that the fishing experience lasted, on average, five hours and required the use of a boat (CEMAGREF, 1986).

11.2.3 *Data processing*

At the end of the fishing season, the fisherman is required to hand over his notebook in order to obtain a new licence. During the first years of recording, a considerable effort was put into data computerization — all statistics were entered into a database (dBASE III). Consequently, over 20 000 records of fishing returns by the recreational fishermen on Lake Geneva are available each year. Similarly, records from each day of fishing for each professional fisherman on both lakes are stored.

11.3   Results

11.3.1   *Annual variations of the recreational fisheries*

To evaluate the annual changes in the fishery the catch declared by each fisherman was plotted in descending order of success as cumulative frequency curves (Fig. 11.1 and Fig. 11.2).

Important variations in the catch from one year to the next were noted on Lake Annecy. Over the six years of the study the catch doubled for each of the species exploited. For the whitefish fishery, the variations were explained by shifts in the recruitment as a result of climatic change and the total number of the cohorts present (Gerdeaux & Dewaele, 1986; Wojtenka *et al.*, 1988).

The relative importance of these variations was the same irrespective of whether the total catch or the catch of the 100 best fishermen was considered. This relationship did not apply to the whitefish on Lake Annecy, where the annual variability was lower for the 100 best fishermen than for all the fishermen.

By comparison, the annual relative variations for the 50 fishermen who declared the highest catch were lower. These 'skilled' fishermen make a constant catch every year, especially of whitefish, on Lake Annecy. This type of fishing requires a precise imitation of the lure, which the skilled fishermen can mimic accurately. With regards the trout fishing in both lakes or char fishing in Lake Geneva, annual variations are also experienced by the fishermen who declare the maximum catch. This type of fishing requires a less elaborate technique and the variations are experienced by all fishermen. Thus the fishing is a better reflection of the stock than the catchability of the fish.

11.3.2   *Distribution of the catch amongst the fishermen*

Closer examination of the data indicated that, on both lakes, the top 100 fishermen caught over 60% of fish each year. On Lake Geneva, the 25 best fishermen declared over 30% of the total catch. However, on average these fishermen declared less than 120 fish, per species, per year. This is well below the authorized annual quota of 250 fish per year or its equivalent daily quota of eight fish per species (Table 11.1). These data are extremely useful in establishing the daily/annual quotas per fishing experience.

11.3.3   *Seasonal variation in catch per unit effort*

The seasonal variation in catch per unit effort was evaluated using a nine-day running mean of daily CPUE. The nine-day running mean was used because it always takes into account a weekend. This analysis was used to compare the daily CPUEs for the professional and the recreational whitefish fisheries in Lake Annecy (Fig. 11.3), and the recreational trout and char fisheries in Lake Geneva (Fig. 11.4).

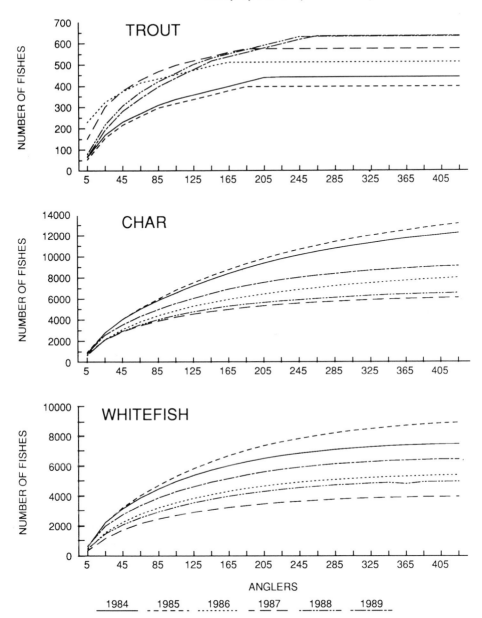

Fig. 11.1 Cumulative frequency polygons of catch classified in descending order of the declared catch of trout, char and whitefish, for the recreational fishing on Lake Annecy 1984–89

The smoothed daily CPUEs for each species and each type of fishing are similar from one year to the next. The professional fishing for whitefish in Lake Annecy shows a general increase in CPUE as the year progesses but peaks in July and August, and again towards mid-October. This general trend is undoubtedly linked to thermal stratification in the lake which becomes well established in July. In the

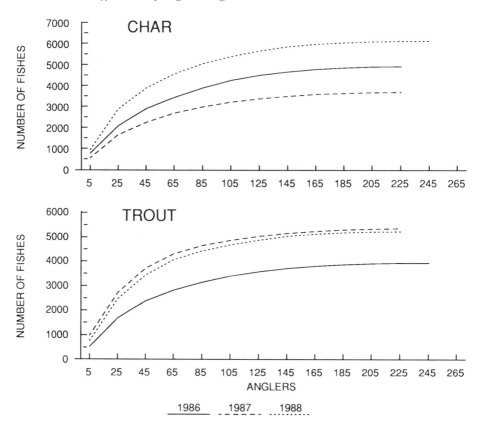

Fig. 11.2 Cumulative frequency polygons of catch classified in descending order of the declared catch of trout and char for the recreational fishing on Lake Geneva 1986–89

Table 11.1 Number of fishing games declared for a considered number of catch in Lake Annecy and Lake Geneva

			Number of fish							
			1	2	3	4	5	6	7	8
Annecy	Char	1987	2137	740	329	168	62	33	18	31
		1988	2300	873	369	136	77	33	23	25
		1989	3203	1191	509	215	119	61	16	28
Annecy	Whitefish	1987	1816	542	197	59	19	11	4	5
		1988	1832	683	248	91	46	27	9	29
		1989	2551	799	296	149	60	42	14	27
Geneva	Trout	1986	1077	423	161	71	37	34	22	71
		1987	1140	576	309	160	73	43	28	80
		1988	909	454	263	138	103	53	55	104
Geneva	Char	1986	1452	652	299	143	54	22	23	6
		1987	1029	485	225	87	54	22	11	6
		1988	1079	618	349	208	125	61	36	22

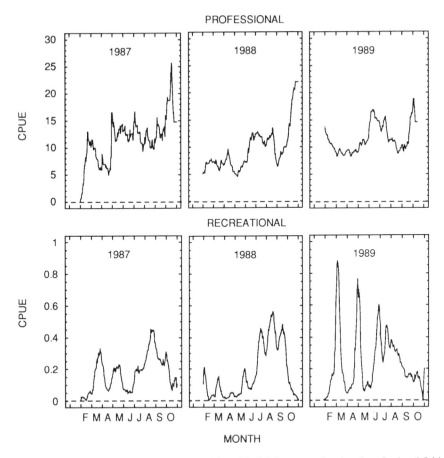

Fig. 11.3 Evolution of the smoothed CPUEs of the whitefish for recreational and professional fishing in Lake Annecy 1987–89. CPUE in fish kg gillnet night^{-1} for professional fishing and CPUE in fish per outing for recreational fishing

recreational fisheries, the annual pattern was also similar from one year to the next. Two peaks in catch occurred whilst a fluctuating but good rate of success was observed over summer. The precise timing of the spring peaks were not always at the same time. In 1989 they were both important and early because of the mild spring climate and a precocious rise in the temperature of the lake. These spring peaks probably correspond to the first hatchings of the chironomid larvae which the hand-line lures imitate. The fluctuations observed in the summer could not be explained.

In Lake Geneva, the catch curves also exhibit similar annual variations (Fig. 11.4). Trout were caught at the beginning of the season, whilst char were more important from June until September every year.

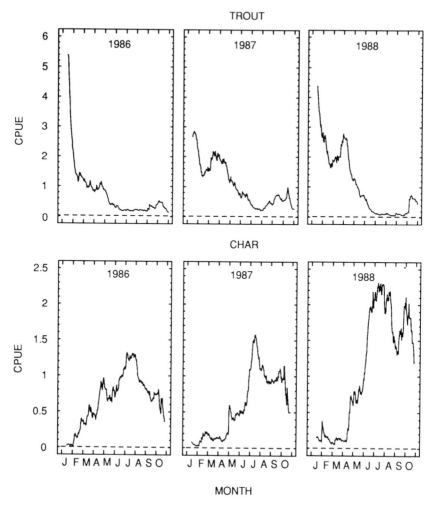

Fig. 11.4 Evolution of the smoothed CPUEs of the trout and char for the recreational fishing on Lake Geneva 1986−88. CPUE in fish kg gillnet night^{-1} for professional fishing and CPUE in fish per outing for recreational fishing

11.3.4   *Daily variability of the CPUEs*

In addition to the seasonal fluctuations in CPUE, there is also considerable variability in the mean daily catch per fisherman (Fig. 11.5). This is particularly prominent when the average catch is over two per fisherman.

11.4   **Discussion**

In the recreational fisheries of both lakes several points important to the management were noted. The CPUEs showed similar trends every year although the timing of the peak catches varied somewhat. Thus if a statistical survey of the data

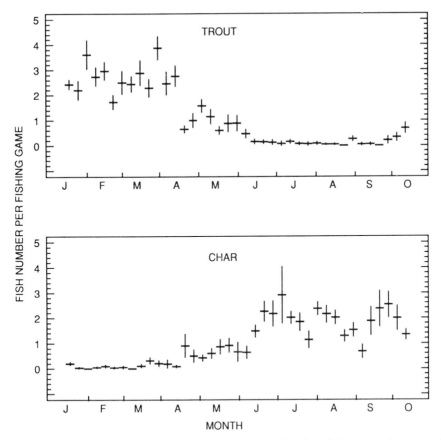

Fig. 11.5 Evolution of the average CPUEs on Sunday and indication of the standard error on this average in 1980 on Lake Geneva

were undertaken these variations would have to be taken into account. Therefore the sampling period would have to be sufficiently long to cover these variabilities. If the fishery is assessed by interviewing fishermen, the daily variability observed amongst the catches indicates that a high number of fishermen must be interviewed each day to obtain a reliable evaluation. For an 80% probability based on 600 fishermen, at least 36 fishermen must be interviewed to reduce the confidence interval about the mean CPUE to 0.4 (STAT-ITCF, 1985). This is excessive compared with the time taken to input data from fishermen's notebooks; the latter method only requires about 60 hours at the end of each year to input the data from all 600 fishermen.

Normally the data collected by an interviewer is of a better quality than the declarations recorded in the notebooks. However, this does not appear to be the case here. Notebook data are inspected by water bailiffs periodically during the year and they appear to be reasonably accurate. They also have the added advantage of evaluating all the data. In addition, the task of data input could be lessened by a stratified sampling of the notebooks. For example, it has been shown that 100 fishermen contribute significantly to the catches and satisfactory results for Lake Annecy would be achieved by inputting data from only 200 notebooks out of the 700 returned.

Statistical analysis also allows the evaluation of management measures which aim to limit the fishing intensity. For example, the annual quota of 250 fish is being reached by only a minority of fishermen, thus a reduction of this quota would have an effect only if it is significantly large. By contrast, the daily quotas are achieved more often and a reduction in the daily quota to a maximum of four fish (instead of eight) would result in a fall in the total catch for the recreational fishery. This is important because this fishery is profitable. In Lake Annecy, this measure would cause a catch reduction of about 15% for char and whitefish, whereas in Lake Geneva a reduction of at least 27% for trout and 13% for char would be observed.

Another management approach is the suppression of fishing activity by reducing the number of fishing opportunities per week. For instance, if fishing is prohibited on Tuesdays in Annecy the reduction of catch varies from 10.5% to 13.5% depending on the year or the species. However, when this restriction is limited to the 20 fishermen who declared over 100 fishing outings in 1987 and 1988, the reduction varied from 14% to 16%. In the total fishery the reduction represents about 2% of the fishing.

The suppression of fishing on one day per week leads to an impact which is equal to or lower than the impact of a lowering of the daily quota from eight down to four fish. This limitation is also easier to implement compared with control of daily quotas.

Acknowledgements

This work could not have been done without the help of the authorities in charge of the fisheries management, particularly Messrs Cassayre, Khal and Michoux, and the water bailiffs, who check the notebooks. J.P. Moille provided help with the database and F. Macchi with the translation.

References

CEMAGREF (1986) Exploitation des déclarations volontaires de captures de corégones. Lac d'Annecy. Années 1984/85/86.

Gerdeaux, D., Dewaele, P. (1986) Effects of the weather and artificial propagation on coregonid catches in Lake Geneva. *Arch. Hydrobiol. Beih. Ergebn. Limnol.* **22**: 343–352.

Wojtenka, J., Gerdeaux, D. and Allardi, J. (1988) Coregonid fishery in Lake Annecy – an example of dual exploitation. *Finnish Fish. Res.* **9**: 389–396.

STAT-ITCF (1985) Logiciel d'analyse statistique. France, Boigneville.

Chapter 12
Comparison of fish population estimates and angling catch estimates in a narrow ship canal in Belgium

P. GERARD and J.A. TIMMERMANS *Government Research Station for Forestry and Hydrobiology, Groenendaal, Hoeilaart, Belgium*

The exploitation rate of anglers was compared with an estimate of the fish population size in a narrow ship canal in Belgium. Catch effort sampling by seine netting was used to estimate the fish population size. Fish biomass ranged from 168 to 832 kg ha^{-1}. Roach and bream represented 76% of the biomass whilst predators comprised only 2% of the community structure.

Angling catch estimate was based on angler counts and creel surveys.

Anglers were counted instantaneously 11 times a month while creel surveys were made two days a month. The importance of the different species in the community was compared for seining and angling. The angling catch rate was estimated and compared with the estimated fish population.

12.1   Introduction

Accurate estimation of fish population size in large water bodies such as ship canals is, in general, problematical. Electric fishing is inefficient and seine netting is often difficult. Nevertheless, seining has proved to be an effective tool under certain circumstances (Penczak & O'Hara, 1983), despite doubts being expressed as to its efficiency for obtaining estimates of different size and species groups (Raleigh & Short, 1981).

Similar difficulties are also encountered when using anglers' catches to evaluate fish stocks. This method requires accurate estimates of the catch per unit effort (CPUE) and the global angling effort (Malvestuto, 1983).

The present study compares angler catch evaluation with fish population estimates in a narrow ship canal in Belgium and attempts to correlate the two methods.

12.2   Study area

The Ath-Blaton ship canal, located in Central Belgium (Province of Hainaut), takes boats up to 300 tonnes. Its depth is constant and does not exceed 3.5 m in the middle of the cross-section; its breadth is about 16 m. The canal (total length 22 km) is separated into 20 reaches by locks. Weed development in the littoral zone is minimal and regular dredging limits the number of obstacles on the bottom, thus allowing relatively easy seining along the whole length of the canal.

Fishing activity (angling), restricted to one bank, is free for every holder of a state licence. The fishing day lasts from sunrise till sunset, whilst the fishing season is closed between 1 May and mid-June. In general, the angler remains relatively static whilst fishing.

Estimation of the fish population size by anglers' catch was conducted in a study area limited to nine reaches of the canal (3 km). Water quality in this area was good and mesotrophic (IHE, 1987). Sampling for fish population estimates by seine netting was conducted in three reaches of the study area.

12.3   Material and methods

12.3.1   *Population estimates by netting*

A total of 12 fish population estimates were made; four per year in each of 1987, 1988 and 1989 (usually in October). For each estimate, three successive fishings were planned, but, depending on working conditions, in practice between one and four fishings were performed. Each fishing was carried out by trawling a 10 mm mesh seine along the study reach. The study reaches varied between 100 to 270 m. At the end of each haul the seine was enclosed by a second net and both seines were retrieved. The fish caught were held in floating cages until the end of the fishing operation. During the fishing operation fish movement was restricted at both extremities of the reach by a fixed net or by a lock.

No significant difference (P > 0.05) was found in the individual weight and relative importance of the different species between successive fishings. Data were therefore combined for the different species of the fish community. Fish density and 95% confidence limits were determined by the 'maximum weighted likelihood' method of Carle and Strub (1978). The probability of capture was also estimated (Cowx, 1983; Gerdeaux, 1987). Biomass was determined from the density and the mean individual weight of fish in each estimate (Penczak & O'Hara, 1983). The mean individual weight was calculated by dividing the total weight by the total number of fish caught. The relative importance of the different species was calculated per reach as a percentage contribution by weight.

12.3.2   *Population estimate by anglers' catch estimate*

The estimate of angling catch was performed during a five-month study period after the opening day of the season in 1988 (i.e. 11 June−30 November). Angling catch per unit effort (CPUE) was based on weight of fish caught during an angling day (for the different species combined). These data were collected during roving surveys made by a clerk at a rate of one week day and one weekend day per month (12 surveys); the days were chosen randomly. A total of 156 anglers were interviewed and the following information was recorded: time spent angling before the interview, presumed time for the whole angling day, catch before the interview. A questionnaire evaluating the real time spent for the completed angling day and catch during this time was given to each angler. A total of 112 anglers (72%) returned their questionnaire.

The data on these questionnaires were used to enhance the incomplete angling information collected during the roving creel surveys. When the number of active anglers during a survey was less than four, the data from the survey was not taken into account for the CPUE estimate.

Angling effort was expressed in terms of number of angling days during the study period and was based on two elements: instantaneous counts of anglers and the pattern of angling activity between sunrise and sunset. Instantaneous counts were made on 50 week days and 14 weekend days (37% of the study period). Preliminary sampling revealed that the period of the day between 2.00 p.m. and 4.30 p.m. would be the most favourable for such counts, as most anglers were present at this time (Fig. 12.1); thus all counts were performed between these hours. The pattern of angling activity during the day was based on the arrival and departure times of each angler, as recorded by the clerks during the surveys (Steinmetz, 1986).

12.4 Results

12.4.1 *Fish population*

The estimated fish density (with 95% confidence interval), fish biomass and the probability of capture for the 12 surveys are given in Table 12.1. Fish density

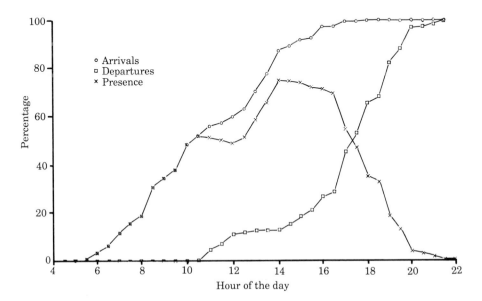

Fig. 12.1 Evolution angling activity (%) during the day

Table 12.1 Fish population estimates based on netting

Estimate	Effort	Density N ha^{-1}	95% CL	Biomass kg ha^{-1}	Probability of capture
1	1	–	–	–	
2	2	11 891	573	202	0.50
3	2	23 150	273	567	0.54
4	2	14 301	73	457	0.90
5	3	24 948	110	679	0.85
6	3	31 165	605	832	0.51
7	3	12 906	17	212	0.90
8	3	1 941	<1	168	0.87
9	3	20 618	696	566	0.45
10	3	29 700	91	284	0.79
11	3	65 476	<1	282	0.60
12	4	10 194	381	284	0.39
Mean		22 390		412	

ranged from $0.2 \, \text{m}^{-2}$ (mean individual weight: 87 g) to $6.5 \, \text{m}^{-2}$ (mean individual weight: 4 g) with an average of 2.2 fish m^{-2}. Mean biomass was estimated at 412 kg ha^{-1} and ranged from 168 to 832 kg ha^{-1}.

The fish population comprised 13 species (Table 12.2) of which roach (*Rutilus rutilus*) and bream (*Abramis brama*) were, by far, the most important species. 85% of the biomass comprised cyprinid fish, whilst the contribution by predators (pike *Esox lucius*, pikeperch *Stizostedion luciopera* and large perch *Perca fluviatilis*) was low at only 1.6%.

12.4.2 *Anglers' catch*

Catch per unit of effort (CPUE) based on the clerk's data was estimated at 240 g day^{-1} (95% Confidence Interval (CI): 103 g day^{-1}). The anglers' questionnaires increased this estimate to 413 g day^{-1} (95% CI: 99 g day^{-1}). The precision of the estimates is not high, but the difference proved to be statistically insignificant (P > 0.05).

The data analysis showed that the difference was neither due to the 28% nonreturn of questionnaires, nor to the fact that the clerk's interviews were based on an incomplete angling day. In fact, the difference seems to result from sampling variability or possible errors in anglers' answers. Consequently, the CPUE estimate of 240 g day^{-1} from the clerks' data was used.

The relative importance of the different species in the catches of anglers is shown in Table 12.2 and compared with the community structure based on netting.

The daily pattern of angling activity is presented in Fig. 12.1. The results calculated for the week days were, in this respect, comparable to the results for the weekend days, thus the two data sets were combined. On average, 69.4% of

Table 12.2 Relative importance (in weight) of the different species in the netting surveys and in anglers' catch

Species		Importance in the population (% of biomass)	Importance in the angling catch (% of the catch)
Roach	*Rutilus rutilus*	51	66
Bream	*Abramis brama*	25	18
Perch	*Perca fluviatilis*	7	15
Ruffe	*Acerina cernua*	7	1
Rudd	*Scardinius erythrophthalmus*	5	(see roach)
Carp	*Cyprinus carpio*	2	–
Pike	*Esox lucius*	1	–
Tench	*Tinca tinca*	1	–
Pikeperch	*Stizostedion lucioperca*	<1	–
Gudgeon	*Gobio gobio*	<1	–
Bleak	*Alburnus alburnus*	<1	–
Leucaspius	*Leucaspius delineatus*	<1	<1
Bitterling	*Rhodeus amarus*	<1	–

the total number of the anglers fishing each day were active between 2.00 p.m. and 4.30 p.m. Consequently, this figure was used to estimate total angling effort over the whole study stretch from instantaneous counts taken at this time of day.

The global angling effort during the study period was estimated as 2259 angling days (95% CI: 340 angling days) during the week (average, 19.3 anglers fishing per day) and 1994 angling days (95% CI: 743) at weekends (average, 35.6 anglers fishing per day).

The total of 4253 angling days (95% CI: 1100 angling days) on the study area corresponds to an angling effort of 945 angling days per hectare during the study period.

12.5   Discussion

During this study the precision in estimating the fish population density, the CPUE and the global angling effort was unequal. Estimates of the density of the fish population proved to be extremely precise, especially if the results are compared with other large water bodies. This could be due to the very high proportion of the population caught by seining (>95% of the estimated number of fish in the population after three successive fishings). There was no indication that seining was size selective and, consequently, it was assumed that the mean individual weight in the catch was a good estimator of the mean individual weight in the population. The CPUE estimate precision was low and increasing the number of surveys to improve precision was considered impractical because of financial and labour restrictions. The global angling effort was estimated rather precisely due to the high frequency of sampling (43% of the week days and 25% of the weekend days in the study).

Table 12.3 Comparison between estimated biomass by netting and anglers' catch exploitation rate (four species)

Species	Biomass (kg ha^{-1})	Angling Catch (kg ha^{-1})	Angling Catch (% of the biomass)
Roach (Rudd)	231	150	65
Bream	103	41	40
Perch	29	34	118
Ruffe	29	2	8

A fish biomass of 412 kg ha^{-1}, dominated by roach and bream, was similar to the situation in an adjacent canal (Gerard & Timmermans, 1985) and, probably, similar to the situation in several other Belgian ship canals where the high density is due to a lack of predatory fish and the fact that only a few roach and bream are removed by anglers. The CPUE of 240 g day^{-1} (40 g h^{-1}; 1.3 fish h^{-1}) compares favourably with similar data from the literature (Cowx *et al.*, 1986). Axford (1979) reported a CPUE of 58 g h^{-1}, Hickley and North (1981) a CPUE of 82–176 g h^{-1} and Steinmetz (1987) from 1.4 to 4.1 fish h^{-1}.

On the basis of the CPUE and the global angling effort, angling catch during the study period was estimated at 227 kg ha^{-1} (about 55% of the estimated biomass). Angling pressure on the most important species shows that angling pressure on perch is extremely high and somewhat lower for roach and bream (Table 12.3).

Cooper and Wheatley (1981) estimated the biomass in a sector of the River Trent as 447 kg ha^{-1} and the annual exploitation rate as 94%. These data are considerably higher than the observations on the Ath-Blaton Canal.

References

Axford, S.N. (1979) Angling returns in fisheries biology. *Proceedings of the 1st British Freshwater Fisheries Conference*, University of Liverpool, 259–271.

Carle, F.L. and Strub, M.R. (1978) A new method for estimating population size from removal data. *Biometrics* **34**: 621–630.

Cooper, M.J. and Wheatley, G.A. (1981) An examination of the fish population in the River Trent, Nottinghamshire, using angler catches. *J. Fish Biol.* **19**: 539–556.

Cowx, I.G. (1983) Review of the methods for estimating fish population size from survey removal data. *Fisheries Management* **14**: 67–82.

Cowx, I.G., Fisher, K.A.M. and Broughton, N.M. (1986) The use of anglers' catches to monitor fish populations in large water bodies, with particular reference to the River Derwent, Derbyshire, England. *Aquaculture and Fisheries Management* **17**: 95–103.

Gerard, P. and Timmermans, J.A. (1985) Inventaires piscicoles dans l'ancien canal Charleroi-Bruxelles. *Trav. Stat. Rech. For. et Hydrob, Groenendaal, SÉrie D* **52**: 28 p.

Gerdeaux, D. (1987) Revue des méthodes d'estimation de l'effectif d'une population par pêches successives avec retrait. Programme d'estimation d'effectif par la méthode de Carle et Strub. *Bulletin Français de pêches et de Pisciculture* **304**: 13–21.

Hickley, P. and North, E. (1981) An appraisal of anglers' catch composition in the barbel reach of the River Severn. *Proceedings of the 2nd British Freshwater Fisheries Conference*, University of Liverpool, 94–100.

I.H.E. (1987) Réseau de mesure de la qualité des eaux superficielles belges en 1987. Institut d'Hygiène et d'Epidémiologie, Brussels. 329 p.

Malvestuto, S.P. (1983) Sampling the recreational fishery. In *Fisheries techniques*. (Ed by L.A. Nielsen and Johnson), American Fisheries Society, Bethesda, Maryland. 468 p.

Penczak, T. and O'Hara, K. (1983) Catch-effort efficiency using three small seine nets. *Fisheries Management* **14**: 83−92.

Raleigh, R.F. and Short, C. (1981) Depletion sampling in stream ecosystems: assumptions and techniques. *Progressive Fish-Culturist* **43**: 115−120.

Steinmetz, B. (1986) Bevissing en vangsten. *Visserij* **39** (2): 58−77.

Steinmetz, B. (1987) Het beheer van de Twenthe kanalen. *Visserij* **40** (2): 113−123.

Chapter 13
Anglers' opinions as to the quality of the fishing and the fishery management in selected Polish waters

A. WOŁOS *Inland Fisheries Institute, Olsztyn, Poland*

Five hundred and thirty anglers' questionnaires were analysed with respect to the quality of angling and assessment of the fisheries management practices. A number of indices were calculated characterizing anglers' catches and their preferred target species. The quality of angling was related to the level of fish catch. Conversely, assessment of the fishery management was not related to the catch; most anglers expressed negative opinion on management irrespective of whether their catch was good or bad. Motives behind the anglers' opinions are presented as well as their suggestions on how to improve the fishery management. Attention was also paid to factors disturbing angling; disturbances of peace and silence, various boats and bad weather conditions were the most important.

13.1   Introduction

Studies carried out in Poland in 1978 and 1979 amongst members of the Polish Anglers Association (PAA), showed that 37.4% of the fishing days spent by anglers was on lakes (Leopold, Bnińska & Hus, 1980). At this time, the number of PAA members was about 800 000, compared with 980 000 in 1987. Thus some 400 000 people participate in recreational lake fishing.

The majority (91%) of Polish lakes are managed by the State Fish Farms (SFF) with a directive aimed at obtaining the highest possible production of market fish. This is not always consistent with anglers' needs and preferences. The remainder of the lakes, about 9% of the total lake area in Poland, belong to the PAA, who under Fishery Law, are obliged to carry out the fishery management so as to fulfil anglers' needs. To this end studies have been undertaken to determine the anglers' needs and preferences, and their opinion on the quality of angling and the management in these fisheries. Attention was also given to the main factors which disturb the anglers and reduce the quality of the fishing. The results will provide information for setting policy to ensure the recreational and commercial lake fisheries in Poland are maintained and improved.

134

13.2   Materials and methods

Studies were based on the questionnaire data collected from 29 lakes managed by the SFF and eight lakes and one reservoir managed by the PAA between 1985 and 1988. Five hundred and thirty questionnaires were collected, 453 from anglers fishing in SFF lakes and 77 from anglers fishing in PAA waters. The questionnaires were completed either during fishing or immediately thereafter. The area of the lakes under study ranged from 7.2 to 3030 ha.

The questionnaire embraced a variety of questions relating to the given water body. The preferred fish was evaluated by ranking such that the species mentioned as the most preferred was ranked as 3, those mentioned as second preference were ranked 2, and of third preference as 1. These data were expressed as a percentage based on the total sum of all ranks. The importance of each species in the catch of each angler was examined by awarding 'points' according to the rank status for each species. For instance, 1 kg of the most preferred species, which embraced 25% of the rank scale, was given 25 'points'. The catch of each angler was also calculated in relation to its monetary value (zl) according to the fish prices of 1987. The following indices were used to estimate the quality of angling and fishery management:

- AC1   annual catch per angler in kg
- AC2   annual catch per angler in zlotys (zl)
- AC3   annual catch per angler in 'points'
- CPUE1   catch per unit of effort in kg man-day^{-1}
- CPUE2   catch per unit of effort in zl man-day^{-1}
- CPUE3   catch per unit of effort in 'points' man-day^{-1}

Chi-squared analysis was used to establish the significance (strength) of the relationship between anglers' desires and their catch. Two categories of anglers, those above and below the mean value of the six indices under study, were used.

13.3   Results

13.3.1   *Fish species preferred by the anglers*

The anglers prefer predatory species (pike, eel, perch and pikeperch) (Table 13.1), because they are attractive both with regards their sporting characteristics (fish size, fighting ability, fishing methods) and consumption value. It should be added that market value of these fish (in zl) does not usually correspond to their sporting value (in 'points'). This is particularly noticeable for perch, which occupies third position in the preference list but has a low market value. Eel is the opposite with a market price almost five times higher than the price of the most preferred species, i.e. pike.

Table 13.1 Preferred fish species (in 'points') and comparison with their market prices (in zl)

Species		'Points'	zl
Pike	*Esox lucius* (L.)	25.00	240
Eel	*Anguilla anguilla* (L.)	21.70	1100
Perch	*Perca fluviatilis* (L.)	18.00	80
Pike-perch	*Stizostedion lucioperca* (L.)	9.00	320
Common bream	*Abramis brama* (L.)	6.40	70
Common carp	*Cyprinus carpio* (L.)	6.30	300
Tench	*Tinca tinca* (L.)	5.80	250
Roach	*Rutilus rutilus* (L.)	5.30	40
Whitefish	*Coregonus lavaretus* (L.)	0.85	250
Grass carp	*Ctenopharyngodon idella* (Val.)	0.60	300
Rudd	*Scardinius erythrophthalmus* (L.)	0.14	40

13.3.2 *Characteristics of anglers' pressure and anglers' catches*

In the population of anglers under study, 499 anglers provided data to characterize fishing intensity (Table 13.2). Anglers fishing SFF lakes usually obtained higher annual (AC1) and daily (CPUE1) catch than anglers fishing in PAA waters. Notwithstanding, the market value of fish (AC2 and CPUE2) caught by the PAA anglers was much higher than that by SFF anglers. On the other hand, catch of the preferred species (AC3 and CPUE3) from SFF lakes was much higher than from PAA waters, confirming their recreational value. Thus intensive commercial fishing in SFF lakes has no negative effect on anglers' catches.

13.3.3 *Anglers' opinion of the quality of angling*

Five hundred and six anglers evaluated the quality of fishing according to a five-score scale used in the questionnaires (Table 13.3). A greater proportion of the waters (53.6%) were considered good or better. The criteria were subsequently simplified into bad (bad and sufficient) and good (the remainder) to assess whether fishing quality was related to the catch (Table 13.4). In general the quality of angling is positively correlated with the catch. The highest percentage

Table 13.2 Average values for the indices characterizing fishing intensity and catches

Index	SFF anglers	PAA anglers	All anglers
Days angler^{-1} year^{-1}	35.90	36.80	36.03
AC1 (kg)	36.61	34.19	36.26
AC2 (zl)	6810.00	8510.00	7050.00
AC3 ('points')	491.02	395.90	477.30
CPUE1 (kg man-day^{-1})	1.02	0.93	1.01
CPUE2 (zl man-day^{-1})	190.00	230.00	200.00
CPUE3 ('points' man-day^{-1})	13.68	10.76	13.25

Table 13.3 Anglers' opinion of the quality of the fishing

Anglers		Bad	Sufficient	Estimate (%) Good	Very good	Excellent
SFF anglers	(N = 429)	11.9	34.0	45.7	7.5	0.9
PAA anglers	(N = 77)	11.7	37.7	40.3	6.5	3.9
All anglers	(N = 506)	11.9	34.6	44.9	7.3	1.4

Table 13.4 Estimates of the condition of angling in relation to the catch level (N = 465)

Index	Anglers' Score	'Bad' (%)	'Good' (%)	χ^2
CPUE3	≤13.25	57.2	42.8	31.17
('points' man-day^{-1})	>13.25	30.8	69.2	p > 0.01
AC3	≤477.3	53.3	46.7	17.23
('points')	>477.3	32.4	67.6	p > 0.01
CPUE2	≤200	53.0	47.0	15.38
(zl man-day^{-1})	>200	33.3	66.7	p > 0.01
CPUE1	≤1.01	53.2	46.8	13.56
(kg man-day^{-1})	>1.01	36.0	64.0	p > 0.01
AC2	≤7050	51.9	48.1	12.11
(zl)	>7050	34.1	65.9	p > 0.01
AC1	≤36.26	52.0	48.0	11.99
(kg)	>36.26	35.7	64.3	p > 0.01

of 'good' estimates (69.2%), and the highest value of χ^2 were found amongst those anglers whose daily catch (CPUE3) exceeded the mean value of this index, i.e. 13.25 'points' man-day^{-1}. The weakest connection between 'good' estimates of the quality of angling and high catch was found in the group of anglers catching more than 36.26 kg (AC1) annually. Hence positive evaluation of the quality of angling was less influenced by the total fish catch or its value (AC1 and AC2) than by the proportion of the preferred species in the catch (AC3 and CPUE3).

Three hundred and eighty four anglers responded to the enquiry about what constituted a good or poor fishery. Altogether, 46 factors were mentioned; the most important of which are presented in Tables 13.5 and 13.6.

These data take no account of the fishery manager (SFF or PAA). In general the responses were similar in both cases. SFF anglers did, however, mention fishery management as a negative factor more frequently (19.9% of the answers) than the PAA anglers (6.1% of the answers).

The opinion that angling quality had deteriorated in recent years was expressed by 64.3% of the anglers, while 33.3% stated that it had not changed. Only 2.4% of the anglers thought that conditions had improved. The main reasons for the deterioration in angling quality are given in Table 13.7

Table 13.5 Factors influencing the judgement poor fishery

Factor	% of answers (n = 209)
Water pollution	26.8
Poor fish stock	20.1
Poor fishery management	17.7
Bad access to water	13.9

Table 13.6 Factors influencing the judgement good fishery

Factor	% of answers (n = 175)
Good fish stock	36.0
Morphometry, surroundings	20.6
Easy access to water	15.5
Clean water	13.7
Silence and peace	7.4

Table 13.7 Reasons for deteriorating quality of angling

Reason	% of answers (n = 427)
Increasing water pollution	38.4
Poor fishery management	27.4
Deteriorating fish stock	15.7
Increasing recreational activity	6.3
Variations in water level	2.6
Increasing number of anglers	1.9
Increased poaching	1.6

13.3.4 *Anglers' opinion of the fishery management*

In Poland, SFF lakes accessible to anglers are usually heavily exploited by commercial fishermen. PAA lakes, by contrast, are usually very small, and subject to no or only sporadic commercial fishing. Anglers' opinions on the fishery management in the two groups of lakes are presented in Table 13.8.

Eliminating 'no opinion' answers, and regrouping bad and sufficient answers as 'bad' and all other answers as 'good' showed that fishery management in SFF lakes was considered 'bad' by 82.2% of the anglers, and in PAA lakes by 71.0%. This negative opinion on the fishery management was irrespective of the success achieved.

13.3.5 *Proposals for changes in fishery management*

In order to improve the quality of the fishing, the anglers proposed a number of changes, the most important of which are listed in Table 13.9.

Table 13.8 Anglers' opinion of the fishery management

Group of anglers	Bad	Sufficient	Good	Very good	No opinion
			Estimate (%)		
SFF anglers (n = 371)	55.0	13.5	13.0	1.9	16.7
PAA anglers (n = 70)	48.6	14.3	24.3	1.4	11.4
All anglers (n = 441)	54.0	13.6	14.7	1.8	15.9

Table 13.9 Changes in the fishery management proposed by the anglers

Proposal to:	% of answers (n = 548)
Increase stocking rates	48.0
Decrease fishing effort	20.4
Forbid electric fishing	10.0
Observe protection during spawning periods	6.6
Displace commercial fishing	2.9
Abolish fishing licences for SFF waters	2.4
Increase selective fishing for roach and bream	2.0
Other (11 proposals)	7.7

The proposal to increase the stocking rates was, in most cases, of a general character. Some anglers, however, asked for increased stocking with selected species, mostly pike. Anglers proposed also a wish to see a decrease in commercial fishing, especially with tow gear. Many anglers are of the opinion that electric fishing is harmful for the fish stocks.

13.3.6   *Additional suggestions for improving the fishery*

Other proposals for improving the fishery, but not directly related to the management, were also given (Table 13.10). Anglers stated that deterioration of the quality of angling was in their opinion mostly caused by water pollution. Hence, they pay a lot of attention to the protection of waters.

Table 13.10 Non-management proposals for improving the fishery

Proposal to:	% of answers (n = 321)
Protect waters (construction of treatment plants, etc.)	26.5
Take more care of the shore	22.8
Fight poaching and anglers not abiding to the rules	22.4
Introduce silence zones (forbid motor-boats)	8.1
Prevent variations in water level	4.7
Limit recreational uses of water	3.7
Other (15 proposals)	11.8

13.3.7   *Proposed changes to angling regulations*

Angling in Polish waters is regulated by the Fishery Law, statutes of the PAA, and some special regulations specifically for different water users. For example, on a number of lakes managed by the SFF, angling is prohibited, night fishing is not allowed, boats can only be used for angling after 1 June. Table 13.11 presents anglers' opinions on these regulations and their proposals for change.

   These results suggest that most anglers view the existing regulations as sufficient. As to the changes proposed, anglers differ in their opinions; no one suggestion predominated. Some anglers proposed increased protection for pike and perch, whilst others asked for earlier use of boats from 1 May, i.e. just after pike spawning but still during the perch spawning period. Requests for night-time fishing are always related to the desire to catch eels.

13.3.8   *Factors disturbing angling activity*

Anglers mentioned 41 factors which they found tiresome and disturbing. Five hundred and thirty anglers gave 925 answers to this question which in itself suggests that the problem is very important. The most frequently mentioned factors (from the most to the least important) are listed below:

(1)   Disturbance of peace and silence.
(2)   Various boats (mainly motor-boats).
(3)   Bad weather conditions.
(4)   Recreational overcrowding.
(5)   Pollution and littering of the environment.
(6)   Overcrowding with anglers.
(7)   Professional and family responsibilities, lack of time.
(8)   Commercial fishing.
(9)   Difficult access to water.
(10) Lack of proper transportation.

Table 13.11   Changes in protective regulations proposed by the anglers

Proposal	% of answers (n = 288)
Regulations are OK	33.0
Permitted angling on more lakes	12.2
SFF should permit night fishing	8.0
Legal size and protective season for perch	7.3
Protection of pike (lower daily limits, increase minimum size)	6.2
Permit angling from boats	5.6
Abolish daily catch limits	3.1
Other (27 proposals)	24.6

According to the anglers, peace is the most important factor associated with angling. Other frequently mentioned factors generally relate to alternative recreational activities and misbehaviour of people engaged in these activities. With regards factors directly related to angling, weather seems quite important.

13.4   Discussion

National studies on anglers have shown that the two most important motives for participating in the practice of fishing are enjoyment derived from being near (or on) water, and the pleasure of fishing *per se* (Leopold, Bnińska & Hus, 1980). Results presented in this paper showed that these two motives were of considerable importance in assessing the quality of angling. Pleasure from being near water is strictly connected with environmental quality which is one of the major factors influencing angling quality.

Pleasure of fishing has a double meaning in Poland. On the one hand it relates directly to sporting values (catching of record fish, fight, etc.), and on the other to fish as a source of food. Almost all species are readily consumed, contrary, for instance, to Great Britain where anglers regard angling as a sport and not a means of obtaining food (Tombleson, 1978). In many west European countries, it is very common to release the fish after capture. In Poland there is no such practice. Hence, assessment of the quality of angling will always be influenced by catch level. Angling is considered high quality by the anglers who obtain a good catch of the preferred species (of high sport as well as consumption quality): pike, eel, perch and pikeperch.

The size of the fish catch was found to have no effect on the assessment of the fishery management. It seems that negative estimates of management by anglers are connected with a decline in the abundance of the most valuable species: pike, perch, tench. This decline has been associated with progressing eutrophication of waters in Poland. Anglers usually connect symptoms of this process with commercial fishing and fishery management. This opinion is far from being true. According to Toews (1985) '... *A common belief by sport anglers is that commercial fishing eliminates high quality angling. This has not only created pressures to eliminate commercial fishing on some lakes but has discouraged the concept of multiple use.*'. Anglers have very little knowledge of the fishery management. For instance, it is generally believed that commercial fishing, and especially tow nets and electric fishing are harmful to the fish stocks. However, comparison of the anglers' catches in lakes managed by SFF (intensive commercial exploitation) and PAA (no commercial exploitation) shows no negative effect of commercial fishing on the angling success.

Alternative recreational use of waters, e.g. boating, motor boats, was found to be one of the most important factors disturbing angling. Conflicts between water users have been documented by many authors (Bielby, 1978; Leopold, Bnińska & Hus, 1980; Parry, 1980; Bnińska, 1984) and are always considered to have a negative effect on angling (Brown, 1978; O'Riordan, 1978) either through the effect they have on water plants (Murphy & Eaton, 1981) and directly on fish. Noise caused by boats also scares the fish (Boussard, 1981). In addition, recreational use of waters also increases the loading of nutrients into the aquatic ecosystem (Bnińska, 1984).

Economic conditions in Poland are now changing rapidly. Irrespective of the possible changes that will take place in the fisheries, or water use in general, there will be a need for more rational lake management.

Acknowledgements

I wish to thank my colleague Katarzyna Grabowska, from the Department of Fishery Economics, Inland Fisheries Institute, for her help in the analyses of the data.

References

Bielby, G.H. (1978) Resolution of conflicts in freshwater recreational fisheries — South West Water Authority. Proc. Conf. Water Research Centre. Medmenham, Stevenage, pp 65−74.

Bnińska, M. (1984) Outdoor recreation — a new factor disturbing lake ecosystems, as exemplified by Lake Gim, Mazurian Lake District. *Vehr. Internat. Verein. Limnol.* **22**: 978−981.

Boussard, A. (1981) The reactions of roach (*Rutilus rutilus*) and rudd (*Scardinius erythrophthalmus*) to noises produced by high speed boating. *Proc. 2nd Brit. Freshw. Fish. Conf.*, Univ. Liverpool, 188−196.

Brown, K.S. (1978) Resolution of conflicts in still waters. Proc. Conf. Water Research Centre. Medmenham, Stevenage, pp. 75−82.

Leopold, M., Bnińska, M. and Hus, M. (1980) Angling, recreation, commercial fisheries and problems of water resources allocation. *Proc. Tech. Cons. Fish. Res. Alloc.*, Auburn University, 212−221.

Murphy, K.J. and Eaton, J.W. (1981) Waterplants, boat traffic and angling in canals. *Proc. 2nd. Brit. Freshw. Fish. Conf.*, Univ. Liverpool, 173−187.

O'Riordan, T. (1978) Angling and boating. Proc. Conf. Water Research Centre. Medmenham, Stevenage, 117−134.

Parry, M.L. (1980) Conflicts between sport fisheries and other recreational uses of water. Synthesis paper. EIFAC Internat. Tech. Cons. Fish. Res. Alloc. Vichy.

Toews, D. (1985) A proposed sport fisheries strategy for Manitoba. Transactions of the 1984 Canadian Sport Fisheries Conference. Ottawa: Department of Fisheries and Oceans, 129−152.

Tombleson, P.H. (1978) The angler's point of view. Proc. Conf. Water Research Centre. Medmenham, Stevenage, 49−60.

Chapter 14
Some factors affecting angling catches in Yorkshire rivers

S. AXFORD *National Rivers Authority, Yorkshire Region, Skeldergate, York, UK*

Collation of records of the catches of individual anglers on particular stretches of river over a period of 24 years, and more general information concerning the results of angling matches at widespread venues on Yorkshire rivers over a period of 19 years, have demonstrated the marked effects on catch rates caused by fish population changes. Examination of angling catch rates has provided a relatively cheap and effective method of monitoring fish populations.

At one venue, where factors affecting angling catches were examined in greater detail, disease outbreaks, strong year-classes and population movements of fish were all observed to have marked effects on catch rates. The effectiveness of some management procedures was also evaluated by this means.

14.1   Introduction

In the 1960s it was recognized that some method of monitoring fisheries was required in order for the organization then responsible for fisheries in the Yorkshire region to be able to judge the success, or otherwise, of its actions to maintain fisheries. Quantitative survey work on fish populations was precluded by cost and equipment efficiency limitations. At that time, the status of fisheries in Yorkshire was judged mainly from comments by anglers concerning their catches, but these were largely subjective views and were affected by such things as recent personal experiences, quirks of memory and forceful expression of opinions by some individuals.

The use of catch effort data for monitoring stock status had been well established for commercial fisheries but little attention had been paid to such data for monitoring recreational freshwater fisheries in Britain. Early creel census work by Lagler and De Roth (1953) in the USA had, however, shown that angling catches could be related to stocks and this suggested that it might be a useful technique to employ for monitoring the fisheries of the major angling areas in the Yorkshire region. In any case, the anglers, who largely paid for the service, would judge the success or otherwise of policies on the basis of the effects on catches rather than any biological statistics of the fish populations.

The larger rivers and lakes of Yorkshire where it was intended to collect such catch effort data are recreational coarse fisheries in which all the fish caught are returned alive to the water during or after the fishing session. This practice prevents catch rate being directly equated with fishing mortality and thus estimation of stock size from an examination of changes in catch rate with cumulative catch. Because the fish are not removed in a creel, this word was avoided when designing the Yorkshire monitoring schemes, which have been termed the angling census and the match return scheme.

14.2   Methods

The first angling census was instituted at a popular fishing venue in 1965 and at a second in 1966. Data have been collected at these two venues until the present time. Censuses have been operated at a number of other venues for varying periods. A clerk is employed to collect from individual anglers details of the species, number and size of fish caught, fishing methods and baits used, the fishing period, and to provide general observations of river and weather conditions. These are entered onto individual cards (Fig. 14.1). The census has generally been taken once a week on either a Saturday or Sunday, when most angling takes place. The individual returns are collated for each day on which the census is taken and the results combined over varying periods. The venues were chosen so as to provide sufficient returns over a period to reduce the effects of the occasional presence of anglers of exceptional skill on overall catch rates. The methods used by anglers and rules governing angling practice have changed very little during the study period and thus should provide data that are comparable between periods.

NATIONAL RIVERS AUTHORITY - YORKSHIRE REGION
Census

	-5"	5"-7"	7"-9"	9"-11"	11"-13"	13"-15"	15"+
TROUT							
GRAYLING							
CHUB							
BARBEL							
BREAM							
PIKE							
DACE							
ROACH							
PERCH							
BLEAK							
GUDGEON							
RUFFE							

Section

Record method in each square and tick other methods used

F - Fly
M - Maggot (float)
B - Maggot (ledger)
W - Worm (float)
L - Worm (ledger)
S - Spinner
A - Other bait
state type below

Any other species here

Tick
Match
Pleasure

Time
Fishing 6...7...8...9...10...11...12...1...2...3...4...5...6...7

NRA 88

Fig. 14.1 Angling census card used by clerks to monitor catches

Angling matches are competitions between individuals or groups of anglers fishing for the same restricted time period at a venue, and which are generally decided on the basis of the total weight of fish of all species captured by each individual or group. Returns of angling match results have been collected since 1971 from a large number of venues throughout the Yorkshire region. The match returns are completed by angling club match secretaries and then submitted to an umbrella angling organization, from where they are collected by the body responsible for fisheries. The details collected are shown in Fig. 14.2 and consist of data on the venue, date, fishing period, weights of catches, number of anglers who caught fish, species caught, and river and weather conditions. These data are codified as necessary for computerized collation. Data comparisons are usually made over a particular stretch of river on an annual basis.

14.3   Results and discussion

The most notable observations from these two main schemes for collecting angling catch data have been the relatively constant species composition at particular venues and the consistent differences in species composition between venues, presumably related largely to their physical characteristics. Nevertheless, there have been notable changes in some instances, which, upon further investigation,

WATER AUTHORITY AND ANGLING ASSOCIATION JOINT MATCH RECORD

Please complete and return both copies (no carbon required) to Match Secretary

Name of Club ....

Date of Match .... Venue ....

Number of Competitors .... Duration of Match .... Hours

Number of anglers .... Winning weight .... lb ..... ozs
weighing in their catch.

                                         2nd   "   .... lb ..... ozs

                                         3rd   "   .... lb ..... ozs

Total weight .... Main species caught: Winner ....

Best Fish ....   Others ....

River Conditions .... Weather Conditions ....

DECLARATION   I hereby certify that at the conclusion of the match the River
              Bank was left free from all litter, and that all gates were
              secured.

SIGNED .... Club Secretary

Fig. 14.2 Match return form to monitor competition results

have indicated significant causal changes in fish populations. Major causes in the Yorkshire region have been identified as disease outbreaks, recruitment of strong year-classes, and habitat alterations. Management procedures, such as stocking, appear to have had little long-term influence.

The angling catch monitoring schemes indicated major effects on fish stocks of the outbreaks of 'roach ulcer disease' in the late 1960s and 'perch ulcer disease' in the early 1970s; both considered, on later investigation, to result from infections of an aberrant form of *Aeromonas salmonicida*, the organism that causes furunculosis in salmonids. Very few dead fish were found during these outbreaks and evidence for their presence came largely from observations of small numbers of ulcerated fish caught by anglers. The effect on catches of these species was very marked, and the spread and recovery from the diseases could be followed easily by this technique. A reduction in roach (*Rutilus rutilus* (L.)) catches in 1968 at Kirk Hammerton was associated with poor catches of some other species that year. The reduction in perch (*Perca fluviatilis* L.) catches in 1972 was accompanied by a low catch rate of dace (*Leuciscus leuciscus* (L.)), but other species had fairly normal catch rates. These reductions in perch and roach catches appeared to have little effect on catches of other species in subsequent years. The fact that catch rates of other species changed little suggests that anglers did not or could not successfully target other species in order to compensate for loss of these species from their catches. Changes, if any, in species vulnerability to angling due, for example, to increased availability of natural food or reduced competition for anglers' baits, did not appear to be significant in years subsequent to those in which the roach or perch declined (Fig. 14.3).

Recruitment of strong year-classes was clearly observed from catch rates of fish in the Kirk Hammerton fishery. In August 1976, large numbers of very small roach and dace appeared in the catches. Examination of some of these fish showed that they were of the 1975 year-class and had just reached a size at which their gape allowed them to ingest the anglers' bait. Over the main angling period of June-September, roach catch rates increased more than six-fold (Fig. 14.4) and dace catch rates more than tripled by comparison with the previous year. Catch rates of all species combined went from about one fish per hour in 1975 to nearly three fish per hour in 1976. Species which were slower to reach a size at which they fed on angler's maggot bait, but which showed a strong year-class in 1975, such as barbel and chub, appeared in subsequent seasons (Fig. 14.5).

There has been little recognition of the possible importance of population movements of coarse fish in British rivers. Again, evidence has come from angling catch data that there are significant seasonal population movements in some rivers, and blocks to such movement may seriously affect stocks present in particular stretches. Erection of a gauging weir on the River Nidd at Skip Bridge in 1978 was associated with marked reductions in the catch rates of small fish upstream at Kirk Hammerton (Fig. 14.6) in subsequent years. This was interpreted as being due to the prevention of upstream migration of small fish in spring from overwintering areas. A subsidiary weir was installed in July 1984 downstream of

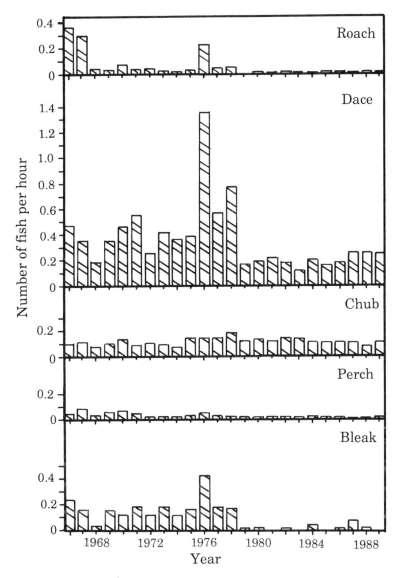

Fig. 14.3 Catch rates (no. hr^{-1}) at Kirk Hammerton for different species of fish

the gauging weir in order to raise the tail water level on the latter and thereby assist fish passage. However, angling results in subsequent years have not revealed any marked benefit from this installation. Since the catches before 1979 comprised a high proportion of small fish, overall catch rates of all species combined have declined, such that anglers now have to fish for about twice as long, on average,

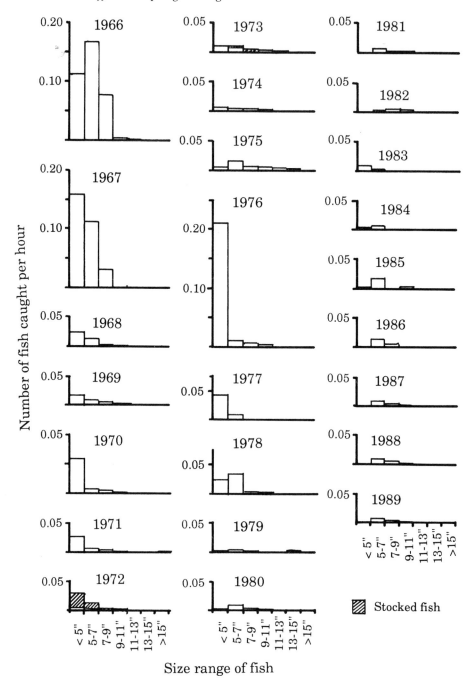

Fig. 14.4 Catch rates of roach by size-classes for the June to September period each year at Kirk Hammerton

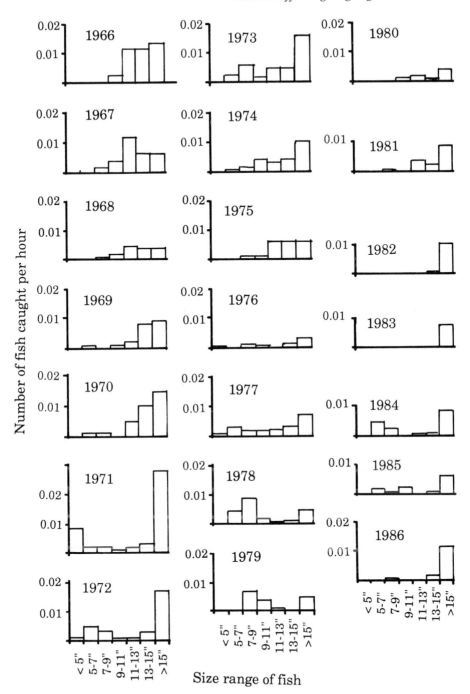

Fig. 14.5 Catch rates of barbel by size-classes for the June to September period each year at Kirk Hammerton

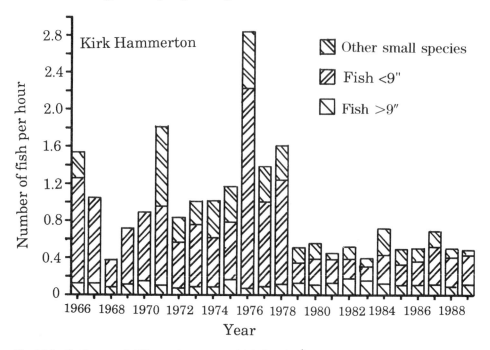

Fig. 14.6   Catch rates of different size groups of fish (no. hr^{-1}) at Kirk Hammerton

to catch a fish as they did before 1979. Angling match return data from up and downstream of the weir provided confirmation of this effect.

The effects of a tidal barrage at the mouth of the River Derwent on catches of flounders can be seen from the percentage of match returns on which this species was recorded among the main species captured. After completion of the barrier in 1975, flounders steadily reduced in catches from being reported on 25% of match returns in 1975 to only 1% in 1978 (Fig. 14.7).

The main remedial management measure suggested by anglers has been re-stocking of rivers with fish from other waters. Nevertheless, angling catch data have generally shown little benefit to fisheries from these introductions. In the winter of 1973–74, 126000 roach weighing 5 tonnes imported from Holland were introduced to the rivers Ure and Ouse, mostly just upstream of York. From Fig. 14.8 it can be seen that there was little immediate effect of these introductions on the relative abundance of roach in match catches and the overall weights of fish caught and the percentage of anglers weighing in their catch in this stretch showed no marked changes (Fig. 14.9). Similarly in the smaller River Nidd, the introduction of 11000 roach in 1972 and a further 1860 roach in 1973 only marginally increased catches in those years and there is a suggestion of some adverse effect on catches of native fish (see Fig. 14.4).

The census at Kirk Hammerton usually operates for about 40 days per year collecting data from about 600 anglers and costs just over £500 at current prices. This is equivalent to the cost of about two days fieldwork for a four-man team

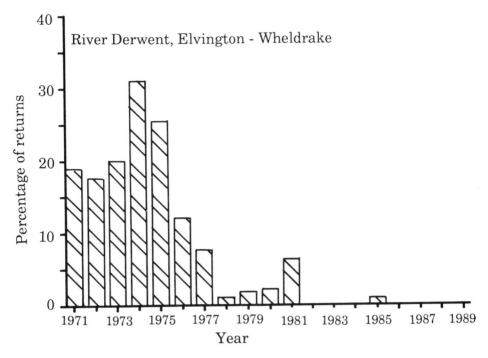

Fig. 14.7 The importance of flounder in the catches between Elvington and Wheldrake on the River Derwent

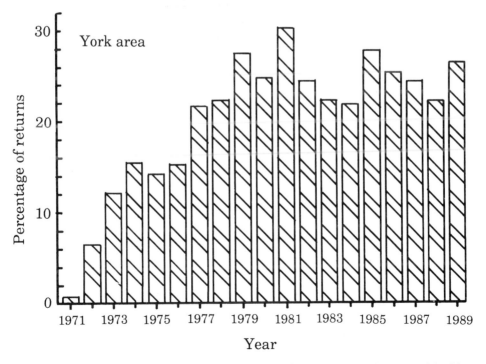

Fig. 14.8 The importance of roach (% frequency of capture) to catch in the York area of the River Ouse

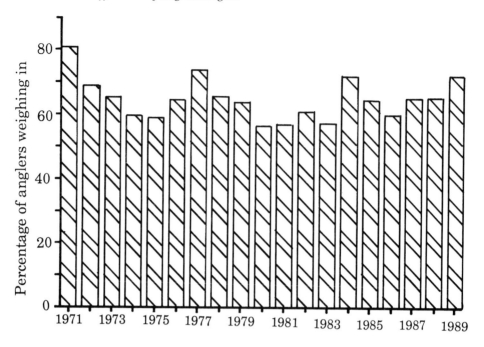

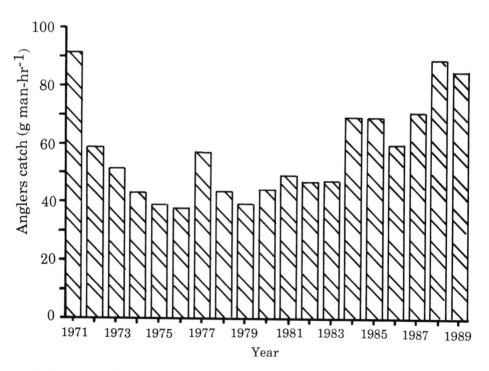

Fig. 14.9 Catch statistics (% anglers weighing in and mean catch weight) in the York area of the River Ouse 1971–89

Table 14.1 Comparison between electric fishing (three runs between nets) and census surveys in the River Nidd. (Note: During the electric fishing survey large shoals of dace and perch were seen but not captured. Good estimates were only made for the chub population.)

	Electric fishing	Angling Census
Man-days	8	23
Cost (£)	520	506
	(staff + equipment)	(contract employee)
Length fished (m)	100	3600
No. spp. captured	8	15
Total no. fish captured	177	4045
No. fish per man-day	22	176

using electric fishing or netting techniques, which have not proved to be very efficient (Table 14.1).

Thus, the collection of angling catch effort data has proved to be a useful, cost-effective means of monitoring the major Yorkshire fisheries. Particular constraints include finding suitable census clerks to operate regularly at venues over several years and in a consistent manner, and getting full completion of individual match returns. Both schemes are in need of evaluation to determine whether more or less detailed data can and should be collected and whether the precision of catch statistics is wastefully great, inadequate, or about right, since this will affect any further geographic expansion of the schemes.

Reference

Lagler, K.F. and De Roth G.C. (1953) Populations and yield to anglers in a fishery for largemouth bass. *Pap. Mich. Acad. Sci.* **38**: 235−253.

Chapter 15
The use of angler catch data to examine potential fishery management problems in the lower reaches of the River Trent, England

I.G. COWX *Humberside International Fisheries Institute, University of Hull, Hull, HU6 7RX, UK*

The catch of each angler participating in fishing matches on a stretch of the River Trent between Stoke Bardolph and Gunthorpe, Nottinghamshire, has been recorded since June 1979. This paper examines these data to identify trends in catch rates between years, times of the year and sections of the stretch which might be used to aid the management of the fishery.

There has been a marked change in the species composition of anglers' catches since 1979. The fishery has changed from a roach/dace to a bream/chub fishery in recent years. This has been coupled with a steady improvement in the catch rates until 1986 when catches started to decline. Catch rates from different parts of the stretch are extremely heterogeneous and reflect the impact of variability in cover and effluent discharge. An allegation that power stations upstream of Nottingham being placed on standby at weekends has a detrimental effect on catch rates was found to hold some credibility. The trend, however, has existed for much longer than anglers have perceived and is also due to heavy fishing intensity at weekends.

15.1 Introduction

In the UK, information relating to the fish stocks of large rivers and lakes is sparse, primarily because of the difficulties in accurately sampling the fisheries quantitatively (EIFAC, 1974). Unfortunately, it is often these fisheries which are the most problematic, usually as a result of man's activities about the water body.

One possible source of fisheries-related data for these waters is anglers. Although in the UK anglers do not usually crop the fish stocks *per se*, they do impose considerable effort to 'exploit' them. This is particularly true during competitions or match angling where there is often considerable financial incentive to record the largest catch in terms of weight.

This source of data has already been adopted by many workers to provide general information about the status of a fishery in rivers (Axford, 1979; North, 1980; Cooper & Wheatley, 1981; Hickley & North, 1981; Pearce, 1983; Cowx &

Broughton, 1986; Cowx *et al.*, 1986; North & Hickley, 1989) but has rarely been used to provide specific answers to management problems (North & Hickley, 1977; Cowx, 1990).

In recent years, anglers fishing the River Trent, England, downstream of Nottingham have made repeated allegations that the quality of fishing in the region has declined. In particular anglers have expressed concern about:

- an overall deterioration in the catches since the mid-1970s;
- a decline in the fishing downstream of the outfall from Stoke Bardolph sewage works (Fig 15.1);
- a decline in catches at weekends associated with power stations upstream of Nottingham, which use the river for cooling purposes, being placed on standby over this period.

With the paucity of data on fish stocks in the river from conventional sampling techniques, in 1979 the Severn-Trent Water Authority, the regulatory body for the river, turned to the anglers for co-operation to provide angler catch data to justify their allegations. This paper reports on the first eight years of this angler catch effort monitoring programme on a stretch of the river below Nottingham controlled by the Nottingham Federation of Anglers, and how it has been used to respond to the anglers' complaints.

15.2   The study site

The River Trent is approximately 280 km long from its source in North Staffordshire (Grid Ref: SJ896548) to its confluence with the Humber Estuary (Grid Ref: SE 865233) (Fig 15.1). The important topographical features of the river with regards fisheries have been described elsewhere (Cowx & Broughton, 1986).

The present study is based on the reach between the 2 m-high weir at Stoke Bardolph (Grid Ref: SK 661407) and Gunthorpe Bridge (Grid Ref: SK682437) 7 km downstream. The stretch is one of the most popular fisheries in the Trent area and includes the 'Golden Mile' length which became nationally famous in the 1960s because of the large catches of roach taken by match and pleasure anglers. The river has an average width of approximately 70 m and a depth of 2−3 m. Along the left bank, 342 permanent 'pegs' (numbered concrete posts) are installed approximately every 12 m and mark the positions from which anglers may fish during angling competitions. Occasionally, where physical obstructions exist, greater distances were found between the pegs.

15.3   Methods

The angling club controlling the left bank of the stretch, the Nottingham Federation of Anglers, agreed to record the catches of match anglers from each peg. These were reported, along with the numbers of hours fished and species caught, on

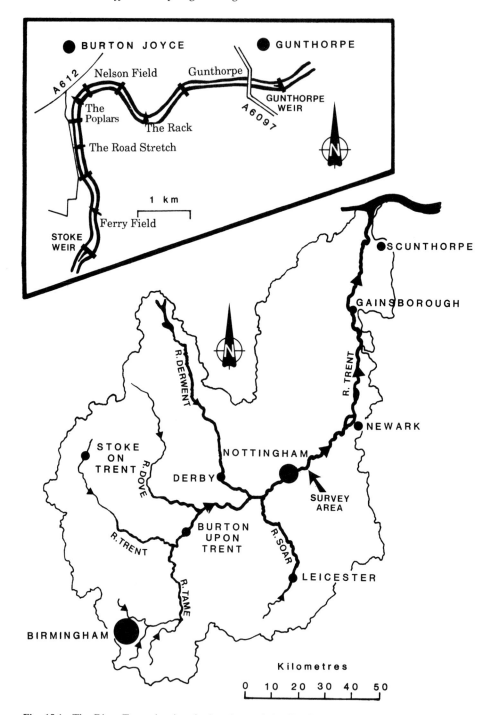

Fig. 15.1 The River Trent showing the location and details of the study reach

catch-return forms (Fig. 15.2) for later analysis. The main statistic compiled from these data is the catch per unit angling effort. The data were combined in various formats to determine the mean annual catch rates, the mean monthly catch rates, the mid-week (Monday to Friday) versus weekend (Saturday, Sunday) catch rates and differences in the catch rate between various parts of the stretch.

In addition, an assessment of species composition of catches in relation to changes in catch rates was undertaken using the index of relative importance (RI) described by Cowx and Broughton (1986), where:

$$RI = (\% \text{ relative abundance} + \% \text{ occurrence}) \times 0.5$$

The percentage abundance of a species was calculated on a points scale according to its recorded importance in the catches. Thus, when a species was recorded as most common it was awarded four points, next most common two points and other captured species one point. Abundance was derived by expressing the total points score for each species as a percentage of the total points awarded for all species. Frequency was determined as the percentage of all matches in which a species was represented in the catches.

15.4   Results

15.4.1   *Annual variation in catch rates*

In match angling, factors such as rumours of a large individual catch or adverse weather conditions are known to influence the number of anglers weighing in their catch (Whiting *et al.*, 1976) and hence depress catch rates. However, in this study there was an overall trend towards an improvement in the mean annual catch rate from $116 \text{ g man hr}^{-1}$ in the 1979−80 season to $220 \text{ g man hr}^{-1}$ in the 1986−87 season (Fig. 15.3). This represents a mean annual increment of $15.5 \text{ g man hr}^{-1} \text{ yr}^1$. The improvement in catch rates observed in this study represents an extrapolation of the trends observed in an earlier study on the fishery by Cowx and Broughton (1986), (Fig. 15.3).

15.4.2   *Monthly and weekly variations in catch rate*

When the seasonal catch rates are broken down into monthly statistics, a clear seasonal pattern in angling success was observed (Fig. 15.4). Early in the season, June−August, the catches were high, but as the season progresses these catch rates steadily decline, only picking up marginally in spring. This pattern was common for all seasons but the comparable monthly catch rates were significantly greater (P < 0.05) in recent years, supporting the evidence that overall catches in the river were not on the decline.

Further subdivision of the data into week-day and weekend catch statistics for each month (Fig. 15.5) showed that weekend anglers have a lower CPUE than their week-day counterparts. This trend was consistent over all seasons examined,

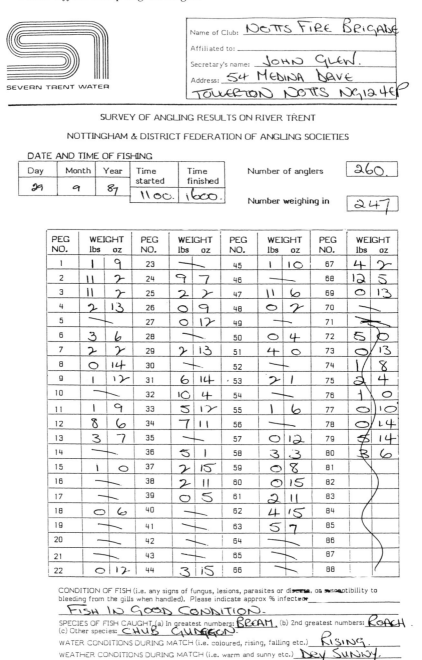

Fig. 15.2 An example of the catch return form

including seasons before the reduction of power station activity, and hence thermal discharge, over weekends. Examination of temperature profiles for the River Trent immediately upstream of the study area (e.g. Fig. 15.6 for November and December 1986) showed a tendency towards lower temperatures at weekdays,

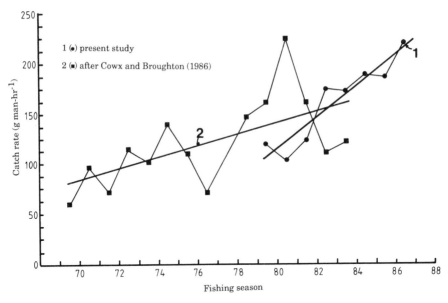

Fig. 15.3 Variation in catch rates between the seasons 1969 and 1987 (2. after Cowx and Broughton, 1986)

possibly as a result of reduction in power station activity, but these differences were not significant (P > 0.1).

15.4.3   *Variations in catch rate within the study site*

To examine whether the discharge from Stoke Bardolph Sewage Works has an impact on the status of the fishery the catch and effort data for different sections of the study zone were combined. The sections were discriminated by sub-division of the study area according to local names for various parts of the stretch (see Fig. 15.1). This was deemed acceptable because the Nottingham Federation of Anglers usually allocate one or more of these sections, depending on the number of anglers participating, to each match.

Although the discharge enters the river below Peg 273 (Fig. 15.7) true mixing is not achieved for some 1 km downstream, at approximately Peg 180. The discharge, however, does have an affect on the fishing. Catch rates were lower for the zone immediately downstream of the sewage outfall compared with above the outfall or at the downstream limit of the study area (Fig. 15.7).

15.4.4   *Variation in species composition of the catches*

Twenty-two species of fish were caught during the study period. Of these, only ten, namely roach, *Rutilus rutilus* (L), dace, *Leuciscus leuciscus* (L), chub, *Leuciscus cephalus* (L), bleak, *Alburnus alburnus* (L), bream, *Abramis brama* (L), gudgeon,

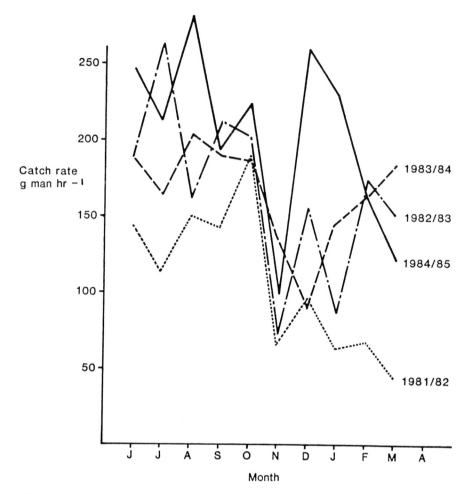

Catch rate
g man hr ⁻¹

250
200
150
100
50

1983/84
1982/83
1984/85
1981/82

J J A S O N D J F M A

Month

Fig. 15.4 Monthly variations in catch rates between different fishing seasons

Gobio gobio (L), perch *Perca fluviatilis* L., tench, *Tinca tinca* (L), carp, *Cyprinus carpio* L., and barbel *Barbus barbus* (L), were caught in sufficient numbers to be of importance to the anglers' catches.

The relative importance of the major angling species varied with fishing season (Fig. 15.8). From 1979 roach showed a gradual decline in importance, although there was a tendency to increase their contribution again in later years. Chub, by contrast, showed a general increase in importance throughout the study period, whilst perch declined. Perhaps the most important changes, however, were barbel and bream which consistently increased their contribution to anglers' catches as the study progressed.

15.5   Discussion

In any study of fish populations as revealed by anglers' catches, several basic assumptions are made. The principal one of these is that catches are representative of

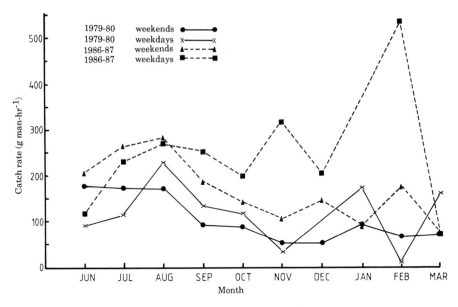

Fig. 15.5 Intra-seasonal catch rates for weekend and week-day fishing matches

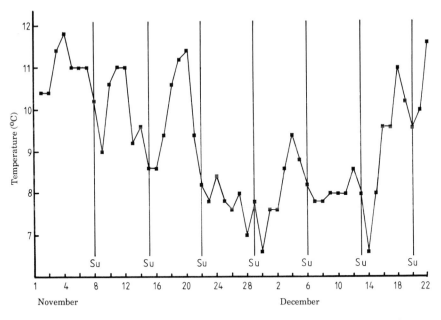

Fig. 15.6 Temperature profile for November and December 1978 (Su indicates Sundays)

the status of the fisheries in terms of species composition and numbers and weight of fish caught. In this present study there is no reason to suggest that this is not the case. Indeed, several authors (Axford, 1979; Cooper & Wheatley, 1981; Cowx & Broughton, 1986) have suggested that creel census techniques are a better

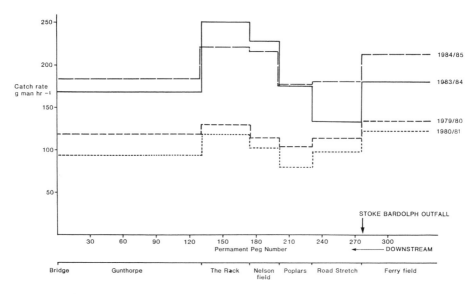

Fig. 15.7 Variation in catch rate with angling season for different zones of the study area

representation of the fish populations than other more conventional and often highly selective fishing methods.

The present investigation shows that many of the anglers allegations are wholly or partially unjustified. A consistent increase in catch rates ($15\,\text{g man-hr}^{-1}\,\text{year}^{-1}$) was observed between 1979 and 1986 (Fig. 15.3), which signifies a marked improvement in the status of the fishery. Intrinsically linked with the 45% increase in catch rates between 1979 and 1986 was a 21% increase in the actual weight of fish caught (5.5–17.0 tonnes) coupled with a 29% fall in angling pressure (44 000–31 000 man-hours). These changes suggest an improvement in the catch per unit effort, and an overall catchability of the stock. With the catch per unit effort reaching $220\,\text{g man-hr}^{-1}$ in 1986, the Nottingham Federation Fishery compares favourably with other large rivers in the UK, e.g. the River Ouse – $158\,\text{g}$ man-hr^{-1} (Axford, 1979), the River Severn – $182–176\,\text{g man-hr}^{-1}$ (Hickley & North, 1981).

These observations prove in real terms that the fishery is not declining as proposed by the anglers. The underlying cause behind the alleged decline in the fishery is probably linked to the change in its status from a principally roach/dace to a chub/bream fishery (Cowx & Broughton, 1986). Anglers favour a mixed catch of fish dominated by a good number of roach. In recent years the catches have been dominated by a few large chub and in later years bream. Thus, it is reasonable to assume that the anglers are confusing the decline in their catches of roach with that of the whole fishery. It should be pointed out that these changes in the fishery are a function of an improvement in water quality in the river since the early 1970s (Cowx & Broughton, 1986) and not a change in the habitat *per se*.

Although the fishery showed an overall tendency towards an improvement

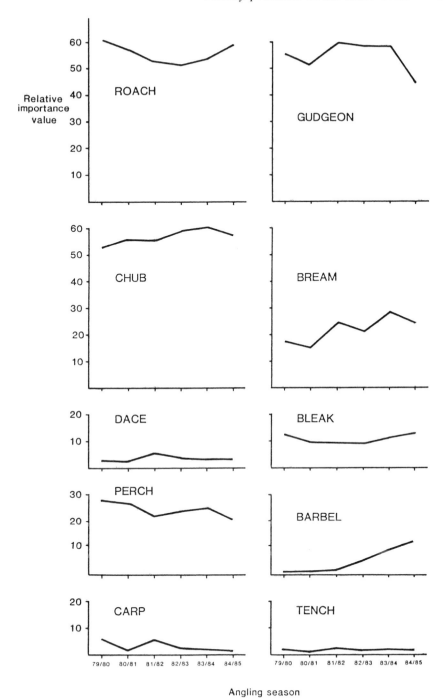

Fig. 15.8 Variation in the relative importance of different fish species to anglers' catches over the study period

between 1979 and 1986, catch rates decline as the season progresses. These results confirm the premise offered by anglers that the River Trent is a 'summer river'.

The allegation that catches deteriorate at weekends was wholly substantiated by this study (Fig. 15.5). However, there is no evidence (Fig. 15.6) to suggest that the reduction in power station activity, and hence depression in water temperature because of the reduction in thermal effluent, is implicated. On the contrary, the most plausible explanation for this disparity in catches is that the fishery is being overexploited in some way.

As coarse fisheries in the UK are traditionally catch and return, the overexploitation is not caused by a reduction in stock density. However, it has been found that high fishing pressures reduces the catchability of fish by inducing temporary hook avoidance (Raat, 1985) and more permanent angler damage (Cooper & Wheatley, 1981). Such overexploitation indirectly reduces the catchable stock size. This mechanism was thought to be initiated in the weekend fishery over July and August when fishing pressure was high. After this time the proportion of stock remaining accessible is sufficient to satisfy the demands of the less intense (in numerical terms) week-day angling pressure (hence their catches were reasonable) but not the large number of weekend anglers.

The pattern of results for angling success in different zones of the study reach (Fig. 15.7) suggest the region downstream of Stoke Bardolph sewage works is not as productive as elsewhere. However, the only season when a dramatic decline in the fishery below Stoke Bardolph was observed was coincident with the dry summer of 1983/84 and may be linked to reduced dilution of the effluent during this period. As complete mixing of the effluent does not occur until the Poplars zone this hypothesis is somewhat suspect, particularly as elevated catch rates were observed further downstream. The change in catch rates is more likely linked to:

(1)   A change in topography of the river. In the Rack and Nelson Field zones, where the catch rates were high, the river forms a narrow channel around the bend of the river and a large shallow area which presents an ideal habitat for chub, bream and roach.

(2)   Upstream of Stoke Bardolph outfall, catch rates are affected by the consistently excellent catches recorded from Pegs 340−342. These are primarily the outcome of large catches of chub taken immediately below Stoke Weir at the upper end of the study area.

The present study demonstrates the value of collecting creel census data as an aid to fisheries management in large water bodies where more conventional sampling gears are inefficient. Without the information provided by the census it would have been impossible to substantiate and/or refute any of the anglers' complaints and formulate a suitable response. In the current study, it was fortunate that the angling federation controlling the stretch of river under examination was willing to co-operate in the work and diligently collect the data from every match. This stems from a great deal of effort being expended, to educate the anglers in the value of such work and how it might help in the long-term, when the project was formulated. Education of anglers/end-users is as much a part of any study as the data collection and processing. To ignore this aspect is usually to the detriment of the project and may ultimately lead to its failure in terms of the quality of data collected. In addition, the results of this study were fed back to the anglers. This has helped to alleviate their fears concerning an overall decline in the fishery and ensure continued co-operation.

Acknowledgements

It is a pleasure to thank all those anglers who co-operated in recording their catch data, the National Rivers Authority − Severn Trent Region and its predecessor Severn Trent Water − for access to the data. I would also like to thank Messrs G.A. Wheatley and J.R. Walker for assistance in analysing the data.

References

Axford, S.N. (1979) Angling returns in fishery biology. *Proc. 1st Brit. Freshwat. Fish Conf.*, Univ. Liverpool, 259−272.
Cooper, M.J. and Wheatley, G.A. (1981) An examination of the fish population in the River Trent, Nottinghamshire, using anglers' catches. *J. Fish Biol.* **19**: 539−556.
Cowx, I.G. (1990) Application of creel census data for the management of fish stocks in large rivers in the United Kingdom. In *The Management of Freshwater Fisheries* (Ed. by W.L.T Van Densen, B. Steinmetz and R. Hughes) Wageningen: PUDOC.
Cowx, I.G. and Broughton, N.M. (1986) Changes in the species composition of angling catches in the River Trent (England) between 1969 and 1984. *J. Fish Biol.* **28**: 625−636.
Cowx, I.G., Fisher, K.A.M. and Broughton, N.M. (1986) The use of anglers' catches to monitor fish populations in larger water bodies with particular reference to the River Derwent, Derbyshire, England. *Aquacult. Fish. Mgmt.* **17**: 95−103.
EIFAC (1974) Symposium of methodology for the survey, monitoring and appraisal of fishery resources in lakes and large rivers. EIFAC Tech Paper **23**, Rome: FAO, UN Publications.
Hickley, P. and North, E. (1981) An appraisal of anglers' catch composition in the barbel reach of the River Severn. *Proc. 2nd Brit. Freshwat. Fish Conf.*, Univ. Liverpool, 94−100.
North, E. (1980) The effects of water temperature and flow upon angling success in the River Severn. *Fish. Mgmt.* **11**: 1−9.
North, E. and Hickley, P. (1977) The effects of reservoir releases upon angling success in the River Severn. *Fish. Mgmt.* **8**: 86−91.
North, E. and Hickley, P. (1989) An appraisal of anglers' catches in the River Severn, England. *J. Fish Biol.* **34**: 299−306.
Pearce, H.G. (1983) Management strategies for British coarse fisheries: the lower Welsh Dee, a case study. *Proc. 2nd Br. Freshwat. Fish Conf.*, Univ. Liverpool, 263−273.
Raat, A.J.P. (1985) Analysis of angling vulnerability of common carp, *Cyprinus carpio* L., in catch and release angling in ponds. *Aquacult. Fish. Mgmt.* **16**: 171−188.

Chapter 16
Anglers' catches as an illustration of the fish community structures, angling pressure and angling regulations, based on inland waters in the Krosno region, Poland

A. WOŁOS and P. PISKORSKI *Inland Fisheries Institute, Olsztyn, Poland*

Analyses were carried out on catch returns collected from 1980 anglers fishing waters of the Krosno region, Poland, during 1986. Annual angling pressure ranged from 0.89 to 249.6 angler-days km^{-1} in zones of the rivers and from 0.47 to 5.96 angler-days ha^{-1} in reservoirs. Anglers caught 28 different fish species. Only part of the river systems were characterized by typical riverine fish zones; the remainder seemed to be strongly disturbed and dominated by less valuable fish species. Anglers' catches appear to be influenced by the creel census and regulatory measures in force in the region under study. The attractiveness of fishery zones in the region is discussed and general guidelines for the fishery management are presented.

16.1   Introduction

Fishery management in waters exploited by anglers and its modifications to meet the needs of anglers is possible only if proper information is available. The information needed includes basic data on the aquatic environment, the fish stocks, data on angling pressure and anglers' preferences, as well as the management measures used to date. A number of authors have advocated that analyses of catch returns are a source of information on the quality of the aquatic environment and its fish stock, and could even be used, frequently in conjunction with other sampling techniques, to study the fish populations. Studies of this type are particularly comprehensive in a number of large British rivers: River Severn (North & Hickley, 1977; North, 1980; North & Hickley, 1989), River Trent (Cooper & Wheatley, 1981; Cowx & Broughton, 1986; Cowx, 1990) and River Derwent (Cowx, Fisher & Broughton, 1986; Cowx, 1990).

In Poland, studies devoted to angling were not initiated until 1978, when questionnaire studies were made on a national scale (Leopold, Bnińska & Hus, 1980; Bnińska & Leopold, 1987; Leopold & Bnińska, 1987). They were of a general character, related to particular regions of Poland and to selected categories of waters. No attention was paid to particular water bodies. These studies showed that rivers were of special significance to anglers; on a national scale 40.6% of all angler-days being spent on rivers. The questionnaires obtained during these studies were also used to estimate the angling pressure and catches on three large rivers: the Biebrza, Bug and Narew (Szlażyńska & Wołos, 1988).

In Poland, most of angling pressure is concentrated on waters used by the State Fish Farms and the Polish Anglers Association (PAA). Angling on PAA waters is permitted for members only (980 000 in 1986). In this context all rivers are managed by the PAA. This association is organized in relation to the administrative situation in Poland, i.e. PAA regional departments are located in each of the 49 voivodeships.

Since 1986, two PAA departments, Słupsk and Krosno, have introduced obligatory registration of anglers' catches. In Krosno voivodeship almost all inland waters (6284 ha) belong to the PAA, of which rivers contribute 3689 ha of water and reservoirs 2560 ha. The region is mountainous and the fisheries are predominantly those associated with trout and grayling zones of rivers, although barbel zones are found in the lower sectors. Consequently, Krosno region is one of the most attractive to anglers in Poland. As a result this region has been used to show how catch returns can be used to formulate management policy and recommendations.

16.2   Material and methods

After the 1986 fishing season, complete catch records were returned by 1376 local anglers (about 20% of all local anglers in the region) and 604 visiting anglers (from other regions) for the Krosno region. These were analysed to assess the status of the fisheries in the region.

Waters of the region were divided into fishery zones as shown in Fig. 16.1. On each day of fishing in a particular zone, anglers recorded their catch, including details of the fish species caught, their numbers, and approximate weight. Altogether, 29 fishery zones were monitored, of which three were reservoirs. Smaller rivers were treated as one fishery zone whilst larger rivers were subdivided. The biggest river of the region, River San, was divided into seven fishery zones, two of which were reservoirs. In eight very small fishery zones only 11 anglers fished during 1986 and they were excluded from the analysis. Zone Ropa-5 was also excluded as most of the catch returns listed fish caught in nearby small pools and not in the river itself.

The following parameters were calculated:

(1)   Angler pressure measured as total number of angler-days per km of a river, or per ha of reservoir.
(2)   Catch per unit of effort (CPUE) expressed as kg angler-day^{-1}. This index was also calculated for the most important species.
(3)   Species composition of anglers' catches and average individual weight of the most important species.

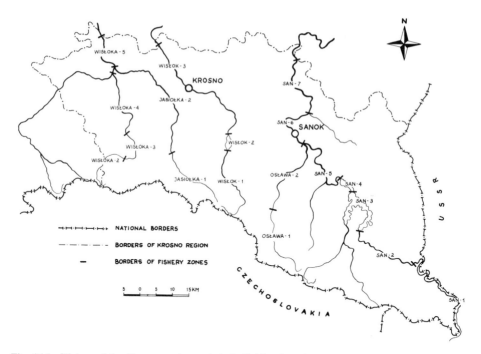

Fig. 16.1   Waters of the Krosno region and their division into fishery zones

16.3   **Results**

16.3.1   *Species of fish caught*

Twenty-eight fish species were caught in the waters of Krosno region. The most important 11 species (see Fig. 16.2) comprised 96.6% of the total catch.

Significant differences were found between catch composition of local and visiting anglers. For instance, grayling (*Thymallus thymallus*) represented 26.8% of total catch of local anglers, but 54.8% of the catch of visiting anglers.

16.3.2   *Angling pressure*

In 1986 registered anglers spent a total of 21 550 days angling in Krosno region, of which 72.3% was on the River San. Local anglers spent less time (66.2% of their angler-days) on this river than visiting anglers (95.3% of fishing effort). Intensive angling pressure on the San River was accompanied by high catches: total registered catch in the region amounted to 21 955 kg of fish, of which 74.6% were caught in the San River.

Angling pressure in the riverine fishery zones was measured as the number of angler-days km^{-1} of river length. This index does not take into account river breadth. Nevertheless, it can be used as a rough estimate of the difference in angling pressure in particular zones (Fig. 16.3). Average angling pressure for the

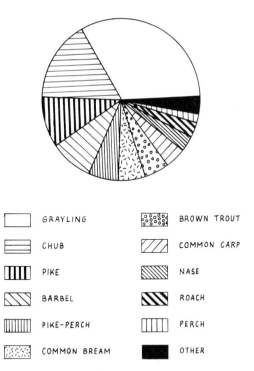

	GRAYLING		BROWN TROUT
	CHUB		COMMON CARP
	PIKE		NASE
	BARBEL		ROACH
	PIKE-PERCH		PERCH
	COMMON BREAM		OTHER

R

Fig. 16.2 Species composition of anglers' catches from waters in the Krosno region

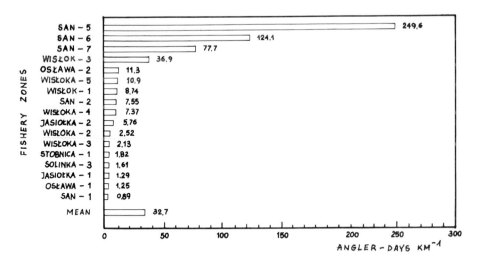

Fig. 16.3 Angling pressure (angler-days km^{-1}) in different fishery zones

whole region amounted to 32.7 angler-days km^{-1}, ranging from 0.89 in zone San-1 to 249.6 in San-5. San-5 (which was almost half the length of San-1) supported 53.1% of the total number of angler-days in this river which was broken down into 7 zones.

16.3.3   *Catch per unit of effort*

Average CPUE for the whole region amounted to 1.02 kg angler-day^{-1}, varying within a wide range from 0.67 kg angler-day^{-1} (Wisłoka-2) to 1.65 kg angler-day^{-1} (San-3) (Fig. 16.4). Correlation between angling pressure and CPUE proved to be statistically insignificant. This is understandable since CPUE is affected by both environmental factors which are not related to angling pressure, such as water quality, river productivity (increasing in lower reaches), and by factors not related to environmental quality, including anglers' preferences and legal restrictions. Anglers' preference is of special importance in Krosno region as most anglers in this region aim to catch brown trout (*Salmo trutta*) and grayling. Figures 16.5 and 16.6 present CPUE for these two species. It was found

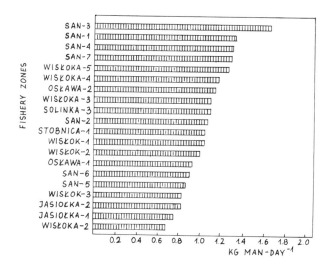

Fig. 16.4 Total catch per unit effort in different fishery zones

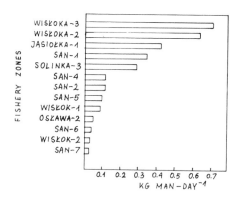

Fig. 16.5 Catch per unit effort for brown trout in different fishery zones

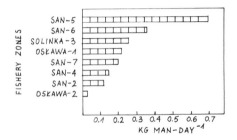

Fig. 16.6 Catch per unit effort for grayling in different fishery zones

that the zones Wisłoka-2 and Jasiołka-1 were characterized by the lowest CPUE in relation to total catch (Fig. 16.4), but the highest for brown trout (Fig. 16.5). Similarly, San-5 and San-6 had very low total CPUE, but high values for grayling (Fig. 16.7).

16.3.4 *Number of fish species caught by anglers*

Significant differences were found in the number of fish species caught in particular zones (Fig. 16.7). The highest number of species caught was recorded in the reservoirs San-3 and Wisłok-2 and in lower parts of large rivers (San-7 and Wisłok-3). The results depended not only on the state of the fish stocks in the particular zones but also on official regulations of the PAA.

Apart from protective regulations (catch limits, legal sizes, protective periods) which are obligatory for the whole country, additional regulations relate to trout and grayling zones. For example, it is not permitted to use natural baits in upper

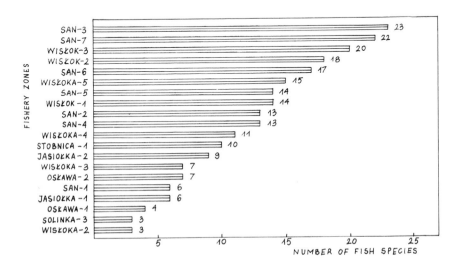

Fig. 16.7 Number of fish species in anglers' catches from different fishery zones

river sections inhabited by brown trout and grayling. In some cases only artificial flies are permitted. Spinning is also forbidden and sometimes angling is completely prohibited during the trout spawning period. As a result of these regulations, the number of species caught in the upper riverine zones is limited and in the case of Wisłoka-2 and Solinka-3 is as little as three. Similarly, no animal baits are allowed in reservoir San-4 and only 13 species were caught which is ten less than in the upper reservoir San-3 where all baits are allowed.

16.3.5   *Species composition of anglers' catches*

The contribution made by the various riverine fishes to anglers' catches in consecutive zones of the San River (Figs 16.8a–g) show a typical picture of natural succession from the trout zone to the barbel zone. The pattern is disturbed by reservoirs in San-3 and San-4. Discharge of cool, well-oxygenated water from Myczkowce Reservoir (San-4) favours the development of the grayling and brown trout. As a result, the trout and grayling zones in the San River are protracted, making the river, especially San-5, attractive to anglers. Catch composition in San-5 is also different from other zones with 80% of the catch comprising big grayling; the total recorded catch of this species reached 5811 kg. In San-7 the catch composition was characteristic of a barbel zone, and this fish represents 36.1% of the catch.

The highest percentage of common freshwater fishes in anglers' catches was recorded from a reservoir, San-3. It was also high in San-4 reservoir and in zone San-2 situated above the two reservoirs. Ichthyofauna in upper sections of the San River was noticeably influenced by the reservoir in San-3.

Of the other fishing zones examined, there was a variation in the composition of the catches. Small streams, such as the Solinka River and the upper reaches of the Wisłok River, have catches dominated by trout and grayling. These sections are extremely attractive to anglers. In other regions chub (*Leuciscus cephalus*) or slow-water, lowland species (e.g. bream, *Abramis brama*) dominate the catches. These zones are often characterized by unfavourable disturbances in the river system and are unattractive to anglers.

16.3.6   *Average fish weight*

Average individual weights of brown trout, grayling, barbel (*Barbus barbus*) and chub (Fig. 16.9) caught by anglers confirmed the attractiveness of the San River. Fishing zones representing other rivers were dominated by relatively small specimens of chub (Wisłok-1 and 2, Osława-1 and 2).

16.4   **Discussion**

Historical data on the fish species of the River San (Rzaczynski, 1721; Nowicki, 1880, 1889, both after Rolik, 1971) indicated the presence of 31 species. Later

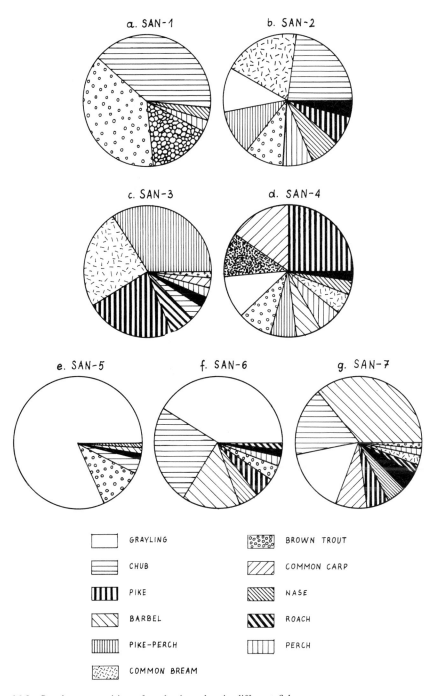

Fig. 16.8 Species composition of anglers' catches in different fishery zones

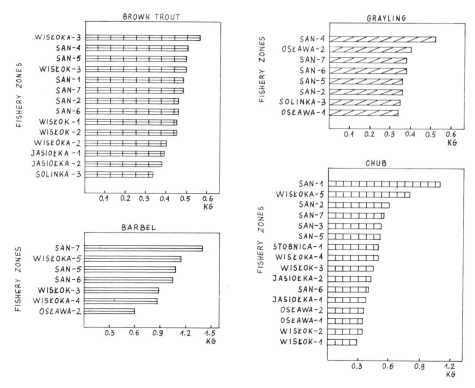

Fig. 16.9 Average weight (kg) of most important fish species caught in different fishery zones

studies of the upper and middle parts of the San River, using electric fishing techniques, revealed the presence of 36 fish species (Rolik, 1971). These studies were made before a large reservoir (San-3) was constructed and filled in 1968. Construction of this reservoir changed the environmental conditions in the river. Anglers' catches records for 1986 revealed a total of 28 species. Five of these were new species introduced to the region when the reservoir San-3 was completed and comprised lake trout (*Salmo trutta morpha fario*), rainbow trout (*Oncorhynchus mykiss*), huchen (*Hucho hucho*), pikeperch (*Stizostedion lucioperca*) and grass carp (*Ctenopharyngodon idella*). The highest number of species (23) recorded for any one zone was in the reservoir San-3. In the fishery zone San-7, which represented a barbel zone, 22 species were recorded. This is as many as were found in the River Trent (Cowx, 1988) and more than the rivers Severn (North, 1980) and Derwent (Cowx, Fisher & Broughton, 1986) where 16 and 19 species respectively were recorded.

In the absence of data on the number of hours spent fishing by the anglers, CPUE was first expressed as kg man-day^{-1}. To recalculate it as g man-h^{-1}, it was assumed, after Leopold *et al.* (1980) that one angler-day was on average equivalent to 6.25 hours. CPUE recalculated for the whole region amounted to 163 g man-h^{-1}, ranging from 107 (Wisłoka-2) to 264 g man-h^{-1} (San-3). The same index in British rivers showed catch rates of 58 g man-h^{-1} in the River Ouse (Axford, 1979), 82 to 176 g man-h^{-1} in the River Severn and 114.7 g man-h^{-1} in the River Trent (Cowx & Broughton, 1986) although the latter increased to over 200 g man-hr^{-1} in recent years (Cowx, 1990). A very high CPUE of over 500 g man-h^{-1} was recorded by Cowx (1990) in the River Derwent.

As catch data were only recorded for one year the possibility of formulating some conclusions, especially those related to the effectiveness of stocking practices, were limited. However, it was still possible to compare particular fishery zones, rivers and reservoirs since data for the whole Krosno region were available. For example, the most attractive zones were identified. Although attractiveness for anglers is not simple to define (anglers' needs and preferences are very variable), three basic measures can be taken, namely, total CPUE, CPUE in relation to the most valuable species and individual weight of the fish caught. In addition, location of the fishing area is also significant in terms of the natural surroundings and distance from home. Using these criteria, the most attractive zones were as follows (in brackets − species determining zone selection): San-1 (brown trout), San-3 (pikeperch, pike), San-4 (pike, grayling), San-5 (grayling, brown trout), San-6 (grayling), Solinka-3 (brown trout, grayling), Wisłoka-2, 3 (brown trout). The least attractive zones (mostly due to a high percentage of small chub and nase, *Chondrostoma nasus*, and low CPUEs) were Wisłok-1 and 2, Jasiolka-2, Stobnica-1, Oslawa-1 and 2.

Angling quality in the whole region can be maintained or improved by the following methods:

(1) Water quality in the whole region should be improved; the River San and its upper tributaries should be protected against pollution.
(2) There should be no minimum size for chub in those zones where stocks of chub are dense. Anglers should also be allowed to use fruit and other plant baits to assist in reducing the density of this fish.
(3) The intense fishing pressure on San-5 should be controlled.
(4) Rates of artificial stocking in the least attractive zones should be reduced. These zones are rarely visited by anglers and resources should be directed towards those zones with the highest angling pressure, especially San-5.

References

Axford, S.N. (1979) Angling returns in fisheries biology. *Proc. 1st Brit. Freshwat. Fish. Conf.*, Univ. Liverpool, 259−272.

Bnińska, M. and Leopold, M. (1987) Analiza ogólnej presji wędkarskiej na poszczególne typy wód. *Rocz. Nauk Rol. H.* **101**, 2: 7−26.

Cooper, M.J. and Wheatley G.A. (1981) An examination of the fish populations in the River Trent, Nottinghamshire, using angler catches. *J. Fish Biol.* **19**: 539−556.

Cowx, I.G. and Broughton, N.M. (1986) Changes in the species composition of angling catches in the River Trent (England) between 1969 and 1984. *J. Fish Biol.* **28**: 625−636.

Cowx, I.G., Fisher K.A.M. and Broughton, N.M. (1986) The use of anglers' catches to monitor fish populations in large water bodies, with particular reference to the River Derwent, Derbyshire, England. *Aquacult. Fish. Mgmt.* **17**: 95−103.

Cowx, I.G. (1990) Application of creel census data for the management of fish stocks in large rivers in the United Kingdom. In *The Management of Freshwater Fisheries* (Ed. by W.L.T Van Densen, B. Steinmetz and R. Hughes). Wagenigen: PUDOC.

Leopold, M., Bnińska, M. and Hus, M. (1980) Angling, recreation, commercial fisheries and problems of water resources allocation. *Proc. Tech. Cons. Fish. Res. Alloc.*, Auburn University, 212–221.

Leopold, M. and Bnińska, M. (1987) Ocena presji połowów wędkarskich na pogłowie poszczególnych gatunków ryb w wodach Polski – konsewencje gospodarcze. *Rocz. Nauk Rol. H.* **101**, 2: 43–69.

North, E. (1980) The effects of water temperature and flow upon angling success in the River Severn. *Fish. Mgmt.* **11**: 1–9.

North, E. and Hickley, P. (1977). The effects of reservoir releases upon angling success in the River Severn. *Fish. Mgmt.* **8**: 86–91.

North, E. and Hickley, P. (1989) An appraisal of anglers' catches in the River Severn, England. *J. Fish Biol.* **34**: 299–306.

Rolik, H. (1971). Ichtiofauna dorzecza górnego i środkowego Sanu. *Fragmenta Faunistica*. **17**: 559–584.

Szlażyńska, K. and Wołos, A. (1988) Ocena presji wędkarskiej na rzeki: Biebrza, Bug, *Narew*. *Rocz. Nauk. PZW.* **1**: 7–22.

Chapter 17
Newspaper information retrieval to assess the movement of valuable sport-fishing stocks in the Parana River Lower Delta, Argentina

P.G. MINOTTI and A.I. MALVAREZ *Ciudad Universitaria Pab.II, 4 piso, lab.57, 1428 Buenos Aires, Argentina*

Weekly newspaper articles on sport-fishing activities in the Parana River Lower Delta region were used to assess the timing and spatial distribution patterns of 11 valuable migrant species. Frequency of species occurrence were used to calculate a monthly index of fish distribution (IC) and possible migratory routes. Autochtonous species showed temporal patterns related to temperature and Parana River water levels. Warm water species, except for the anadromous sea catfish (*Netuma barbus*), appeared from upstream reaches after flow reduction. Temperate species were also anadromous but entered the delta during autumn, when the waters were rising. Common carp (*Cyprinus carpio*), the only exotic species, showed two periods of movement that may be related to past and present reproductive seasons.

17.1   Introduction

Argentina's main inland fishery is situated in the De la Plata River Basin and, as in other parts of Latin America, it is based on migrant species (Welcomme, 1985). The main channel in this watershed is the Parana River, a $11\,000\,km^2$ freshwater tidal delta (Lower Delta) which is formed below the confluence of the Uruguay River and De la Plata Estuary (Fig. 17.1). Fish movements between these three water courses have been studied by means of tagging experiments (Bonito & Pignalberi, 1984; Delfino & Baigun, 1985; Sverlij & Espinach Ros, 1988) but there is no information with regard to the Lower Delta *per se*, although it is implicitly assumed that fish travel across this region in their up or downstream migration.

Recently, flow regulation projects in the upper and middle course of the Parana and recreation projects in the Lower Delta have warranted the need for information on the timing and main routes of migratory species that use the Lower Delta. Sampling and tagging methods were considered inappropriate because of the area's large dimension, its poor accessibility from upland roads and restricted funding. Catch statistics from landings were available from only one of the Parana River's tributaries (Parana de las Palmas) and these were incomplete or non-existent for many species. An alternative source of data was newspaper reports

177

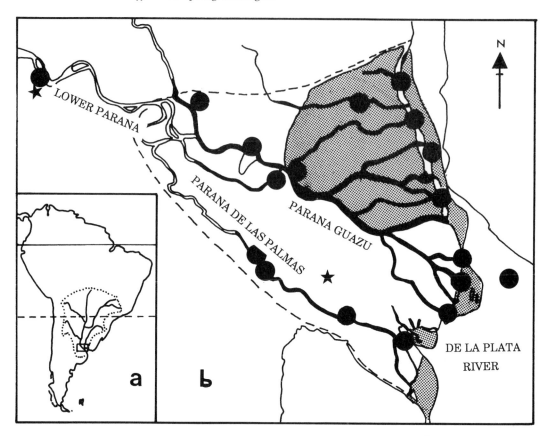

Fig. 17.1 (a) Situation of the Parana River Lower Delta in South America. De la Plata River watershed boundary is represented by the dotted line. (b) Lower Delta region with reported fishing localities. ▓ area site, ⟋ reach site and ● point site. Localization of hydrologic and meteorological stations

of sport-fishing activities (see Table 17.1). In these reports 15 migrant species were frequently mentioned (at least once a month) as being caught by anglers at a number of fishing localities which covered the main drainage system of the Lower Delta. The present paper attempts to use this information to assess the movements of the more valuable migratory species across the Lower Delta and thus provide baseline data on which management decisions can be formulated.

17.2   Materials and methods

A database was created from articles on sport-fishing activities that appeared weekly in newspapers from Buenos Aires City between March 1987 and April 1988. These articles gave information about angling success at different fishing localities, and included at least one of the following descriptors: species caught, relative size, approximate weight, catch per unit effort (fish per angler) and

Table 17.1 List of species reported in the newspaper and known habits (from Ringuelet *et al*. 1967): R − resident; M − migratory; ? − unknown

Scientific name	Common name	Habits
Ageneiosus brevifilis	Manduva	M
Ageneiosus valenciennesi	Manduvi	M
Astyanax (A.) eigenmaniorum	Mojarra	?
Astyanax (A.) fasciatus fasciatus	Mojarra	?
Astyanax (P.) bimaculatus	Mojarra	?
Brevoortia pectinata	Saraca	?
Ilisha flavipinnis	Saraca	?
Cyprinus carpio	Carpa	?
Gymnotus carapo	Morena	R
Hoplias malabaricus malabaricus	Tararira	R
Leporinus obtusidens	Boga	M
Luciopimelodus pati	Pati	M
Netuma barbus	Mimoso	M
Odonthestes bonariensis	Pejerrey	M
Odonthestes perugiai	Juncalero	R
Pterodoras granulosus	Armado	M
Rhinodoras d'orbigni	Armado	M
Oxydoras kneri	Armado	M
Parapimelodus valennviennesi	Porteñito	?
Pimelodus albicans	Bagre blanco	R
Pimelodus clarias maculatus	Bagre amarillo	R
Pseudoplatystoma coruscans	Surubi, cachorro	M
Rhapiodon vulpinus	Chafalote	M
Roeboides bonariensis	Dientudo	?
Oligosarcus jenynsi	Dientudo	?
Charax gibbosus	Dientudo	?
Salminus maxillosus	Dorado	M
Sorubim lima	Cucharon	?

relative fishing quality. No reference was made to the number of fishermen that visited each reported site.

Records were based on 55 fishing localities within the Lower Delta of the Buenos Aires province and were classified according to their geographic dimensions, i.e. point, reach or area (Fig. 17.1). A regional zonation pattern, based on landscape and flooding characteristics (Minutia & Kandus, 1988), was used to pool information from adjacent sites (Harding, 1982). Occurrence was the only species descriptor that was reported in high frequency for all the species and was consequently used for the migration analyses.

Temporal patterns of migration through the Lower Delta as a whole were based on species occurrence. A monthly index of occurrence (IC_{it}) was calculated for each species according to:

$$IC_{it} = \frac{\Sigma_j \overline{IC}_{ijt}}{n_t} = \frac{\Sigma_j (x_{ijt}/y_{jt})}{n_t}$$

where $\overline{IC_{jt}}$ is the mean number of weeks that species i was reported in zone j during month t, x_{ijt} is the number of weeks that species i was reported in zone j during month t, y_{jt} is the number of weeks that fishing sites were reported for zone j during month t and n_t is the number of zones reported in month t. $IC_{it} = 1$ when the species was present the whole month in all reported zones. $IC_{it} = 0$ when it was absent in the whole area during that month. IC_{it} values were compared with mean monthly water levels of Parana River at the Delta apex and mean monthly air temperatures.

Probable migration routes for each species were determined by joining sites where fish were first reported (first month with IC_{it} different from 0) to the last week of the month with maximum IC_{it}. Sites were joined according to the main drainage system and the resulting network was considered the dispersion area.

Species were classified as upstream or downstream migrants depending on their trend in site occurrence and their probable stream habitat origin given by Ringuelet *et al.* (1967).

17.3   Results

During the study period mean daily air temperature was inversely related to discharge regime (Fig. 17.2). Three different fish migration patterns were identified according to changes in IC_{it} in relation to these environmental characteristics. The first group, comprising eight species, was positively correlated to air temperature and hence negatively to water levels (Fig. 17.2a). Conversely, *Odonthestes bonariensis* (silverside) and *Lycengraulis olidus* (river anchovy) were positively related to water levels but negatively to temperature (Fig. 17.2c). Common carp (*Cyprinus carpio*) by contrast showed no relationship to either temperature or water levels (Fig. 17.2b).

Seven species were classified as downstream migrants. *Leporinus obtusidens* was the only species to disperse throughout the drainage system (Fig. 17.3a), whilst *Rhapiodon vulpinus* showed the most restricted spatial utilization (Fig. 17.3g). *Salminus maxillosus, Pseudoplatystoma coruscans, Ageneiosus brevifilis, Ageneiosus valenciennesi, Pterodorus granulosus, Rhinodoras d'orbigni* and *Oxydoras kneri* all used an intermediate number of water courses (Fig. 17.3b–17.3e). The remaining species could be classified as upstream migrants (Fig. 17.4). Silverside used the entire drainage network (Fig. 17.4a) while *Netuma barbus* exhibited the most restricted distribution (Fig. 17.4d).

17.4   Discussion

Those migratory species whose presence is positively related to elevated temperatures are all considered of tropical/sub-tropical, freshwater lineage (Ringuelet *et al.*, 1967). They are warm water species that appear from upstream reaches during the dry season. Within this group only *Netuma barbus* is an upstream migrant. From a biogeographical standpoint, the brackish water contribution to

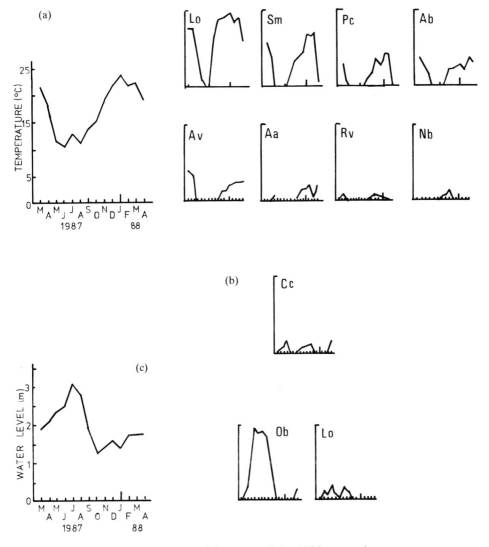

Fig. 17.2 Environmental parameters and Occurrence Index (IC_{it}) temporal patterns:
(a) Mean monthly air temperatures from March 1987 to April 1988 with IC_{it} plots for species with positive correlation with this environmental factor. (Lo – *Leporinus obtusidens*; Sm – *Salminus maxillosus*; Pc – *Pseudoplatystoma coruscans*; Ab – *Ageneiosus brevifilis*; Av – *Ageneiosus valenciennesi*; Aa – armoured catfish (armados); Rv – *Rhapiodon vulpinus*; Nb – *Netuma barbus*)
(b) IC_{it} plot for *Cyprinus carpio* (Cc) which has negative correlation coefficients both with air temperature and with water levels.
(c) Mean monthly water levels of Parana River at San Pedro from March 1987 to April 1988 with IC_{it} plots for species with positive correlation with this factor. (Ob – *Odonthestes bonariensis*; Lo – *Lycengraulis olidus*)

the life-cycle of this species (Ringuelet *et al.*, 1967) is only of secondary importance and its IC_{it} pattern is possibly indicative of its former fluvial origin.

By contrast, the discharge related species, *Odonthestes bonariensis* and *Lycengraulis olidus* are of temperate origin. They enter the delta when the waters are

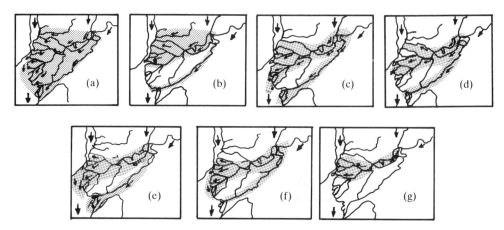

Fig. 17.3 Dispersion patterns and main routes used by downstream migrants: (a) *Leporinus obtusidens*; (b) armoured catfish (armados); (c) *Salminus maxillosus*; (d) *Pseudoplatystoma coruscans*; (e) *Ageneiosus brevifilis*; (f) *Ageneiosus valenciennesi*; (g) *Rhapiodon vulpinus*

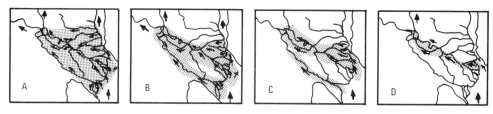

Fig. 17.4 Dispersion patterns and main routes used by upstream migrants: (a) *Odonthestes bonariensis*; (b) *Lycengraulis olidus*; (c) *Cyprinus carpio*; (d) *Netuma barbus*

rising and the weather is cooling; both have marine lineage but live in brackish waters. Silverside is also well adapted to temperate freshwater, having important populations in pampean lakes and rivers (Ringuelet *et al.*, 1967).

Common carp is an exotic species that portrays two periods of upstream movement each year. The first in April–June can be related to its ancestral spawning season in the northern hemisphere (Breder & Rosen, 1966) while the other, during spring in South America, could be related to adaptation to this new environment. This observation represents the first evidence of carp migration within the Lower Delta.

The remaining upstream migrants coming from the De la Plata estuary are known to be anadromous species (Ringuelet *et al.*, 1967).

If IC_{it} index values could be used as predictors of reach utilization, species with low mean annual values would be expected to use the Lower Delta only as a corridor. This is true at least for *Netuma barbus*, which mate during upstream migration and come down carrying the developing eggs and larvae in their mouths, travelling at least 600 km in a two-month period. Conversely, fish with relatively high index values probably use the Delta extensively. To this end, *Leporinus obtusidens, Salminus maxillosus, Pseudoplatystoma coruscans, Ageneiosus brevifilis, Odonthestes bonariensis* which display high values have feeding areas within this region (Ringuelet *et al.*, 1967).

Sport-fishing data from newspapers suggest a seasonal replacement pattern of migrant species that can be related to hydrology and climatic variations and the lineage of fish. In early spring, when Parana water levels are lowering, warm-water species enter the Lower Delta from the sub-tropical freshwater side while temperate species move towards the estuary. During early autumn, with rising waters, the reverse trend occurs.

Finally, although IC_{it} has not been correlated to biomass or fish densities it may be used to assess the relative importance of each species in the sport-fishery, particularly as a species with a high frequency of occurrence over a large area is probably indicative of high density.

These results show that data retrieved from newspaper articles provide a useful source of information on which management decisions can be based.

Acknowledgements

This study was carried out with funds from Buenos Aires University grant 135/ 87−88 to A.M. while P.M. was on a research fellowship. We would like to acknowledge Papel Prensa S.A., the National Direction of Navigation and Harbour Constructions and the National Meteorological Service for providing the environmental data.

References

Bonetto, A. and Pignalberi, C. (1984) Nuevos aportes al conocimiento conocimiento de las migraciones de los peces en los rios mesopotamicos de la Republica Argentina. *Comun. Inst. Nac. Limnol. Sto. Tome.* **1**: 1−14.

Breder, C. and Rosen, E. (1966) *Modes of reproduction in fishes. How fishes breed*. T.F.H. Publications.

Delfino, R. and Baigun, C. (1985) Marcaciones de peces en el embalse de Salto Grande, Rio Uruguay (Argentina-Uruguay). *Rev. Asoc. Cienc. Nat. Litoral.* **16**: 85−93.

Harding, E. (1982) Landform sub-divisions as an interpretive tool for stream and fish habitat assessment. In *Acquisition and utilization of aquatic habitat inventory information* (Ed. by E.N. Armandtrout). Proceedings of a Symposium held 28−30 October 1981, Portland, Oregon, 41−53.

Minotti, P. and Kandus, P. (1988) *III*. Zonificacion de patrones de paisaje en el Bajo Delta. In *Condicionantes ambientales y bases ecologicas para la formulacion de alternativas productivas en el Delta del Parana*. Annual report for Grant-contract 143/88 UBACyT program, Buenos Aires University, 15−20.

Ringuelet, R., Aramburu, R. and Alonso de Aramburu, A. (1967) *Los peces argentinos de agua dulce*. Comision de Investigaciones Cientificas de la Provincia de Buenos Aires.

Sverlij, S. and Espinach Ros, A. (1988) El Dorado, *Salminus maxillosus*, en el Rio de la Plata y Rio Uruguay inferior. *Rev. Invest. y Des. Pesquero*, **6**: 57−75.

Welcomme, R. (1985) River Fisheries. *FAO Fish. Tech. Pap.* **262**: 330p.

Chapter 18
A comparison of methods used for coarse fish population estimation

L. KELL *Freshwater Unit, University of Liverpool. Current address: Ministry of Agriculture, Fisheries and Food Fisheries Laboratory, Lowestoft, NR33 OHT, UK*

Estimation of fish population for fisheries management is often based on the quantitative survey of a limited number of sites. The results are then used to determine appropriate management action. Data from a study of the impact of the zander (*Stizostedion lucioperca* L.) on a lowland freshwater fish community are presented. Routine monitoring has shown that fisheries containing the zander had low cyprinid stocks. The Sixteen Foot Drain, a 16 km stretch of Fenland drain in Cambridgeshire, was used as an indicator of changes in the fish populations following a cull of zander and pike (*Esox lucius* L.) in the Middle Level System, of which it forms a part. Data obtained by successive seinings between stop nets are compared with data collected during an angling match, trawling and micromesh seine netting. These methods are discussed in relation to the distribution of common bream and computer modelling is used to investigate sampling strategies.

18.1 Introduction

Population surveys are an important tool in fisheries management where estimates of population size are often used to determine appropriate management action. Unfortunately little is known about the suitability of sampling methods for providing accurate data on which management decisions are made. This paper presents population data collected during the study of the management of the fishery of the Middle Level Drain, England, following the identification of a predator prey imbalance by monitoring of the fish populations. Four survey methods (quantitative seine netting, single fishing of a micromesh seine net, trawling and recording catches during an angling match) were used and their suitability for routine monitoring in relation to survey design is discussed. These results are important in establishing the most cost-effective method of sampling but provide the most suitable information for management decision making.

18.2 The study site

The study area, the Sixteen Foot Drain, is part of a system of interconnected rivers and artificial channels known as the Middle Level System. They were constructed in the 17th Century to drain the southern half of the Great Fenland occupying parts of Cambridgeshire, Northamptonshire, Norfolk and Lincolnshire in eastern England and have traditionally been an important roach (*Rutilis rutilis* L.), common bream (*Abramis brama* L.) and pike (*Esox lucius* L.) fishery.

The Sixteen Foot Drain, in common with other rivers in the area, has a uniform trapezoidal cross-section and poor macrophyte growth due to land drainage and irrigation management. It varies in depth in the centre from 2.5 m to 3.0 m and is just over 20 m wide.

18.3 Management of the fishery

A series of quantitative seine net surveys in the Middle Level System (covering 38 sites from 145 km of river) were conducted by Anglian Water between 1979 and 1981 (Table 18.1). These gave low estimates of fish biomass and pointed to an imbalance between piscivores and their prey species which was attributed (Linfield, 1985) to the introduction of the zander (*Stizostedion lucioperca* L.). Anglian Water, in an attempt to bring about a recovery of fish stocks, initiated a cull of piscivores by asking anglers not to return zander and pike caught in the affected waters. Between August 1980 and February 1981, it was estimated that 1415 kg of zander and 2917 kg of pike were removed. This was followed by the release of 1188 kg of roach and bream at various points in the system between April and May 1981. A further 1987 kg of roach and bream were stocked in the following February (Linfield, 1985).

18.4 Methods

The survey area was sampled either by seine netting or trawling. All areas of the Sixteen Foot Drain were accessible and the sites were selected randomly for the seine and trawl surveys.

Table 18.1 Summary of fish biomasses ($kg\,ha^{-1}$) in the Middle Level System

	Anglian Water	Sixteen Foot Drain	
	up to 1980	1981	1983
Total biomass	35	23.9	182.0
Zander	2	0.0	16.0
Pike	10	0.1	19.0
No. of sites	38	4	10
Range of site lengths	60–380 m	200 m	150 m
Author	Linfield, 1985	Kell, 1985	Kell, 1985

18.4.1   *Seine netting*

Quantitative seine netting techniques for the routine monitoring of fish stocks in the slow-flowing rivers of the region were pioneered by Anglian Water Authority (AWA) (Coles *et al.*, 1985). All seine data presented were obtained by, or in conjunction with, AWA teams. Sections of river were isolated by stop nets and two beach seines were used to sample the area between them. The upstream seine was pulled slowly downstream until it could be encircled by the downstream net which was laid along one bank and next to the downstream stop net. The downstream seine was then pulled into the bank. Site length generally varied between 150 and 200 m and two successive removals allowed depletion estimates of the fish populations to be made.

   To sample the juvenile stocks micromesh seine netting was also carried out. The nets used were made from 2.5 mm knotless netting and were 14 m long and 4 m deep (Penczak & O'Hara, 1983). The net was laid from a small boat parallel to the bank, 5 m out; each end was then hauled in towards the landing beach.

18.4.2   *Trawling*

A small otter trawl 0.5 m deep by 3 m wide was fished in the centre of the river from a boat with a 15 h.p. outboard motor. Trawling was carried out over 150 m lengths of river. Catches were recorded as number or gram per metre trawled.

18.4.3   *Match fishing*

The Great Ouse Championship was held on the Sixteen Foot Drain on the first weekend of the 1983/84 season. The anglers fished pegs at intervals of 40 yards over a distance of 6 km on the northern end of the Sixteen Foot Drain. Observers recorded the catches of the anglers as they weighed in.

18.5   **Results**

Quantitative seine netting estimates of fish populations in the Sixteen Foot Drain in 1981 and 1983 are compared in Fig. 18.1. The biomass of all species except zander increased, with the largest change being seen in the roach population. The concurrent changes in age structure of roach is shown in Fig. 18.2 and common bream in Fig. 18.3. Populations were characterized by few fish from year classes prior to 1979 with subsequent good recruitment to the cyprinid stocks in 1979, 1981 and 1982, particularly to the roach population (Kell, 1985).

   Single fishings of the micromesh seine net in the margins, however, gave estimates of pike densities comparable to, or higher than, those obtained by quantitative seine netting, although the primary use of micromesh seine nets was to sample fry populations. For example, during fry surveys in October 1981 after the zander cull, it was calculated that pike were present at a density of 36 kg per

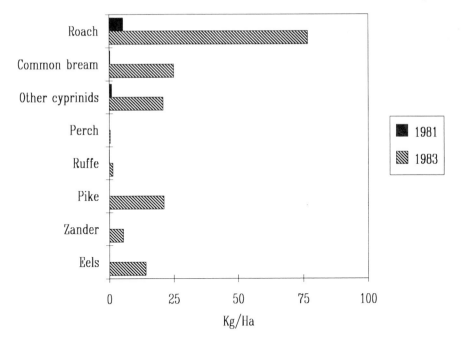

Fig. 18.1 Comparison of seine net biomass estimates in 1981 and 1983

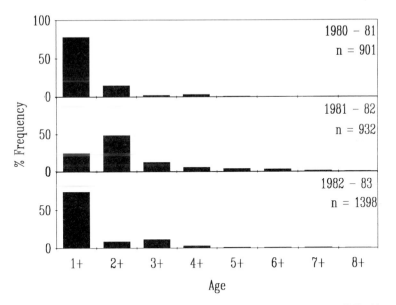

Fig. 18.2 Age composition (percentage frequency) of roach from trawl catches, 1980–83

hectare (16 sites) (cf. the estimate of 10 and 0.1 kg per hectare from the pre-1980 and 1981 seine surveys, Table 18.1), although in a further micromesh seine net estimate in September 1982 this had dropped to 10 kg per hectare (40 sites).

Large variances in population size, calculated using the variance formulae of

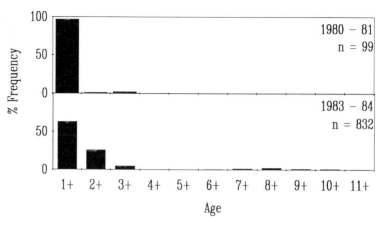

Fig. 18.3   Age composition (percentage frequency) of common bream from seine catches, 1980–83

Bohlin (1981) (e.g. 100 000 and 350 000 per hectare for common bream in 1981 and 1983 respectively), were found due to the great variation in the number and size of fish at each site. The variance, and hence the precision, of a population estimate depends on the spatial distribution of the population. The relationship between the variance (σ^2) and the arithmetic mean ($\bar{x}$) can be used to detect a non-random distribution within a population. Figures 18.4a–d show the log of variance plotted against log of mean for roach, common bream, perch and zander catches (numbers per metre trawled) for each quarter, respectively. The data for each quarter in a year are plotted in the form of a kite representing the 95% confidence intervals for the variances and means. The diagonal is the line log (variance) = log (mean). Values above the diagonal represent ratios >1 (contagious distributions) and those below <1 (uniform distributions). For all species the variance : mean ratio is greater than 1 in the final quarter due to the fish becoming more aggregated at this time of year. The increase in the mean suggests increased efficiency of the trawl, presumably due to fish being more lethargic and concentrated in the centre of the drain where the trawl operates.

The negative binomial probability distribution was used to model the data on common bream greater than 200 mm in length from both trawl and seine catches and to test for a contagious distribution. The negative binomial frequency distribution is given by the expansion of $(q - p)^{-1}$ multiplied by the sample size, where μ is the arithmetic mean of the population, $p = \mu/k$ and $q = 1 + p$, and the reciprocal of the exponent k, $(1/k)$, is a measure of the clumping of the individuals in a population (Anscombe, 1949; Bliss & Fisher, 1953; Debauche, 1962; Elliott, 1977). The χ^2 test of goodness of fit was used to test departure from randomness by comparing the observed values with the terms of the negative binomial.

The distribution of bream >200 mm for both trawl and seine catches (Tables 18.2 and 18.3 respectively) was found to agree with a negative binomial, suggesting a contagious distribution.

Table 18.2 Trawl catch of common bream >200 mm (1982), χ^2 test for agreement with a negative binomial distribution. (Mean catch per site – 0.74; Variance – 6.31; k – 0.18). * Agreement with the negative binomial accepted at the 95% level

Catch (number)	Obs.	Exp.	Obs. – Exp.	χ^2
0	65	64.15	0.805	0.01
1	10	9.23	0.77	0.06
2	4	4.39	−0.39	0.03
3	3	2.57	0.43	0.07
4	1	1.65	−0.065	0.26
5	1	1.12	−0.12	0.01
6−7	1	1.33	−0.33	0.08
>7	1	1.52	−0.52	0.18
Total	86	85.96		0.71*

Table 18.3 Seine catch of common bream >200 mm (1982), χ^2 test for agreement with a negative binomial distribution. (Mean catch per site – 3.70; Variance – 20.46; k – 0.47). * Agreement with the negative binomial accepted at the 95% level.

Catch (number)	Obs.	Exp.	Obs. – Exp.	χ^2
0	4	3.57	0.43	0.05
1	1	1.50	−0.50	0.17
2	1	0.63	0.37	0.22
3−9	2	1.77	0.23	0.03
>10	2	2.53	−0.53	0.11
Total	10	10.00		0.58*

Table 18.4 Summary of trawl catches

Year	No. of Trawls	Roach	Bream	(No. m^{-1} trawled) Zander	Pike	Perch	Ruffe
1981	11	0.2681	0.0101	0.0007	0.0023	0.0556	0.7272
1982	82	0.2667	0.0570	0.0051	0.0019	0.0181	0.1733
1983	47	0.0759	0.1520	0.0018	0.0018	0.0073	0.0163
Significance at 95% level		82 > 83	83 > 81 & 82			81 > 82 & 83	82 > 83

Tests of significance by ANOVA and Tukey-Kramer method (Sokal & Rolf, 1969).

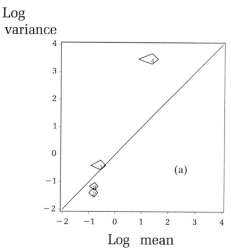
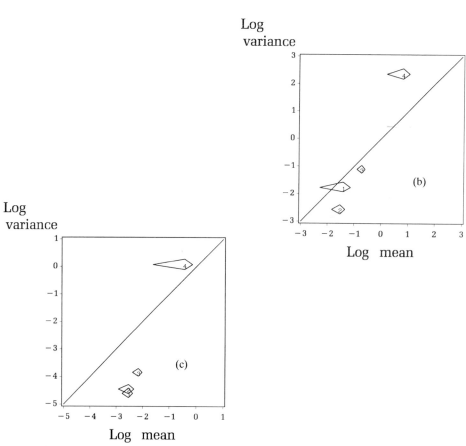

Fig. 18.4 Log plots of variance against mean number per metre trawled by quarter data from 1980–83: (a) roach; (b) common bream; (c) perch; (d) zander. Confidence limits on mean and variance are represented as kites

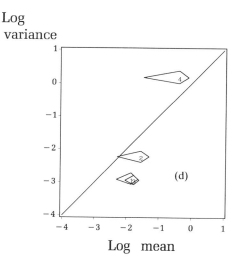

Log
variance

Log   mean

Fig. 18.4 Continued

The trawl data (summarized as number per metre trawled, Table 18.4) were used to investigate variation in population size after excluding the data from the fourth quarter, when fish distributions differed from the rest of the year. Bream catches increased through the years of the study but only in 1982 were any significant differences seen in the roach population when the strong 1981 year class was present at its greatest abundance in catches (Fig. 18.2).

Data are not presented on fish biomass as no significant differences were found between years. The cyprinid populations were characterized by large numbers of juveniles and relatively few older fish. This produced large variances in catch data expressed as gram per metre trawled.

The Great Ouse Championship presented an opportunity to compare survey methods with anglers' catches and the match data are summarized in Tables 18.5 and 18.6. The main species caught was common bream; catches of bream are plotted by peg number in Fig. 18.5. Bream were specifically targeted by anglers as they knew that winning weights would be produced by catches of bream. The anglers catch of bream during the match (equivalent to 13 kg per hectare) was a third of the estimated biomass given by quantitative seining (Table 18.5).

The catches of bream during the match (Table 18.7) were fitted by a negative binomial as were seine and trawl catches >200 mm (Tables 18.2 and 18.3). The trawl and match data have similar values for mean, variance and k, because the frequency distribution of both match and trawl catches were very similar. This suggests that the likelihood of capturing large bream is the same for both trawl (one haul of 150 m) and match units (one angler fishing for four hours). The relationship between trawl and angler efficiency is complicated by variations in

Table 18.5 Summary by weight of species caught in Great Ouse Championship

Species	Kg
Common Bream	93.27
Eels	3.84
Roach	0.13
Rudd	0.30
Tench	0.41
Pike	0.20
Zander	1.75
Perch	0.10

Table 18.6 Summary of catches during the Great Ouse Championships 1983 (* population estimates for northern 6 km only)

No of pegs		150
No of pegs fished		120
No of anglers weighing in		74
Length of a peg	(m)	40
Total weight caught	(kg)	194.5
Biomass of fish caught	(kg/ha)	12.97
Biomass of bream catch	(kg/ha)	12.1
Seine estimate of fish biomass	(kg/ha)	124.6*
Seine estimate of bream biomass	(kg/ha)	37.42*
Mean weight caught per man	(kg)	1.62
Mean weight caught per man catching	(kg)	2.63

Table 18.7 Match catch of common bream (1983), χ^2 test for agreement with a negative binomial distribution. (Mean catch per site − 0.92; Variance − 7.48; k − 0.13). * Agreement with the negative binomial accepted at the 95% level

Catch (number)	Obs.	Exp.	Obs. − Exp.	χ^2
0	91	93.35	−2.35	0.06
1	14	9.95	4.05	1.65
2	3	4.92	−1.92	0.75
3	4	3.07	0.93	0.28
4	4	2.18	1.82	1.52
5−9	1	4.87	−3.87	3.08
10−14	2	1.12	0.88	0.69
>15	2	1.56	0.45	0.13
Total	121	121.01		8.16*

area and time fished. In angling the timescale is such that fish are able to move between sites, whilst for trawling the area covered can vary to increase the chance of encountering concentrations of fish.

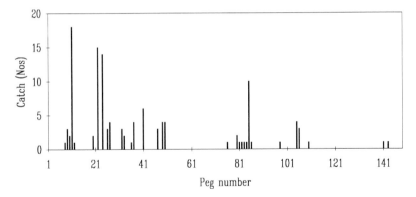

Fig. 18.5 Catches (number) of common bream by peg during the Great Ouse Championship

The observed frequency distribution of angling catches is likely to be determined by two main factors, angler skill and fish availability. The position of an individual angler is determined by a draw at the beginning of the match and so the location of anglers is random. If catch is related to availability, however, and concentrations of bream are localized, then areas of good and poor catches should be apparent. This theory was tested by the variance : mean ratio method.

The variance in catch per peg was calculated after grouping adjacent pegs, for a range of groups from 1 to 20. The variance was then plotted against the number

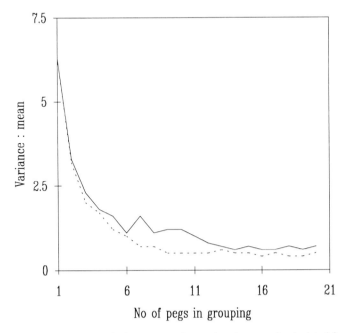

Fig. 18.6 Plots of variance: mean ratio for common bream (catch per peg) calculated for groupings of 1−20 pegs, data from the Great Ouse Championship. Solid line calculated for groups of adjacent pegs, dotted line for groups of pegs selected at random

of pegs in a particular grouping (Fig. 18.6) enabling the scale of pattern to be detected (Kershaw, 1973). The scale of pattern reflects the area within which fish are concentrated. The dotted line in Fig. 18.6 was produced by assigning pegs to a group at random and repeating the analysis. A peak is shown at a peg grouping of seven, suggesting that there is a scale of pattern based on 300 m. However, it is not possible to determine the significance of this peak because under non-randomness the use of a variance ratio test is invalidated. Repeated analysis of match data would have to be used to determine whether such peaks are due to chance effects or reflect real variations in distributions.

Catches do seem to be concentrated in certain areas, consistent with a contagious distribution for large bream. Bream >200 mm (the size range featuring in anglers' catches) were shown to be contagiously distributed by seining and trawling. Within areas, there is considerable variation in catches consistent with variations in angling skill. These are likely to be manifest through the ability to get fish feeding when they are present and subsequently to keep them at a peg.

18.6   Computer simulation

18.6.1   *Methods*

The SAS System was used for modelling because of its ability to handle complex data sets (Kell, 1989). Two hypothetical fish populations occupying a river of two hundred sites were constructed (Figs. 18.7(a) and (b)) and randomly sampled by a variety of strategies. The populations represented uniform and complex spatial distributions where fish are highly clumped.

The uniform distribution was composed of ten groupings of fish each of a hundred individuals (Fig. 18.7(a)). The position of fish within a grouping was calculated using a normal distribution, the site location having a standard deviation of five. The second population representing a contagious distribution (Fig. 18.7(b)) was again composed of ten groupings of fish of a hundred individuals but with a standard deviation of 1.5 sites (the median standard deviation of bream location calculated from the match data). The location of each grouping of fish was determined by dividing 200 sites (representing the river) into 10 sections of 20 sites each and randomly assigning a rank to each section. Sites were selected at random from the sections, 10 from the section with rank 1 and nine from the section with rank 2, down to one from the section with rank 10. The groupings of fish were centred on 10 sites randomly chosen from the 45 sites selected. The resulting distribution thus models a heterogeneous fish population where there are preferred areas within a river.

A range of sampling strategies were investigated for the two populations. Each sampling programme was run a thousand times and the areas sampled in any run were selected at random. Sample areas were composed of adjacent sites with no two sample areas overlapping during an iteration. The relative size of sample areas ranged from 1 to 20 and the number of areas sampled from 1 to 50. The

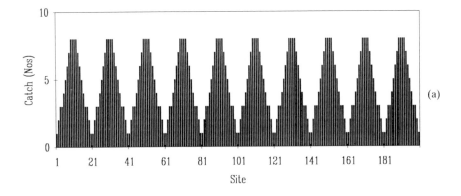

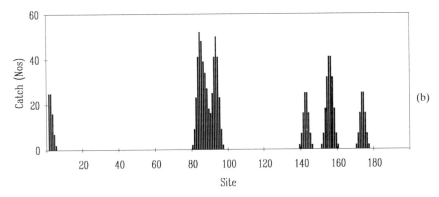

Fig. 18.7 Computer generated fish distributions: (a) uniform distribution; (b) contagious distribution

calculation of an estimate for a given number and area of samples was performed by an iteration which corresponded to a single survey. The cumulative probability of obtaining any particular estimate was plotted against its value as a percentage of actual population density for each sampling strategy.

18.6.2   *Results*

The effects of varying sample size are shown in Figs. 18.8(a) and 18.8(b) and number in Figs. 18.9(a) and 18.9(b). The steeper the gradient of the curve around the true population density the greater is the probability of getting an estimate close to that value.

As expected, estimates are more accurate for the uniform distribution. However, for a site of unit size, the probability of getting a more accurate estimate increases as the number of sites sampled increases (Figs. 18.8(a) and 18.8(b)). Increasing site area (Figs. 18.9(a) and 18.9(b)), however, makes little difference to the probability of getting an accurate estimate. For a given total area sampled, the best strategy appears to be, therefore, to maximize the number of sites rather than the area of individual sites covered. This is due to the increased probability of encountering a shoal and because a large number of small sites will cover a wider range of habitat types than a small number of larger sites:

Quantitative seine netting is often used to produce precise population estimates with 95% confidence limits. The accuracy of estimates of total population size, however, also depend on the spatial distribution of a population. Bohlin *et al.* (1983) state that '... the possibility of making useful estimates about the entire stock ... is negatively influenced by the spatial variation in population density ...'. Contagious distributions mean that the variance of the total population size will be so large that estimates based on small numbers of samples will be unreliable.

The distribution and size of shoals and the size of samples can sometimes obscure contagious frequency distributions. The observed level of contagion may be reduced where there are only a few individuals in a clump, causing the dispersion of a population to appear more random. Similarly, when sample units are large enough always to include a shoal, the variance of the estimate will decrease. This effect was shown for the trawl data. The apparent distribution of bream is uniform in the first three quarters (Fig. 18.4(b)); however, when only fish over 200 mm (Table 18.7) were included in the analysis the distribution was highly clumped. The trawl data are biased towards the recruiting year-classes and although the distribution of juveniles will actually be contagious due to shoaling behaviour the small size and large number of shoals will make their distribution appear less contagious.

The difficulty in obtaining representative biomass estimates when the populations are contagiously distributed is illustrated by the survey results. The increase in the bream seine estimate in the 1983−84 season (Table 18.1) was due to the number of large fish caught (Fig. 18.3), not the number of juveniles. These larger individuals must have been present in the system during the earlier survey. The estimate of bream of 27.8 kg per hectare in 1983−84 would have represented a biomass of 11.6 kg per hectare in 1980−81 (calculating the weight of the same number of fish three years previously) and 53.6 kg per hectare if adjustment is also made for a mortality of 40%. This contrasts markedly with the actual estimate of 0.4 kg per hectare even when allowance was made for the 3 kg per hectare of roach and bream stocked in the 1980−81 season.

Similar discrepancies arise in the estimates of zander biomass. In August 1980 it was estimated that 629 kg of zander were present in the Middle Level System (obtained by multiplying the seine estimates by the area of the Middle Level System). By February of the following year, however, 1415 kg of zander had actually been removed by anglers during the cull.

The survey results suggest, therefore, that quantitative seining may not always be the ideal method for following detailed changes in a fish population.

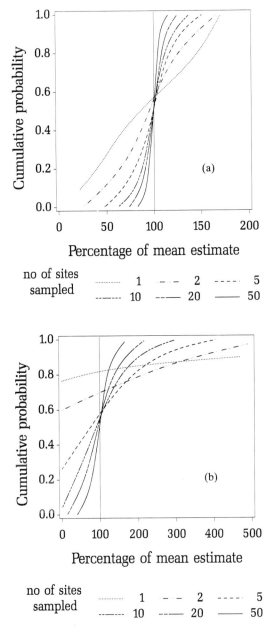

Fig. 18.8 Plots of cumulative probability of obtaining a particular estimate against percentage of its true value, for sampling strategies where the number of sites in a sample are varied: (a) samples from a uniform distribution; (b) samples from a contagious distribution

Population estimates can be improved by increasing the proportion of a population covered. The computer simulation showed that assessment of changes in total population size based upon a randomly selected large number of small sites are more reliable than those from a limited number of larger sites. It is neither

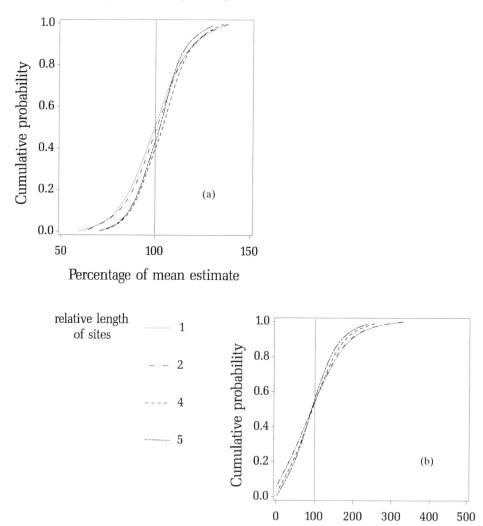

Fig. 18.9 Plots of cumulative probability of obtaining a particular estimate against percentage of its true value, for sampling strategies where the area of sample sites are varied: (a) samples from a uniform distribution; (b) samples from a contagious distribution

practical nor desirable, therefore, to increase the area of samples sufficiently to reduce the apparent contagion.

This has important implications for the choice of appropriate sampling methods. Quantitative seine netting is very labour intensive. Halving the size of sample sites, for instance, will not double the number of sites covered because of the fixed costs of travelling to sites and setting up. Only the effort due to catching and processing fish varies depending on the area of sites. Methods such as simplified seine netting or trawling, although giving much lower levels of precision within sites, may give better estimates of total population size, as they are able to cover the large number of sites necessary to account for fish distribution.

While the latter methods do not produce absolute density estimates, they can produce indices (e.g. catch per unit length or site) that reflect changes in populations. Data on spatial and temporal distributions will also be collected and any survey monitoring fish populations must take into account, for both biological and statistical reasons, variations due to fish behaviour and habitat characteristics. Without an understanding of the factors influencing fish distribution and abundance meaningful interpretation of survey results is difficult. Hankin (1984) recommended that the size of sites used to sample salmonid populations in small streams, should vary according to natural habitat units. This helps to reduce the variance due to variations in relative frequency of catches between habitat types. The division of a relatively homogeneous environment like the Middle Level System into habitat units, presents more difficulties than the division of salmonid habitats into riffle and pool areas. It is therefore extremely important to ensure that any sampling strategy also takes fish distribution into account.

Population surveys are likely to become increasingly important in assessing environmental changes, as well as for the monitoring of fisheries. It is important that if changes in populations are to be followed survey techniques must provide accurate information on both total population size and community structure. Methods producing precise estimates for a limited number of sites might not always be the best strategy due to the complex distributions of fish populations.

References

Anscombe, F.J. (1949) The statistical analysis of insect counts based on the negative binomial distribution. *Biometrics* **5**: 165–173.

Bliss, C.I. and Fisher, R.A. (1953) Fitting the binomial distribution to biological data and a note on the efficient fitting of the negative binomial. *Biometrics* **9**: 176–200.

Bohlin, T. (1981) Methods of estimating total stock, smolt output and survival of salmonids using electrofishing. *Rep. Inst. Freshw. Res., Drottningholm* **59**: 5–14.

Bohlin, T., Dellefors, C. and Faremo, U. (1982) Electro-fishing for salmonids in small streams – aspects of the sampling design. *Rep. Inst. Freshw. Res., Drottningholm* **60**: 19–24.

Coles, T.F., Wortley, J.S. and Noble, P. (1985) Survey methodology for fish population assessment within Anglian water. *J. Fish. Biol.* **27** (Suppl. A): 175–186.

Debauche, H.R. (1962) The structural analysis of animal communities in the soil. In *Progress in soil Zoology* (Ed. by P.W. Murphy). London: pp. 10–25.

Elliott, J.M. (1977) *Some methods for the statistical analysis of samples of benthic invertebrates*. Freshwater Biological Association. Scientific Publication No. 25.

Greig-Smith, P. (1952) The use of random and continuous quadrats in the study of the structure of plant communities. *Ann. Bot., Lond., N.S.* **16**: 293–316.

Kell, L.T. (1985) The impact of an alien piscivore the zander (*Stizostedion lucioperca* L.) on a freshwater fish community. Ph.D. Thesis, University of Liverpool.

Kell, L.T. (1989) Automated data analysis using the Macro language. *Proceedings of the SAS Users Group Britain and Ireland*, 1989. SAS Software Ltd. UK 104–121.

Kershaw, K.A. (1973) *Quantitative and dynamic plant ecology*. London: Edward Arnold.

Linfield, R.S.J. (1985) The impact of zander (*Stizostedion lucioperca* L.) in the United Kingdom and the future management of affected fisheries in the Anglian region. *EIFAC Technical Paper* **42** (Suppl. 2): 353–362.

Penczak, T. and O'Hara, K. (1983) Catch effort efficiency using three small seine nets. *Fish Management* **14**: 83–92.

Sokal, R.R. and Rohlf, F.J. (1969) *Biometry: The principles and practices of Statistical and Biological Research*. 2nd edition. San Francisco: W.H. Freeman.

Chapter 19
The customization of recreational fishery surveys for management purposes in the United States

STEPHEN P. MALVESTUTO *Fishery Information Management Systems,*
500 Dumas Drive, Auburn, Alabama, 36830 USA

In the United States, recreational fishery surveys are the most common and widely applied sampling technique employed by fish and wildlife agencies. It is estimated that $20 million are spent each year by fishery agencies for recreational fishery surveys. However, there is little accountability with respect to the quality of the information collected relative to money spent. This paper presents a four-stage planning process for the customization of recreational fishery survey designs that begins with setting survey objectives (stage 1). The survey objectives form the foundation for collection of the data (stage 2) which leads to synthesis of the information collected (stage 3). Once data have been collected and synthesized, the efficiency of the design can be evaluated for more optimal data collection in the future (stage 4). Based on the initial survey objectives of estimation of fishing effort and catch rate, each stage in the planning process is discussed with particular reference to critical design features that should be considered to effectively customize survey designs for specific recreational fisheries.

19.1   Introduction

In the United States, recreational fishery surveys, or more traditionally, creel surveys, may be the most common and widely applied sampling technique employed by fish and wildlife agencies. It is the only sampling method that deals directly with assessment of the fishery and additionally is the only method that puts agency employees directly in contact with anglers on a regular basis. Creel surveys thus are important not only from a fishery management point of view, but also in terms of public relations and information transfer between agencies and clientele.

Given the widespread use of recreational fishery surveys with the associated costs of collecting this kind of information, it is important that surveys be customized for the particular circumstances at hand. Here, customization refers to the design and conduct of a survey such that the objectives are satisfied at minimum cost. Thus, customization represents the development of a particular methodology to obtain specific information. In essence, we are asking ourselves what aspects of the survey are under our control and how should we vary these aspects to collect needed information from anglers in an efficient manner.

As a first step toward effective customization of a recreational fishery survey, it is useful to visualize a sequence of events, or steps, that address several areas where particular survey technique decisions must be made. This sequence might be termed a survey planning process (Fig. 19.1). The planning process begins with the definition of the objectives of the survey. The objectives should be sufficiently detailed so that the specific variables to be measured are identified, as well as the comparisons to which the variables will be subjected and the accuracy to which statistical differences must be documented.

The second step in the process is the collection of the data in a manner that will satisfy the survey objectives. This can be adequately accomplished only if the objectives are stated in sufficient detail. This step not only represents the customization of a survey sampling design to the objectives of the survey, but also the development of an interview schedule for verbal acquisition of information from anglers. The design customization must, of necessity, take into account the physical and temporal aspects of the environment, fishing modes and angler activity patterns, and the money, manpower and equipment available for the survey. This step in the planning process encompasses most of the customization decisions.

After the data have been collected, they will be subjected to analysis. The particular analytical procedures should follow largely from the way in which the data were collected, but there may be certain choices that can be made within the defined sampling structure. Thus, there may be several estimators of a particular variable (catch rate for example), or several comparative approaches that might be appropriate to the situation, so that some customization of analytical method, over and above field survey technique, may be possible. These decisions usually will be directed at obtaining the most precise estimates possible, or at using the most powerful comparative approaches for statistical documentation of changes in the fishery.

The final step in the planning process is one of evaluation of the method used so that the survey process can be streamlined for efficient collection of information. This step can take place only after the survey has been implemented. Gross inadequacies in field procedure or in the interview schedule may be spotted early through implementation of a pilot, or pre-test, survey over a short period of time (one month). However, statistical evaluation of the utility of certain design decisions, such as stratification, definition of sampling units and sample size, can only be made after the survey has been conducted for a full cycle of time, be it a fishing season or an entire year.

The approach taken here will be to briefly address various aspects of the customization of recreational fishery surveys by following the planning process presented in Fig. 19.1. The discussion will attempt to provide basic guidelines that might be used by others to more effectively define their survey methodologies. Examples of decisions that might be made in particular situations are taken from the author's own experience in the design of angler surveys, as well as from information contained in the literature.

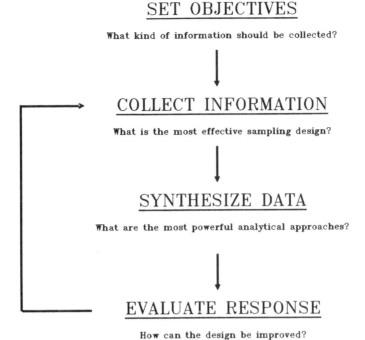

SET OBJECTIVES

What kind of information should be collected?

COLLECT INFORMATION

What is the most effective sampling design?

SYNTHESIZE DATA

What are the most powerful analytical approaches?

EVALUATE RESPONSE

How can the design be improved?

Fig. 19.1 Survey design planning process

19.2 Planning step 1: setting survey objectives

The objectives of the survey provide the foundation on which the other steps of the planning process depend. Two traditional objectives of creel surveys are estimates of fishing effort and catch rate, which are important for recreational fishery management in the United States. Fishing effort is a measure of the use of the resource with a direct bearing on the economic value of the fishery (Palm & Malvestuto, 1982) and harvest (or catch) rate is a measure of fishing success which is a primary benefit derived by anglers through the fishing experience (Hudgins, 1983). It is the product of these two variables that provides an estimate of catch; thus, the validity of catch values rests solely on the validity of estimates of effort and catch rate. Catch and release fishing, as a benefit associated with the implementation of length limits as a management option, may necessitate enumeration of the number of fish caught and returned to the water, as well as those caught and kept.

Under circumstances where bag or length limits restrict harvest, or where catch and release fishing is voluntarily practiced, catch rate is likely to be the appropriate measure of fishing success, as well as the most appropriate index of stock density. Current recreational fishery management concerns in the United States go beyond the measurement of traditional catch and effort descriptors, so that collection of particular kinds of socio-economic data may be included as survey objectives.

These objectives might include the estimation of fishing trip expenditures, investments in durable equipment items, willingness to pay, fishing satisfaction, fishing trip quality, attitudes concerning benefits provided by the recreational site, or opinions on planned implementation of management policy.

To simplify the situation this paper will focus on the measurements of fishing effort and catch rate as primary survey objectives. Fishing effort is typically measured in angler-hours or angler-trips and catch rate is expressed as number or weight of fish caught per hour (CPUE). Number per hour is probably more useful as an index of fishing success because it is difficult to estimate the biomass of fish caught and released, and additionally, creel and length limits are conceptualized on a number basis. In keeping with the premise that the stated survey objectives should contain sufficient detail to allow the sampling methodology to be adequately defined, it will be assumed here that the primary comparisons to which the data will be subjected involve documenting annual changes in the two descriptors, and there is a desire to statistically document relative changes of more than 25% and to be at least 80% certain that these changes truly occurred. This is a statement of the desired accuracy to which the data must be collected.

19.3   Planning stage 2: data collection

Given the above stated objectives, it is now possible to customize the survey method. What general approaches to obtaining fishing effort and catch rate information are available? In general, either physically intercepting anglers, or obtaining information by mail or phone, are the only options. If information specific to a particular fishery is desired and unbiased estimates of catch characteristics is a primary survey objective, then interception of anglers at the fishing site is the most appropriate alternative. Attention will be focused on on-site angler-intercept approaches, which can either be classified as roving intercept surveys or access point intercept surveys.

Usually, if access points are numerous, or if fishing pressure is relatively low, such that few interviews would be available through any given access point, then the roving survey would be the desired approach. However, assumptions inherent to the roving intercept method must be carefully considered with respect to the objectives of the survey. Of special concern is the assumption that catch rate is independent of fishing time. This is critical, because as a consequence of the roving survey, anglers are interviewed in the act of fishing so that catch rate data are based on incomplete fishing trips, rather than on completed trips as would be obtained by sampling access points. This assumption has been tested in only a few cases and for the most part has been found to be valid (Malvestuto *et al.* 1978; Malvestuto 1983). There are circumstances, however, that might dictate that the assumption would not hold. For example, when low creel limits exist, experienced anglers may catch their limit and terminate their trips early such that the majority of anglers intercepted on the water are the relatively unsuccessful ones, producing negatively biased estimates of catch rate. Thus, if it is difficult to assume that

CPUE is independent of the time spent fishing, then it might be better to use an access point approach. The primary benefits of the roving technique are that a much larger portion of the fishery can be sampled during any period of time and also that anglers are actively contacted for interview, usually ensuring that more interviews are obtained per unit time spent sampling. However, if critical information is likely to be biased, then the increased efficiency of the roving approach does not provide clear justification for its use.

Effective customization of on-site intercept survey designs for the collection of both effort and CPUE information rests on consideration of the fundamental difference in the basis for the generation of these estimates. Fishing effort is generally based on counts of anglers taken within a certain time period such that the number of angler-hours expended equals the angler-count times the number of hours in the time period, i.e. a count value times a constant. Catch rate, on the other hand, is based on interviews of anglers and depends on anglers' recollections of how long they fished and also on how accurately the creel clerk can obtain the catch data. The estimate of CPUE represents a ratio of estimated fishing effort over a count of fish caught. Ideally, the clerk should view and measure caught fish, rather than depend on anglers' verbal responses. In any case, it is the count process that is critical for valid estimates of effort, whereas the interview process is critical for valid estimates of catch rate. If only estimates of effort are required to meet the survey objectives, then interviews can be eliminated.

19.3.1 *Estimation of fishing effort*

Theoretically, to estimate fishing effort the best way to obtain counts of anglers is via a vantage point so that an 'instantaneous count' can be obtained. A mean of several instantaneous counts taken during each hour of the sampling period and then averaged over the total time period would provide the most representative count for the estimation of effort. In most cases, suitable vantage points are not available and a progressive count of anglers must be made as they leave from access points, or as they actively fish, by making a circuit of the sample area by boat or car (roving survey). The shorter the progressive count period, the closer the count will be to a theoretical instantaneous count. Counting anglers leaving through access points is relatively straightforward, but roving counts can pose problems.

It is critical that the creel clerk makes a complete circuit of the sampling area within the allotted time, or estimates of effort will be negatively biased. If sampling periods are relatively long, i.e. longer than the average length of a fishing trip, then it will be better to make two circuits of the sample section, one for counts and the other for interviews. In this way, counts can be conducted as quickly as possible. If sampling periods are relatively short, then counts and interviews can be conducted simultaneously using some method to ensure that the clerk stays on schedule and still makes a complete circuit of the sample section. The creel clerk can skip angling parties to be interviewed in some objective manner, e.g. every other party or every fifth party, etc., depending on the intensity of angling activity on the sample day (Malvestuto, 1983); or, the clerk can establish check points that must be reached at particular times during the sample period to ensure that a complete circuit is made (Wade *et al.*, in press).

If counts and interviews are conducted concurrently, then length of stay bias will enter into the effort estimate, i.e. anglers will be missed for counting purposes depending on how long and how many interviews are taken within the sampling period (Wade *et al.*, in press). Staying on schedule during the circuit will minimize this bias, the magnitude of which also is dependent on the particular daily fishing patterns exhibited by the fishery. Short sampling periods within a day also ensure that this bias is minimized because the probability is less that angler numbers will change substantially during the sampling period. Malvestuto (1983) found that circuits of four hours in duration where anglers were interviewed and counted concurrently were similar to those obtained from one-hour count circuits.

19.3.2   *Estimation of catch rate*

The interview process is inherently more complex than the count process. If interviews are to provide catch rate data that is representative of the area of the lake or river being surveyed, then the creel clerk should endeavour to conduct interviews proportional to the spatial occurrence of anglers on the water. This depends, again, on the circuit methodology used with roving surveys while interviews are being conducted and it is critical that a complete circuit of the sample area be made. With access points, peak exit periods may dictate that not all anglers can be interviewed and the creel clerk should ensure that interviews are spaced over the entire period of time that anglers are leaving through the access point.

Given that in the interview process the creel clerk can view and count fish caught by anglers, the critical piece of information needed for the calculation of CPUE is the amount of time spent catching these fish. There are two possible alternatives. One is simply to ask anglers how long they have been fishing. The other is to ask anglers what time they began to fish and to also record the time of interview, where the difference in the two times provides an estimate of the time spent fishing. The second alternative may provide the most accurate data because anglers need only recall when they started fishing, which most seem to know without much hesitation. The first alternative forces anglers into a more complex mental recall process which may increase memory and digit bias, although there is no evidence to support this.

19.3.3   *Definition of sampling units*

The temporal and spatial components of a fishery survey that are subjected to random sampling are the sampling units for the survey. The logical primary sampling units are fishing days, usually defined in terms of the daylight hours within which anglers are normally active, although night-time fisheries are not uncommon and may need to be addressed in particular circumstances. When fishing days are divided into shorter time periods and the water area is divided into smaller spatial sections for sampling purposes, these units become secondary or sub-sampling units which would be randomly chosen within any given sample day. These types of survey designs are designated as multi-stage random sampling designs which typically are applied to inland recreational fishery surveys in the United States.

For customization purposes, the above considerations usually dictate that large systems be divided into sample sections that can be thoroughly covered in short time periods, e.g. not more than four hours using roving surveys in the south-eastern United States. Given that counts and interviews can be conducted concurrently under these conditions, money and manpower can be reduced by elimination of the independent count circuit. The number of secondary sampling units is not particularly critical and will depend on the surface area and physiography of lake basins, and primarily on the length of river basins.

Experience suggests that it is better to collect good data within short time periods over smaller areas than to collect questionable information by trying to sample too much of the fishery on any given day. It is possible by using small time periods to choose more than one period to sample within any given day, thus providing replicate estimates of effort and CPUE which can be averaged to give more representative values for that day. Smaller temporal and spatial secondary sampling units also reduce the heterogeneity of fishing activity and catch rates within the unit, reducing the probability of bias and increasing the precision of the estimates. From a practical standpoint, shorter sampling periods lessen creel clerk fatigue and boredom leading to more conscientious information collection and ultimately to better data on which to base management decisions.

It should be noted that these considerations apply more to roving than to access point surveys. Where access points are numerous, it may be difficult to get adequate numbers of interviews through any given access point, even if the creel clerk spends the entire day at the site. Thus, dividing the day into short time periods would simply lessen the chances of intercepting anglers as they terminate their fishing trips. Currently, research is underway to define methodology for roving access point surveys (bus route method) where the creel clerk moves from one access point to another under a rigidly prescribed timetable, which would increase the efficiency of this approach in terms of the interception of anglers and coverage of the fishery within any given sample day (Robson & Jones, 1989).

19.3.4 *Stratification of the fishery*

The activity of anglers on the water is the most visible aspect of a fishery and for this reason many survey design decisions are based on temporal and spatial patterns in fishing effort. Thus, common temporal stratifications such as dividing the year into time blocks (months or seasons) and dividing time blocks into day types (week days and weekend days), are justified by the expectation that fishing effort will vary significantly and in a systematic way across the defined strata. If this is so, then the precision of the population estimates can be improved. It has been shown recently (Malvestuto & Knight, in press) that stratification by day type will generally improve the precision of estimates of fishing effort, but will not improve estimates of catch rates or catch for recreational lake fisheries. Meredith and Malvestuto (1991) found the same results for the recreational fishery on the Tombigbee River in Alabama. Thus, design specifications based on temporal patterns in fishing effort may not provide effective sampling structures for estimates of catch rate.

The precision of estimates of both effort and catch rate can be improved by the creation of monthly time blocks, rather than by using larger seasonal blocks. Meredith and Malvestuto (in press) showed that residual variance associated with estimates of effort, catch rate and catch can be reduced by 10−50% by sub-dividing the year into months rather than into seasons of three to four months each. Given the relatively high variances associated with creel survey estimates (generally CV's range from 70−120% for estimates of effort and CPUE for river and reservoir fisheries in Alabama), it is wise to stratify the temporal dimension of the survey period into the shortest time blocks possible, while still maintaining adequate sample sizes within the blocks.

To control known or perceived variability in descriptors of interest through stratification, it is relatively easy to create too many strata such that sample sizes within strata, given limited money and manpower, may be too small to obtain estimates of desired precision. In this case, a viable alternative to stratification may be to employ non-uniform probability sampling of secondary sampling units (Malvestuto *et al.*, 1978). As an example, a survey year may be stratified into 12 monthly time blocks of one month and each month also stratified into day types (week days and weekend days). If, additionally, fishing effort or catch rate systematically change from the morning to the afternoon, two additional temporal strata may be appropriate, giving a total of $12 \times 2 \times 2 = 48$ temporal strata. If spatial heterogeneity also exists, as on a large lake or river, then spatial strata may help to increase the precision of estimates as well. Only two spatial strata would increase the total to 96 strata. Even the minimum sample size of n = 2 per strata, which certainly would be inadequate, would imply 192 samples within the year. It is obvious that creation of strata can quickly force sample size requirements upward to prohibitive levels.

It would be possible, however, in the above instance to treat the two temporal strata within each day, as well as the two spatial strata, as sampling units which could be chosen at random (one time-space combination each day) by assigning sampling probabilities proportional to the magnitude of a particular variable of interest, e.g. fishing effort. Thus, if 20% of the fishing effort typically occurred in the morning and the other 80% in the afternoon, then sampling probabilities become 0.20 and 0.80 respectively for the two time periods. If time periods were sampled at random in accordance with these non-uniform probabilites, then on average 20% of the sampling effort would be expended in the morning period and 80% in the afternoon. This would allocate sampling effort proportional to the amount of fishing occurring on the system and additionally would increase the number of interviews taken for the purpose of estimating catch characteristics.

Each sample estimate of fishing effort could be raised to an estimate of fishing effort for the entire day by dividing by the appropriate sampling probability so that day-to-day variability in the total estimate of effort for the time block would be minimized. The same advantages would apply to the spatial sections if non-uniform probabilities were applied. Thus, the assignment of non-uniform probabilities to sampling units can increase precision without the creation of strata, serving to keep sample sizes within practical bounds.

19.3.5   *Sample size considerations*

The logical primary sampling unit for recreational fishery surveys is the fishing day. It is common practice to generate estimates of survey statistics for each day sampled and to subject the daily estimates to the appropriate survey design formulae, which typically involve some type of stratification as discussed earlier. Thus, sample size is generally viewed in terms of the number of fishing days that must be randomly chosen to generate estimates of the desired precision. The situation is really more complex, particularly with respect to the estimation of catch rates, because interviews must be taken so that there is also a sample of fishing parties to be considered.

However, it is extremely difficult to control the number of interviews taken on any given day and the general approach is to take as many interviews as possible within the allotted time, either at an access point or by making a complete circuit of the sample area. On less intense fishing days, perhaps 100% of the active anglers can be interviewed, but on heavy fishing days, only a portion of the population can be intercepted. Regardless of the situation, the daily estimate of catch rate can be based only on the interviews taken, whatever the proportion of the total population the sample may represent on that day, and this estimate is taken as representative of the sample day. Thus, the overall consideration with respect to sample size becomes the number of days sampled.

Essentially, sample size requirements depend on the stated accuracy to which the data must be collected, which should be a part of the objectives of the survey as per the example given earlier. Given a statement of accuracy, a simple sample size formula can be used to generate a first level estimate of the sample size needed. Thus, for simple random sampling:

$$n = t^2 \, (CV)^2/d^2$$

where n = required sample size, t = the student's t-value for n given the desired confidence associated with the estimate, CV = the coefficient of variation (standard deviation/mean) associated with the response variable in question, and d = the error around the true mean which can be tolerated based on statistical comparisons to which the data may be subjected. In essence, the statement of accuracy in the study objectives provides a definition of d. The previously stated requirements to estimate catch rate within plus or minus 25% of the true mean and to be 80% certain of this result implies that $d = 0.25$ and that the t-value is based on an alpha level of 0.20.

The critical aspect of generating sample sizes for customization purposes over and above the statement of desired accuracy, is the CV associated with the response variable in question. The CV can change, based on the fishery being surveyed and on the variable to be estimated. It is typical in the south-eastern United States to obtain CV's for fishing effort of between 35% and 80%, based on monthly sampling and depending on the type of/particular fishery. Therefore, sample sizes, measured as randomly chosen days, would range from 5 to 19 days per month to obtain estimates within the previously stated limits of tolerable error. However, CV's for catch rate typically range from 70% to 120% which would demand sample sizes of 15 to 41 days per month. Thus, to maintain an effective sampling program for estimates of catch rate might be impossible on a monthly basis at the upper end of the range of variability, and would certainly imply excess sampling for the estimation of fishing effort. Sample sizes based on the estimation of effort generally would be inadequate for the estimation of CPUE.

The above example is somewhat simplistic because survey design components such as stratification and non-uniform probability sampling allow more precise estimates than using simple random sampling, but initial sample size goals can be established in this manner. Given inherent differences in the variability of fishing effort and catch rates, it always will require more samples to meet accuracy objectives for estimates of catch rate than for estimates of effort.

A related aspect of sample size determination is the allocation of sampling effort among strata and also among secondary sampling units within fishing days. The rules for allocation of samples among strata dictate that a larger sample should be taken if the stratum is larger, or more variable internally, or less expensive to sample. Typically, stratum size is measured in terms of the fishing effort exerted within the strata. Knight and Malvestuto (in press) attemped to generate an optimum monthly allocation scheme for the estimation of fishing effort using stratum size and predicted values of internal monthly variability (CV) to obtain the most precise annual estimates possible. It was found, however, that because of natural fluctuations in the monthly CV's associated with fishing effort from year to year, the optimum allocation did not produce better estimates than using a constant monthly sample size. Thus, it may be difficult to obtain consistent results with respect to the quality of data based on the expected behaviour of stratum characteristics.

With respect to trade-offs in the allocation of sampling effort among primary sampling units (days) versus secondary sampling units (within days), i.e. multi-stage sampling designs, Malvestuto and Knight (in press) found for south-eastern reservoirs in the United States that within-day variance increased as the year progressed from the spring to the fall. Thus, to obtain the most precise estimates of fishing effort more sampling effort should be devoted to taking sub-samples within days at the expense of sampling more days as the year progresses. This was a consistent finding across several fisheries analysed. However, for catch rate and total catch, within-day and among-day variance components behaved on a fishery-specific basis, and there was no overall rule that could be established for the allocation of sampling effort with respect to the estimation of catch characteristics.

19.4   Step 3: data synthesis

A primary reason for the use of statistical survey designs is to provide a valid measure of the error associated with estimates of fishery descriptors. The error term for a particular variable is critical with respect to documenting changes in the fishery through time and space. The analytical methods established for the estimation of fishing effort and its variance are straightforward, where, as pointed out earlier, daily effort is the product of an angler-count times a constant (hours in the work period) and the variance is based on the calculation of a mean across sample days using standard formulae for multi-stage and stratified random sampling.

The estimation of catch rate, however, poses unique analytical problems because there are several ways in which the rate estimate might be generated. Crone and Malvestuto (in press) have shown that the rate estimates and their variances can change substantially depending on the particular estimator used. For example, regression estimators of catch rate always provided lower estimates than those based on taking means across fishing parties or using total ratio estimators. Rate estimates based on means across individual fishing parties always gave inflated values relative to other approaches. Estimators which utilized rate data summarized on a daily basis provided the most precise estimates of catch rate. Based on the premise that a fishing day is the logical primary sampling unit, it is likely to be most appropriate to summarize rate data on a daily basis and to subject the daily estimates to the survey sampling formulae appropriate to the design used.

Given the use of appropriate design structures, ANOVA is the most powerful analytical technique available to compare annual changes in the fishery. Thus, sources of variance associated with strata or with sampling stages of multi-stage designs, can be partitioned from the total variance of the response variable to reduce the residual variance and thus increase the power of the analysis. It also is possible to use co-variables within the design model to reduce residual variance for comparative purposes. This will be particularly valuable for comparisons involving catch rate, given the inherently high variances associated with the rate estimates. Experience suggests that by using ANOVA design models (without co-variables) for the comparison of annual catch rates for lake fisheries, sample sizes of 15−20 days per month are usually adequate to document annual differences in accordance with the tolerable error objective stated earlier.

19.5   Step 4: evaluation of design response

Initial design structures should be evaluated for efficiency so that optimally customized designs will be forthcoming. Existing arrangements of strata and sampling units may not be effective in reducing residual variance and improvements may be possible. ANOVA allows these components of the design to be statistically evaluated. This kind of evaluation necessitates the availability of previous survey data, preferably over the entire survey period. For example, as stated earlier, the customary division of the temporal dimension into day type strata (week days and weekend days) has not been found to be beneficial for improvement of precision for catch rate or catch. Post-stratification applied to previously collected data allows the likely effects of various stratification strategies to be evaluated using ANOVA.

Currently in the United States, thousands of sets of survey data on recreational fisheries exist; however, rarely have these data been evaluated for survey design customization purposes. Fishery biologists primarily are interested in using information for management purposes and are not particularly interested in whether the collection of the data could have been improved. Fishery management agencies become satisfied with the status quo and are afraid to vary from their traditional ways, as inefficient as they may be. It has been estimated by the United States Fish and Wildlife Service that $20 million are spent each year on creel and angler surveys in the United States and there has been little accountability with respect to the quality of the information collected for the money spent. It is hoped that the concept outlined will aid fishery biologists to approach the customization of on-site intercept survey designs for the collection of information for recreational fishery management more effectively.

References

Crone, P.R. and Malvestuto, S.P. (in press) A comparison of five estimators of fishing success from creel survey data on three Alabama reservoirs. *International Symposium and Workshop on Creel and Angler Surveys in Fisheries Management*, Houston, Texas, 26–31 March 1990.

Hudgins, M.D. (1984) Structure of the angling experience. *Transactions of the American Fisheries Society* 113: 750–759.

Knight, S.S. and Malvestuto, S.P. (in press) Three approaches for the allocation of monthly sampling effort for the roving creel survey on West Point Lake. *International Symposium and Workshop on Creel and Angler Surveys in Fisheries Management*, Houston, Texas, 26–31 March 1990.

Malvestuto, S.P. (1983) Sampling the recreational fishery. In *Fisheries Techniques*. (Ed. by L.A. Nielsen and D.L. Johnson). American Fisheries Society: Bethesda, Maryland, USA. pp 397–419.

Malvestuto, S.P. and Knight, S.S. (in press) Evaluation of components of variance for a stratified two-stage roving creel survey design with implications for sample size allocation. *International Symposium and Workshop on Creel and Angler Surveys in Fisheries Management*, Houston, Texas, 26–31 March 1990.

Malvestuto, S.P., Davies, W.D. and Shelton, W.L. (1978) An evaluation of the roving creel survey with non-uniform probability sampling. *Transactions of the American Fisheries Society* 107: 255–262.

Meredith, E.K. and Malvestuto, S.P. (1991) An evaluation of survey designs for the assessment of effort, catch rate and catch for two river fisheries. In *Catch Effort Sampling Strategies: their*

Application in Freshwater Fisheries Management, (Ed. by I.G. Cowx) Fishing News Books, Blackwell Scientific Publications Ltd.

Palm, R.C. and Malvestuto, S.P. (1983) Relationships between economic benefit and sport-fishing effort on West Point Reservoir, Georgia-Alabama. *Transactions of the American Fisheries Society* **112**: 71–78.

Robson, D. and Jones, C.D. (1989) The theoretical basis of an access site angler survey design. *Biometrics* **45**: 83–98.

Wade, D.L., Jones, C.M., Robson, D.S. and Pollock, K.M. (in press) Computer simulation techniques to assess bias in the roving creel survey estimator. *International Symposium and Workshop on Creel and Angler Surveys in Fisheries Management*, Houston, Texas, 26–31 March 1990.

Chapter 20
Analysis of catches from the British National Angling Championships

K. O'HARA and T.R. WILLIAMS *Department of Environmental and Evolutionary Biology, University of Liverpool, P.O. Box 147, Liverpool, L69 3BX, UK*

Data obtained from the National Angling Championships are presented to show how the distributions of catches vary between different water bodies in relation to the community structures. A method of estimating the concentration of the distribution of weight of fish caught within a competition is presented and these results are correlated with the quality of the fishery as perceived by the anglers. Observations on the use of angling statistics as a sampling technique and a management decision tool are made.

20.1 Introduction

The use of data derived from angling competitions to monitor coarse fisheries has become more widespread in Britain in recent years (Cowx *et al.*, 1986). Where detailed species catches from anglers are available, such as those collected by Severn Trent Water Authority (Cooper & Wheatley, 1982; Cowx *et al.*, 1986; Hickley & North, 1981), a detailed appraisal can be made. This is labour intensive, however, and some Water Authorities who utilize competition data for management purposes only record, by questionnaire, the catches of the winning few individuals (Axford, 1979; Banks, 1981). While this latter approach may detect large-scale trends in community structure and species changes, it does not allow detailed interpretation of the population dynamics of recreational fisheries.

Data are available on a long-term basis from the British National Angling Championships which are annual competitions in which the fish captured are non-salmonid (coarse fish) species. These championships have a number of features that make them potentially an excellent source of material for analysis and they are readily available and collected without a large commitment of time or labour. The positive aspects are:

(1) All anglers must record the weight of fish they catch (even if this is zero).
(2) Since 1972 there have been at least two championships a year with at least 936 anglers fishing in each competition.
(3) All the competitions are based on linear systems (rivers or canals), with anglers usually being spaced at a distance of 20 metres apart. Thus the effective length angled in each competition is approximately 18.7 km.

214

(4) The competitions are fished by teams and, because they are based on a points system with each member of a team contributing, all anglers are attempting to catch fish. This is not the case in other coarse fishing competitions where the total weight caught by a team represents the winning result.

The questions therefore posed were:

(1) Are these data capable of giving some indication of fish community structure?
(2) On waters where several years data are available are the patterns of catches consistent or do they indicate trends that could be assessed in the light of known changes in species composition?
(3) Do anglers perceive different water bodies as better in competition angling terms than others?

20.2   Methods

All data were obtained either from results made available from the National Federation of Anglers or from results printed in the national angling weekly newspaper, *The Angling Times*. These data were computed to provide weight-frequency histograms and estimates of the mean weight. One problem of interpretation with weight-frequency distributions is that they are difficult to analyse by normal statistical techniques, in that the distribution that is of interest is the concentration of weights, and the mean of deviations from the mean do not give a measure that is meaningful when a large proportion of the total weight caught may be contributed by a relatively few anglers. Skewness could be one measure of this, but skewness and weight dominance by a few large catches are different properties of a weight distribution. Essentially the analysis required is one that gives a measure of how evenly the weights in a competition are distributed. The method adopted was one usually used in business or economic analysis known as the Lorenz curve and Gini coefficient. To examine whether the Gini coefficient gave an indication of the quality of the fishing as perceived by the anglers, the National Federation of Anglers (hereafter referred to as the NFA) ranked their fishing matches from 1972 to 1982 in the order that they considered were the most successful fishing matches. To avoid any bias, this information was sought without providing the NFA with any information on why the ranking was required.

20.3   Results and discussion

The results from the following three potentially different fisheries were examined (Fig. 20.1):

(1) The River Ancholme, a roach/bream community.
(2) The Oxford Canal, a gudgeon and stunted/small roach dominated community.
(3) The Leeds Liverpool Canal, a general mixed fishery comprising mainly roach and tench.

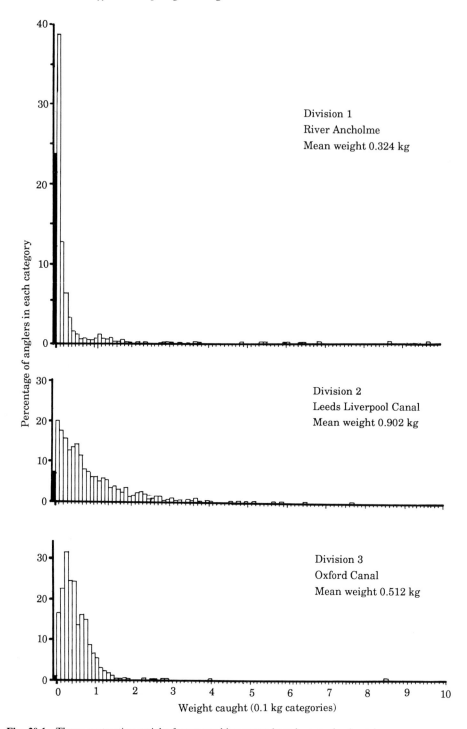

Fig. 20.1   Three contrasting weight frequency histograms based on anglers' catches

These fisheries have been broadly categorized on the basis of competition angling reports or Water Authority survey data to give an indication of their fish community structure. It should be noted that the average maximum size that competition-caught fish usually attain is in the order of approximately 2 kg for bream and 0.5 kg for roach. Gudgeon, on the other hand, only attain a weight of perhaps 20−30 g and tench can grow up to approximately 2 kg.

The shape of the catch curves is broadly consistent with the anticipated patterns. Those fisheries which contain strongly shoaling fish of a large size, for example bream, showed the most skewed distributions. Further examination of the results from the Oxford Canal Division 3 and River Ancholme Division 1 show that both are extremes of catches, the Ancholme registered a large number of anglers who failed to catch any fish whereas virtually all anglers caught fish in the Oxford Canal. This result confirms the difference in the fish community structure because if ability of angler were considered influential better results should have been obtained by the Division 1 anglers who generally are more skilled (National Federation of Anglers, *pers. comm.*). The visual interpretations of these results are supported by the values of the Gini coefficients and the shape of the Lorenz curves (Fig. 20.2).

The results of the ranking presented by the National Federation of Anglers and the Gini coefficients are shown in Table 20.1. The statistical significance of these results was tested for all divisions' competitions, although Division 4 was considered to be rather low in number of years to give reliable results. One problem of interpretation with the lower divisions is that the anglers are not as skilled as those fishing in Divisions 1 and 2. The correlations between the ranking of the Gini coefficient and those provided by the National Federation of Anglers were tested using a non-parametric test, the Spearman-Rank Correlations Test. For Divisions 1, 2 and 3 a significant correlation, ($P < 0.05$), was obtained. A non-significant result was obtained from Division 4. This correlation tends to indicate that fisheries giving consistent catches in terms of reliability are rated more highly than those in which some high individual catches are found but in which a large number of anglers fail to catch any fish. It should be noted that these results may not be entirely representative of other fishing competitions in which a team result may not be as important as individual results. Nevertheless, the method does demonstrate that the perception of anglers can be reliable when compared with a statistical appraisal of the catches.

The results can also be used to obtain some information on the level of exploitation in systems. This can be very high: in the Leeds Liverpool Canal Division 2, the 1982 catch average was 0.9 kg per angler; taking the width of the canal as 12 metres this indicates that approximately 37 kg/hectare were caught in this fishing competition. If a biomass of 400 kg per hectare is assumed as reasonable for this type of water body, it can be seen that there is a very high exploitation rate. This tends to indicate that it would be inadvisable to remove fish from water bodies after fishing matches, as has been advocated by some

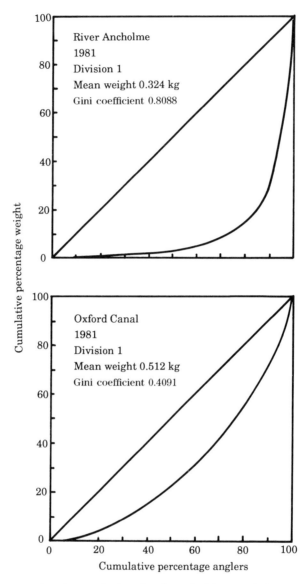

Fig. 20.2 Lorenz curves based on anglers' catches from the River Ancholme and Oxford Canal

managers. Exploitation at this rate could potentially cause drastic reductions in population densities.

The data from the River Trent (Fig. 20.3) are in general agreement with the results presented by Cowx *et al.*, (1986), who have reported a change in the community structure from roach to one which is dominated by chub. This species grows to a larger size than roach and tends to be much more contagious in their distribution because of shoaling.

The results of some poor competitions, in terms of both average weight and the number of anglers failing to catch fish, are consistent with results produced by the

Table 20.1 Relationship between Gini coefficients and NFA ranking of Championship matches

Year	Gini Coefficient	NFA Ranking	Venue
Division 1			
1980	0.4992	1	River Trent
1978	0.5325	2	Bristol Avon
1976	0.5598	3	River Trent
1973	0.6270	4	River Witham
1972	0.6065	5	Bristol Avon
1975	0.5849	6	River Nene
1977	0.5595	7	River Welland
1974	0.8191	8	River Welland
1982	0.6257	9	River Huntspill
1981	0.8088	10	River Ancholme
1979	0.7094	11	Rivers Cam & Ouse
Division 2			
1975	0.4150	1	River Trent
1974	0.5710	2	Avon (Warks)
1980	0.4890	3	River Witham
1981	0.5200	4	Leeds Liverpool Canal
1982	0.5270	5	Bristol Avon
1976	0.5130	6	River Witham
1979	0.4840	7	River Witham
1973	0.7130	8	Great Ouse Relief Channel
1978	0.6090	9	River Trent
1972	0.6860	10	River Welland
1977	0.5320	11	Great Ouse
Division 3			
1982	0.5392	1	River Trent
1981	0.4091	2	Oxford Canal
1977	0.6203	3	River Trent
1978	0.5279	4	River Huntspill
1975	0.7333	5	River Welland
1979	0.6863	6	River Witham
1980	0.7401	7	River Ancholme
1976	0.7267	8	River Huntspill
Division 4			
1981	0.6484	1	River Thames
1982	0.5634	2	Leeds Liverpool Canal
1979	0.6359	3	River Trent
1978	0.7267	4	River Huntspill
1980	0.7707	5	Yorkshire Ouse

Anglian Water Authority from their detailed biomass surveys. Examination of two fisheries, the Rivers Ancholme and the Cam, which were ranked badly by the National Federation of Anglers in this study, had in the case of the Cam a large part of the river ranked as a Class D system by Anglian Water, that is, a river with a biomass of only $0-5\,\mathrm{gm}^{-2}$ and the lower half of the Ancholme as a Class C, which has only $5-10\,\mathrm{gm}^{-2}$.

Water bodies where a large component of the competition weights are contributed by a few large catches appear to be vulnerable to producing bad competition results as perceived by anglers. Similarly, those waters where the catch is composed

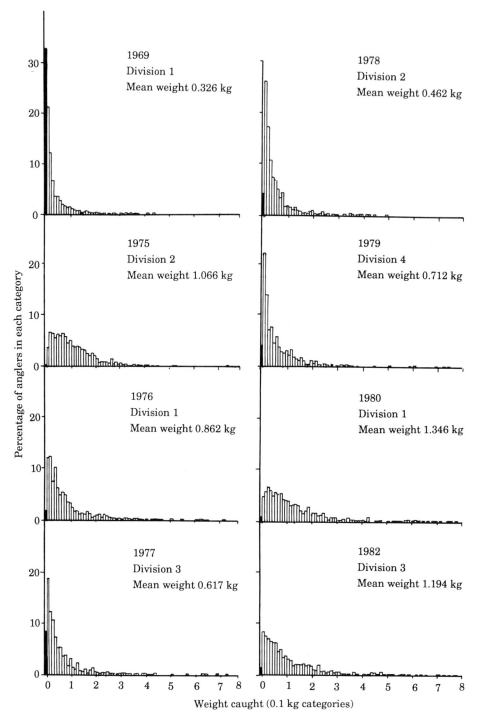

Fig. 20.3   Weight-frequency distributions for championship matches in the River Trent 1969–82

of short-lived species such as gudgeon could decline rapidly, possibly in some instances within a year.

Where the distribution of weights deviates markedly from a normal pattern the average or mean weight may be some measure of the behaviour of the competition recreational fishery but does not necessarily reflect the perception of anglers of the quality of the fishing. The Lorenz curve and Gini coefficient give some indication of the concentration of weights in a competition and, therefore, are a useful quantitative alternative.

Some criticisms of the data used in this study are that only the total weight of fish captured by each angler is recorded, no individual sizes of fish are obtained, and the species composition is not measured, the only assessment available being that reported for the winning individuals in angling journal reports. It could be further argued that the results of individual competitions are subject to the vagaries of weather and other variables that may produce a poor result on a water body on a particular day. Nevertheless, the comparability demonstrated with other source material on particular water bodies could justify the use of these data for management interpretation purposes and as an aid to other sampling techniques. A strong skew in angling catches suggests that many of the sampling programmes which only sample a short section of river or canal, are unlikely to detect major components of the biomass that are recreationally exploited. Only by using a very intensive sampling effort over long sections can this problem be overcome. When angling data are combined with other approaches, for example species and age composition data, it may be possible to predict future catches.

A further use of competition angling results may be in the formulation of stocking policies. In particular, it is envisaged that where a large component of the competition catches rely on the capture of relatively few large fish, there may be the basis for supplemental stocking of a few large fish. There is some indication that such an approach can be used in practice, as shown by the results of Fisher *et al.* (1984), who were able to demonstrate successful stocking of chub into a fishery. The use of competition data for interpretation and management purposes has been advocated previously by several authors (Cowx *et al.*, 1986; Hickley & North, 1981; Pearce, 1983) and the data presented in this paper tend to support their opinions that fishing competition results provide a useful and readily available method of sampling large rivers in an intensive way that is not quantitatively or qualitatively achieved easily by any other method.

Acknowledgements

The authors would like to thank the National Federation of Anglers for their support and financial assistance in this project and their permission to use their competition data is gratefully acknowledged.

References

Axford, S.N. (1979) Angling returns in fishery biology. *Proc. 1st Br. Freshwat. Fish Conf.*, Univ. Liverpool, pp 259–272.

Banks, J.W. (1981) Approaches to stock assessment in the Thames area. In *Assessment of freshwater fish stocks*. London: Water Space Amenity Commission, pp 1–15.

Cooper, M.J. and Wheatley, G.A. (1981) An examination of the fish population in the River Trent, Nottinghamshire, using anglers' catches. *J. Fish Biol.* **19**: 539–556.

Cowx, I.G., Fisher, K.A.M. and Broughton, N.M. (1986) The use of anglers' catches to monitor fish populations in larger water bodies with particular reference to the River Derwent, Derbyshire, England. *Aquacult. Fish. Mgmt* **17**: 95–103.

Fisher, K.A.M. and Broughton, N.M. (1984) The effect of cyprinid introductions on angling success in the River Derwent, Derbyshire. *Fish. Mgmt* **15**: 35–40.

Hickley, P. and North, E. (1981) An appraisal of anglers' catch composition in the barbel reach of the River Severn. *Proc. 2nd Br. Freshwat. Fish Conf.*, Univ. Liverpool, pp 94–100.

Pearce, H.G. (1983) Management strategies for British coarse fisheries: the lower Welsh Dee, a case study. *Proc. 2nd Br. Freshwat. Fish Conf.*, Univ. Liverpool, pp 263–273.

Chapter 21
An evaluation of survey designs for the assessment of effort, catch rate and catch for two contrasting river fisheries

E.K. MEREDITH and S.P. MALVESTUTO *Fishery Information Management Systems, Inc., 500 Dumas Drive, Auburn, AL 36830, USA*

A landing point catch assessment survey was conducted on the Niger River artisanal fishery in Niger, West Africa, from 1983 to 1985. The sources of variability associated with geographical and temporal strata were evaluated using ANOVA. In the south-eastern United States, a roving creel survey was conducted on the Tombigbee River recreational fishery in 1981. The sources of variability associated with a two-stage temporally stratified survey design were evaluated in the same fashion as for the Niger River data. Results showed that in both cases, stratification seasonally was not significantly ($p < .10$) beneficial. Additionally, temporal stratification by week days and weekend days for the Tombigbee River recreational fishery was only beneficial for the estimation of fishing effort. Comparisons with tailrace and reservoir survey designs showed that monthly stratification may improve the precision of the designs relative to the seasonal stratification used. As with the river fisheries, the reservoir and tailrace fisheries showed that temporal stratification by week days and weekend days was beneficial for estimation of fishing effort, but not for catch or catch rate. In all cases, catch rates and catch were inherently more variable than effort. Recommendations for improving riverine fishery survey designs are suggested.

21.1   Introduction

Evaluation of sport fisheries using creel surveys and artisanal and small-scale commercial fisheries using catch assessment surveys are well documented. The theory, advantages and problems associated with creel survey techniques were discussed by Carlander *et al.* (1958), Lambou (1961), Malvestuto *et al.* (1978), Malvestuto (1983), Crone and Malvestuto (in press), and Malvestuto and Knight (in press). Bazigos (1974), Bazigos *et al.* (1975), Malvestuto *et al.* (1980) and Malvestuto and Meredith (1989) discussed catch assessment surveys for small-scale commercial and artisanal fisheries.

Stratification is commonly used in conjunction with random sampling to obtain survey data. The temporal dimension typically is divided into monthly or seasonal strata (time blocks) and, for recreational fisheries, cross stratification by day type (week days *vs.* weekend days) is also commonly practised. These temporal stratification schemes recognize expected systematic variability in the fisheries being sampled to create more homogeneous temporal strata within which random sampling is conducted. Geographical or environmental heterogeneity may demand spatial stratification as well. This is likely for river and large lake fisheries.

A sampling design which incorporated temporal (seasonal) and geographical stratification was used to evaluate the artisanal fishery on the Niger River in Niger, West Africa. By comparison, the Tombigbee River sport fishery, Alabama, USA, was surveyed using a two-stage temporal stratification scheme where the year was stratified into seasons and also into day types. The objective of the analysis presented here was to evaluate the overall utility of the survey designs used based on their ability to reduce residual variability in estimates of fishing effort, catch rate and catch for the fisheries surveyed.

21.2   Methods

21.2.1   *Characterization of survey areas*

Niger River

The Niger River is the longest river in West Africa (4200 km) and the 14th longest in the world (Fig. 21.1). Its source, at an elevation of 800 m in the highlands of Fouta Djallon in Guinea, is approximately 250 km from the Atlantic coast of West Africa. It flows through the arid region of West Africa known as the Sahel, traversing Mali, then turns south and flows through Niger and Nigeria to empty into the Gulf of Guinea. The Niger River is classified as Sudanian by Daget and Illis (1965), as it contains a largely Nilotic fish fauna (Beadle, 1974) and drains an arid savanna region with no fringing forests.

The 575 km of the Niger River in Niger have been classified into three ecological zones (Malvestuto & Meredith, 1989). The river from the Mali border south to the confluence with the Sirba River contains a broad, internally braided floodplain. A central section downstream from the Sirba River south to the confluence of the Mekrou River is primarily a main channel with little or no fringing floodplain. The third zone is the stretch of river running from the Mekrou River to the Nigeria border which has a broad fringing floodplain primarily along the southeast border with Benin (Fig. 21.1).

Malvestuto and Meredith (1989) have described the conditions that existed in the Niger River during the study period. Since the late 1960s, the annual cycle of the flood in the Niger River has been progressively changing, and the peak and duration of the flood was much depressed as a result of the drought across the Sahelian region of West Africa during the study period from 1983–85.

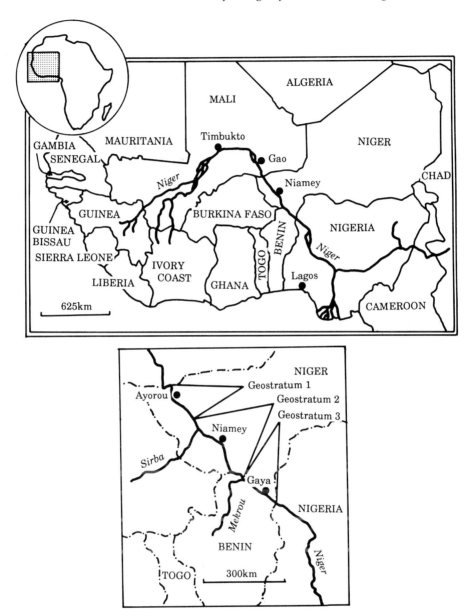

Fig. 21.1 Regional location of Niger River and locations of ecological zones and sampling strata for the river in Niger (as per Malvestuto & Meredith, 1989)

Tombigbee River

The Tombigbee River originates in north-eastern Mississippi at an approximate elevation of 230 m in the southern foothills of the Appalachian Mountains and flows approximately 709 km in a south-easterly direction (Shell, 1980). The study area included the section of the Black Warrior-Tombigbee Waterway in Alabama, south of the Demopolis Lock and Dam and north of the confluence of the Alabama River and Tombigbee River (Fig. 21.2).

The Tombigbee River in this area is primarily a main channel with little fringing floodplain except areas where tributaries flow into the main river. In these areas, there are inundated backwaters that serve as nursery and feeding grounds for fish. Both main river and backwater areas of the river were surveyed. The inclusion of backwater areas was dependent on ease of accessibility. Some areas were accessible during certain seasons and not others; thus, their inclusion changed seasonally.

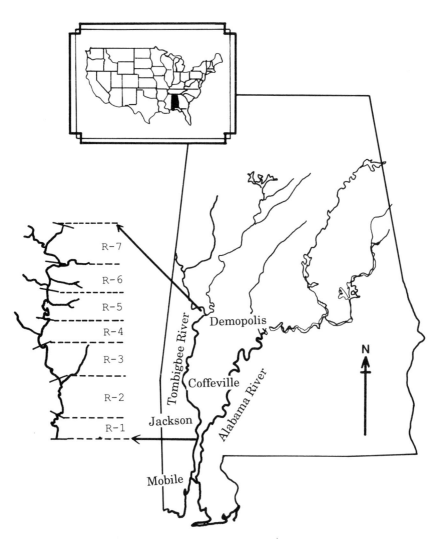

Fig. 21.2 Location of the Tombigbee River study area in Alabama, USA, and sampling areas. The seven spatial sampling units are designated R-1 to R-7

21.2.2   *Survey designs*

Niger River

Malvestuto and Meredith (1989) described the access point catch assessment survey (CAS) which was used to assess the artisanal fishery on the Niger River from March 1983 to December 1985. The survey design was based on geographical and temporal stratification. Spatially, the 575 km of river in Niger was stratified from north to south into four geographical sampling strata (geostrata). Later, the four strata were re-defined into three strata which coincided with the ecological zones of the river described earlier (Fig. 21.1).

The seasonal stratification was based on normal river flow data. The hydrological year was divided into four temporal strata (hydrostrata) as follows:

Falling water: 24 Feb–6 May = 72 days
Low water: 7 May–17 Jul = 72 days
Rising Water: 18 Jul–27 Sep = 72 days
High Water: 28 Sep–23 Feb = 144 days

Simple random sampling was used to choose the days during which interviews would be taken within any given geo-hydrostratum unit. An average of 22 days was sampled within each unit for a total of 1069 sampling days over the three-year study.

A fish landing was defined as any fishing village or camp where fishermen were actively fishing during any part of a hydrostratum. The survey teams censused each geostratum during a two- to three-week period before the start of each hydrostratum to determine which villages or camps would remain active during that hydrostratum. If inactive, a village was excluded from the sampling frame.

Fish landings were randomly chosen with uniform probabilities and without replacement so that any one landing could not be sampled twice within any one hydrostratum. Socio-cultural constraints precluded the use of non-uniform probability sampling because if larger villages were imposed upon too many times during the sampling process the survey teams would not remain socially acceptable to the village.

At the randomly chosen landing sites, all active fishing canoes were counted to provide an estimate of total fishing effort at the chosen landing during the sample day. Interviews were taken using a standard interview schedule and each fisherman's catch was enumerated down to the species level. Fish were weighed and measured according to a standardized procedure detailed by Malvestuto and Meredith (1989). If fishermen were too numerous for total enumeration, then they were interviewed as time and manpower permitted, but all fishermen actively fishing were counted. Canoe catches were pooled to give an estimate of catch rate (kilograms per canoe) for the landing for that day which was taken to be representative of the catch rate for the entire geographical stratum.

Effort at each landing was multiplied by catch rate to give an estimate of the catch at that landing on that day. Both effort and catch were expanded upward using the appropriate sampling probabilities to give estimates of total effort and total catch for the entire geostratum for that day. Sampling probabilities were adjusted seasonally based on the number of landings active at the start of each hydrostratum.

Tombigbee River

A roving creel survey was used to assess the sport fishery on the Tombigbee River. The survey design incorporated a two-stage temporal stratification scheme. The temporal dimension was stratified into seasonal time blocks and these blocks were further divided into day type strata, i.e. week days and weekend days. The seasonal time blocks, each covering three months, were defined as winter (December 1980 through February 1981), spring (March through May 1981) and summer (June through August 1981).

The river was divided into spatial sampling sections and each sample day was divided into three four-hour time periods. An actual sample during any day consisted of taking counts and interviews of anglers in a randomly chosen river section during one of the randomly chosen time periods. Effort within the sampling unit was estimated as the angler-count times the number of hours in the time period, and then expanded for the entire day using the appropriate sampling probabilities.

Interviews were taken using a standardized interview form and each angler's catch was enumerated down to species level. If anglers were too numerous, only a portion were interviewed as time permitted, but all anglers active in the sample section were counted. Angler catches were pooled to give a weighted mean estimate of catch rate (pounds per hour) for the time period, which was taken to be representative of the catch rate for the entire day. Catch for each sample day was calculated by multiplying the expanded effort estimate by the catch rate.

The benefits associated with temporal and spatial stratification of the fisheries surveyed were statistically evaluated using ANOVA by testing the significance of these sources of variability at alpha = 0.10 using the Statistical Analysis System (SAS Institute Inc, 1988).

21.3   Results and discussion

21.3.1   *Niger river*

The ANOVA for each response variable showed that a relatively small amount of the total variation was explained by the sources described in the model (Table 21.1). For effort, the unexplained variance ranged from 79−83% of the total variance. Catch rates were more variable with residual variation ranging from 85−90% of the total. Catch was similar to catch rate with residual variation ranging from 86−92% of the total.

Table 21.1 The residual variance associated with estimates of effort, catch rate (CPUE) and catch, expressed as a percentage of total variance, for the Niger River fishery in Niger (1983–85) and for the fishery on the Tombigbee River in Alabama (1981)

Year	Effort	CPUE	Catch
Niger			
1983	83	85	86
1984	81	95	92
1985	79	90	92
Mean	81	90	90
Tombigbee			
1981	74	89	86

Although the sources of variability associated with seasonal and geographical strata were largely significant (p < .10) because of the relatively large annual sample sizes, these sources described, on the average, only 3–11% of the total variation for effort, 3–8% for catch rate, and 2–9% for catch. Thus, the design model was relatively inefficient with respect to the temporal and geographical stratification scheme used for the survey.

Annual coefficients of variation (CV = standard deviation/mean) for effort, catch rate and catch for the Niger River CAS are listed in Table 21.2. From 1983 to 1984, the CV's were probably more representative of normal conditions. The low-water conditions in 1985 forced a more sporadic fishery as adequate economic returns from fishing became questionable and fishing activity and catch became more inconsistent on a daily basis. Thus, variability in that year increased over previous years.

21.3.2   *Tombigbee River*

The design model for the Tombigbee River was slightly more efficient in accounting for variability than that for the Niger River design. The residual variation for effort, catch rate and catch was 74%, 89% and 86% of the total, respectively (Table 21.1). For effort, season and day type strata as sources of variability in the design model were significant (p < .10). Neither season nor day type strata represented significant (p > .10) sources of variability for catch rate or catch. The CV's for effort, catch rate and catch for the Tombigbee study (Table 21.2) were greater than those for the Niger River study. This indicates that, relative to the magnitudes of the variables estimated, the design for the Niger River fishery was more efficient than that for the Tombigbee River fishery.

21.3.3   *Comparison with other fisheries*

It is informative to compare these results with various other creel surveys that have been conducted on tailrace and reservoir recreational fisheries in the south-eastern United States. From June 1982 to October 1983 a roving creel survey was

Table 21.2 Annual coefficients of variation (%) for effort, catch rate (CPUE) and catch for the catch assessment survey on the Niger River in Niger (1983–85) and the sport fishery creel survey on the Tombigbee River in Alabama (1981)

Year	Effort	CPUE	Catch
Niger			
1983	70	76	115
1984	77	87	166
1985	105	118	352
Mean	84	94	211
Tombigbee			
1981	116	141	167

conducted on West Point Reservoir in Alabama/Georgia. The survey design was a two-stage temporal stratification scheme similar to that used for the Tombigbee survey except that the year was divided into monthly time blocks, rather than into seasonal blocks. This design was substantially more effective in explaining total variation than were the river fishery designs, giving a residual variance equal to 36% of the total variance. Both monthly time blocks and day type sources of variation were highly significant ($p < 0.001$). Stratification by months explained 46% of the total variation and day type strata explained 12% of the total variation. An additional 6% of the total variation was explained by the interaction between month and day type.

When the same survey data were post-stratified into seasonal time blocks and the day type strata were retained, the precision of the model for effort decreased by approximately 50%, i.e. residual variation = 68% of the total. The seasonal time blocks only described 17% of the total variation compared to 46% with monthly blocking.

A roving creel survey was conducted concurrently on three systems of the Tallapoosa River in Alabama from May 1988 to April 1989. Yates and Thurlow Reservoirs and the Thurlow Dam tailrace recreational fisheries were surveyed using the same design model as used with the West Point study. For fishing effort, the tailrace fishery acted much like the reservoir fisheries and the total variability explained by the model using monthly time blocks was relatively high (mean for all systems = 55% of the total). When seasonal post-stratification was applied to the data, the average residual variation for the three systems only increased by about 10% relative to monthly stratification.

For the Tallapoosa River fisheries, the total unexplained variance for catch rate (number of fish caught per hour) was higher for seasonal than for monthly stratification. Using monthly strata, the mean residual variation for catch rate on all three systems was 79% of the total. Using seasons, mean residual variation increased to 98% of the total. Thus, an improvement of approximately 20% was realized using monthly *vs.* seasonal stratification.

Similar patterns existed when catch rate measured as biomass per hour and catch were analysed. Catch rate in biomass and catch in number and biomass improved by 10−15% when monthly strata were used relative to seasonal strata.

21.4   Conclusions

These results indicate that current survey designs for assessment of riverine fisheries are relatively inefficient with respect to accounting for variation in the basic fishery descriptors of effort, catch rate and catch. Relative to reservoir fisheries in the south-eastern United States, the riverine fisheries surveyed appeared to be much more inherently variable in time and space and it is necessary to identify factors associated with this variation in order to improve the efficiency of survey designs for riverine fisheries, whether they be small-scale commercial or sport fisheries.

Based on comparisons concerning temporal stratification, it appears that the creation of monthly strata relative to seasonal strata of three to four months each, generally will improve the precision of survey estimates. The reduction in residual variance ranged from 10−50% when monthly strata were used. The benefit of monthly stratification was best seen in a separate study on a large lake fishery (West Point Lake). This implies that monthly stratification would have improved the fishery estimates substantially for the riverine fisheries surveyed where seasonal stratification was used.

Another characteristic of the data analysed was that estimates of fishing effort were always more precise than estimates of catch rate or catch. This appears to be an inherent aspect of fishery survey data. Generally in the south-eastern United States, estimates of fishing effort that are precise enough for most comparative purposes for lake fisheries can easily be obtained. Given sample sizes of 6 to 15 days per month, relative standard errors (RSE = standard error/mean) of less than 15−20% for large lake fisheries and less than 10% for smaller fisheries can be consistently obtained, where the entire body of water can be sampled within a single time period on a sample day. Estimates of catch rate and catch, however, are typically 20−100% more variable than estimates of fishing effort and these levels of precision are unsatisfactory for management purposes. Thus it is necessary for better survey designs to be established for the estimation of catch statistics for both riverine and lake fisheries.

Acknowledgements

The data used for this study were collected while the authors were employed by the Department of Fisheries and Allied Aquacultures at Auburn University. We express our thanks to the Food and Agricultural Organization of the United Nations, the United States Peace Corps and the United States Agency for International Development for their contributions to the Niger River project. Primary funding for the Tombigbee River project was provided by the United States Army Corps of Engineers and we thank Mr Garry Lucas for collection of the creel data.

References

Bazigos, G.P. (1974) The design of fisheries statistical surveys – inland waters. *FAO Technical Paper* No. 133, 122 pp.

Bazigos, G.P., Kapetsky, J.M., and Grandos, J. (1975) Integrated sampling designs for the complex inland fishery of the Magdalena River Basin, Colombia. FI:DP/COL/72 552 Working Paper No. 4, 55 pp.

Beadle, L.C. (1974) *The inland waters of tropical Africa.* Longman: New York, 365 pp.

Carlander, K.D., DiCostanzo, C.J. and Jessen, R.J. (1958) Sampling Problems in creel census. *Progressive Fish-Culturist* 20: 73–81.

Crone, P.R. and Malvestuto, S.P. (In press) A comparison of five estimators of fishing success from creel survey data on three Alabama reservoirs. *International Symposium and Workshop on Creel and Angler Surveys in Fisheries Management*, Houston, Texas, 26–31 March 1990.

Daget, J. and Illis, A. (1965) Poissons de Cote d'Ivoire et de la Basse Guinee. *Mem. I.F.A.N.* 65: 1–207.

Lambou, V.W. (1961). Determination of fishing pressure from fishermen or party counts with a discussion of sampling problems. *Proceedings of the Annual Conference of the Southeastern Association of Game and Fish Commissioners* 15: 380–401.

Malvestuto, S.P. (1983) Sampling the recreational fishery. Pages 397–419. In *Fisheries Techniques*. (Ed. by A. Nielsen and D.L. Johnson). American Fisheries Society. Bethesda, Maryland, USA.

Malvestuto, S.P. and Meredith, E.K. (1989) Assessment of the Niger River fishery in Niger (1983–85) with implications for management. In *Proceedings of the International Large River Symposium*. (Ed. by D.P. Dodge). Canadian Special Publication of Fisheries and Aquatic Sciences 106 pp. 533–544.

Malvestuto, S.P. and Knight, S. (in press) Evaluation of components of variance for a stratified two-stage roving creel survey design with implications for sample size allocation. *International Symposium and Workshop on Creel and Angler Surveys in Fisheries Management*, Houston, Texas, 26–31 March 1990.

Malvestuto, S.P., Davies, W.D. and Shelton, W.L. (1978) An evaluation of the roving creel survey with non-uniform probability sampling. *Transactions of the American Fisheries Society* 107: 255–262.

Malvestuto, S.P., Scully, R.J. and Garzon, F.F. (1980) Catch assessment survey design for monitoring the upper Meta River fishery, Columbia, South America. *International Center for Aquaculture, Agricultural Experiment Station, Research and Development Series* No. 27, 15 pp.

SAS Institute Inc. (1988) *SAS/STAT$_{tm}$ Users Guide, Release 6.03 Edition.* Cary, North Carolina, SAS Institute Inc. 1028 pp.

Shell, J.D. (1980) Drainage areas in the upper Tombigbee River Basin, Mississippi-Alabama. Jackson, Mississippi, U.S. Geological Survey, 305 pp.

Chapter 22
Assessment of the historical downfall of the IJsselmeer fisheries using anonymous inquiries for effort data

WILLEM DEKKER *Netherlands Institute for Fishery Investigations, P.O. Box 68, 1970 AB IJmuiden, The Netherlands*

The fisheries for eel, perch and pikeperch in the IJsselmeer (Netherlands, $\pm 180\,000$ ha) have shown a tremendous decline during the 1970s and 1980s (eel: $2-5\,\text{kg ha}^{-1}\,\text{yr}^{-1}$; perch: $3-4\,\text{kg ha}^1\,\text{yr}^{-1}$; pikeperch: $0.5-3\,\text{kg ha}^{-1}\,\text{yr}^{-1}$), which has tentatively been attributed to the unrestricted increase in fishing effort (eel boxes, fyke nets and gill nets). Until the mid-1980s no measures were taken to determine and/or restrict the fishing effort, although from 1985 the number of fyke nets was fixed at the existing level; other gears were left unrestricted. Conversely, catch figures have been routinely collected from the auctions by the authorities since 1978.

In 1988, all 100 fishermen were asked to report their annual catches and effort on anonymous forms. Nearly 50 replies of effort were received; complete and true catch figures were almost never reported. A small minority of the responses made their name known. In order to analyse these effort data in relation to the corresponding catch data, names need to be assigned to the anonymous replies; a statistical model, based on the assumption that catches of every fisherman should be related to their effort by a log-linear relationship, was developed.

22.1 Introduction

The fisheries for eel, perch and pikeperch in the IJsselmeer have shown a tremendous decline during the 1970s and 1980s, which has tentatively been attributed to the unrestricted increase in fishing effort (eel boxes, fyke nets and gill nets). Until the mid-1980s no measures were taken to determine and/or restrict the fishing effort. In 1988, however, all the fishermen were asked to report their annual catches and effort in order to assess the status of the fisheries. This paper presents the findings of this survey and describes the selection of an appropriate model relating catch to effort.

22.2 Description of the fisheries

The IJsselmeer (sometimes translated as 'Lake IJssel') is a former part of the Waddensea, but separated from it by a 32 km dam, built in 1932. This former

233

brackish estuary has turned into a highly productive freshwater lake, with a fish fauna consisting of eel (*Anguilla anguilla*), perch (*Perca fluviatilis*), pikeperch (*Stizostedion lucioperca*), roach (*Rutilus rutilus*), bream (*Abramis brama*), smelt (*Osmerus eperlanus*), ruffe (*Gymnocephalus cernua*) and several other species. Its total area is about 180 000 ha, with an average depth of 2–4 m, and a sandy to muddy bottom. The River Rhine discharges about 10% of its water into the lake via a branch known as the River IJssel. In 1932, about 1000 fishing vessels re-orientated their fisheries from estuarine species to freshwater species, mainly eel, perch and pikeperch and, despite their low price, smelt and ruffe. The government deliberately planned to reduce the fishing industry, in order to reduce the costs of buying out fishing vessels which was necessary for the planned land reclamation programmes. This policy, together with an increasing trend towards overfishing of the stocks, has now reduced the number of fishing companies to less than 100.

The following fisheries are practised:

(1) Fyke net fishery with fyke nets fixed to poles along the shore. This fishery catches eel (including silver eel), and a small proportion of the perch. Additionally, small perch and pikeperch are caught and discarded. The activity has been shown to exert a discard mortality of 50–90% per year for these species. In spring, ripe smelt are caught on their spawning migration towards the rocky dikes.

(2) Fyke net fishery with fykes anchored to the bottom; so-called shoot-fyke nets. This fishery catches the same range of species as the previous one, except that silver eel are of minor importance. The anchoring to the bottom allows for a much more flexible fishing strategy, yielding more fish in the currently overexploited stage.

(3) Eel boxes catching eel only.

(4) Long lines catching mostly eel, with some perch and pikeperch.

(5) Gill nets, catching perch, pikeperch, bream and roach. This fishery is mainly directed towards perch and pikeperch; the price of bream and roach is too low to make it profitable.

(6) Trawling. This fishery was aimed at (valuable) eel, but also takes large quantities of smelt and ruffe. However, because of the by-catch of small perch and pikeperch, it has been banned since 1970, and only one small trawl is allowed to catch bait for the eel boxes and long line fishery. The trawl is restricted in size.

Annual total catches of eel, perch, pikeperch and smelt are given in Fig. 22.1. Since the Second World War, catches have declined steadily. In 1970, the ban on trawling reduced the catch of smelt and eel, but improved the catch of perch and pikeperch. However, the subsequent introduction of the shoot-fyke nets in the mid-1970s again improved the eel catches temporarily, but also lowered the catches of perch and pikeperch to their former levels. The restoration of the smelt fishery in the 1980s was due to the development of new markets abroad, and the

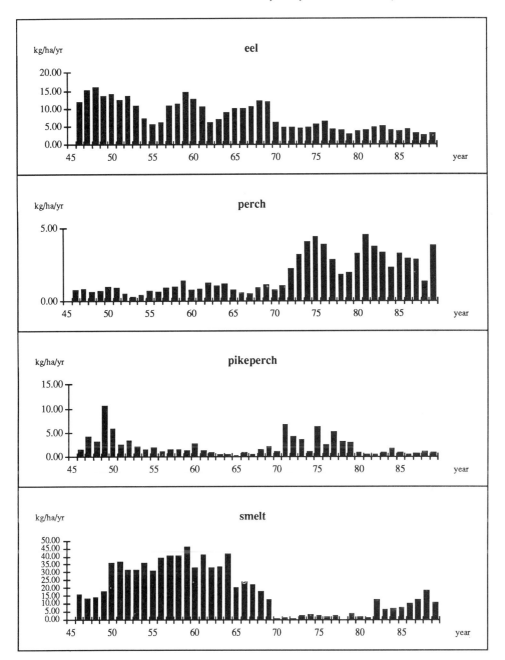

Fig. 22.1 Annual catches (kg ha^{-1}) of the main fish species in the IJsselmeer

use of shoot-fyke nets. Renewal of the trend towards increasing the fishing intensity during the mid-1970s seems to have brought back the overfishing problem of earlier decades. In the mid-1980s, the number of fyke nets was frozen at the existing level, following an enforced but unconfirmed reporting of the current number of fyke nets by each company. This freeze was motivated by the high by-catch of perch and pikeperch, and was not directed at an improvement in the eel fishery. Finally, following the work described in this paper, the government reduced the fishing effort in almost all fisheries by approximately 50%, on the grounds that all fisheries (except the smelt fishery) were heavily overexploited.

22.3   Catch effort enquiry

Preliminary research on the size compositions of the commercial catches (Dekker, 1987; Willemse, 1977) had shown the stocks of eel and pikeperch were heavily, and increasingly, overexploited. Market landings had been recorded for many years, but they did not account for fish sold directly from the vessels, and basic information on a per ship basis was lost for the years prior to 1978. Fishing effort had never been recorded. In the absence of reliable data on fishing effort and yield, it was not possible to decide whether the downward trend in the total production was caused by failing natural productivity or simply overfishing. Further research was more or less blocked by the absence of data on the fishery. Thus the need to restore the basic information on fishing effort, or at least start a new data series, was identified. Since no information was available, interviewing individual fishermen was considered the best source of data, but approaching all 100 would be extremely time consuming. As an alternative, an enquiry was devised which would overcome this problem, but which could be supplemented by additional detailed information when required.

The enquiry involved several letters to the fishermen's organizations, and meetings with their foremen. Details of the need for an enquiry, the method to be used and exact enquiry forms were discussed. Finally, the enquiry was presented at local meetings in every harbour, and anonymous enquiry forms were sent to each fisherman. Both fishing effort and fishing yield for each species were requested. Additionally, a second meeting was organized, stressing the need for the enquiry data, and replenishing any lost forms. Finally, a new set of anonymous forms were sent to each fisherman, asking them to forget about the fishing yield data, but only provide effort data. Every form was accompanied by a stamped return-envelope.

Unfortunately, during the course of the enquiry, fierce political debates broke out on the need for considerable reduction in fishing effort. Many fishermen suspected the enquiry was contributing to the reduction plans. The anonymity of the forms was stressed repeatedly. On the other hand, discussing the need for effort reductions also implied stressing the need for a better knowledge of the fisheries, and thus a better knowledge of the fishing effort.

In the course of the three months following the first dispatch of enquiry forms, nearly 50% of the fishermen replied. Nearly half made their name known on the form or the return-envelope.

Directly following the return of the forms, preliminary results were compiled (unchecked sums for all replies, divided by the fraction of replies), see Figs. 22.2

CPUE
kg/net/yr

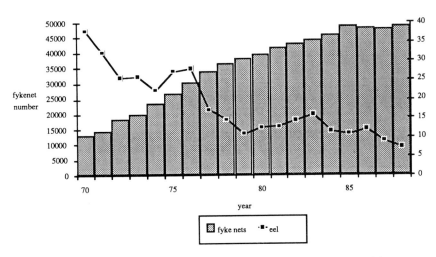

Fig. 22.2 Preliminary estimates of the total number of fyke nets and eel catch per fyke net per year, based on results from the anonymous enquiry

and 22.3. Following the presentation of these preliminary results, a few extra replies were received.

22.4   Modelling a partially answered fishing effort enquiry

One of the most difficult problems in analysing enquiry results is judging the representativeness of the reply. Nearly 50% replied, which is a reasonable response. Nevertheless, the replies were not considered representative; small and old-fashioned 'honest' companies and large ambitious developing companies were overrepresented, whilst the intermediate status companies were underrepresented.

Considering the fishing process from a system analysts point of view, effort has several prerequisites (inputs) and a clear product (output): fishing requires a licence, netting material, fuel and labour time, etc., and produces landed fish (output). Although some information was present on some inputs (namely licences), this information was inadequate: some fishermen provided anonymous statistics on illegal fishing mixed with legal operations. On the output side, the market statistics were known to represent only a (large) share of the total catch. From personal communication with a majority of the fishermen, a strong impression was gained that this misreporting of the market statistics varied from year to year and from company to company. However, variations in both characteristics were more or less consistent: strong misreporters have always misreported, and misreporting rose and fell for almost all companies simultaneously. Therefore, the following statistical model was set up:

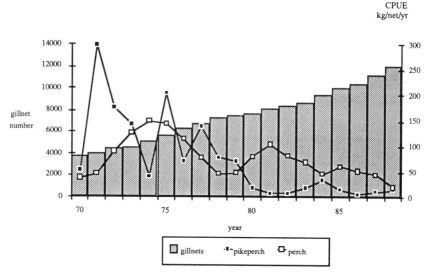

Fig. 22.3 Preliminary estimates of the total number of gill nets and perch and pikeperch catch per gill net per year, based on results from the anonymous enquiry

$$market\text{-}catch_{year,\ boat,\ species} = effort_{year,\ applicant,\ species,\ gear} \times$$
$$year_i \times species_j \times applicant_k$$

with i, j and k being indices over the range of years, species and enquiry applicants respectively, and market-catch taking only the catch passing through the official auctions into account.

This model conforms with a general multiple analysis of variance (ANOVA) model, with *market-catch* as the dependent variable and four explanatory variables (year, gear, boat, species). The interaction terms: *year × species, year × applicant, applicant × species, applicant × gear* and *species × gear* might have a sensible interpretation; others (*year × gear*) would deny the constant efficiency of an otherwise unchanged gear.

One important deviation from the standard model is that for the *market-catch* the boat is known with certainty, whilst on the right-hand side the applicant is unknown. Thus, on top of this ANOVA model, a *who-is-who* pattern matching should be applied. Fitting the general ANOVA model is straightforward, but a complete pattern matching is practically unattainable: a full search of all possible combinations would require more than 10^{29} matchings, with a separate fit of the ANOVA model for every matching. Thus, in analogy with a stepwise ANOVA, the following local search procedure is proposed:

(1)   Assign an initial *who-is-who* pattern for at least two enquiry applicants, and use their effort data as an initial data set for the ANOVA fitting.
(2)   Fit the ANOVA model to the current data set.

(3) Add at least one applicant, choosing the one with the best fit to the current ANOVA model.

(4) Optionally delete one or more applicant(s) from the data set, choosing the ones with the worst fit to the current ANOVA model.

(5) Repeat step (2) to (4) until all applicants have been assigned a boat name.

Note that the criterion for adding applicants to the data set used in step (3) should be less critical than the criterion for deleting applicants in step (4); during implementation, step (3) always adds one applicant, while step (4) deletes any applicant with a worse fit than the fit of the added applicant.

Whether this procedure finds the global optimum or just locates a local best fit, is difficult to assess. If different initial data sets find the same optimum, it will be a convincing procedure. Two widely differing initial sets are proposed: the first being a random allocation of boat names to the enquiry applicants, the second being the subset of all applicants for which the boat name is known for sure. Secondly, the usefulness of the proposed procedure might be judged by the degree of misfitting names to the known applicants.

The proposed procedure is currently being implemented in actual computer programs.

22.5   Discussion

The fish stocks of the IJsselmeer, except for the smelt, have been overfished for at least 20 years, and perhaps since the Second World War. Until recently, this statement was not accepted by the government, or at least judged insufficiently proven. Following the anonymous enquiry described in this paper, no-one questions the cause of the declining production: overfishing is the cause, or at least it prevents recovery. Thus, from a political standpoint, the enquiry has proven to be successful. In hindsight, the following points were essential for this success:

(1) The enquiry was broadcast verbally on several occasions.
(2) The enquiry was anonymous.
(3) The agency responsible for the enquiry (the research institute) is independent of the licensing agency.
(4) The results were reported quickly.
(5) The results of the enquiry do not affect any individual applicant.

From a scientific standpoint, the enquiry elucidated the fishing pattern in a qualitative way. Whether the quantitative statistical analysis proves to be useful, remains to be seen.

The approximate costs of the enquiry were:

• discussions and preparations, circa 30 man-days in total;
• 4 visits to (foremen of) the fishermen's organisations;

- 4 days talking for 3 staff members: 12 man-days;
- 2 visits to each of the 6 primary harbours for 3 staff members: 36 man-days;
- 3 sets of forms for each of the 100 fishermen−1200 pages in total, and 300 stamped return-envelopes;
- 4000 km travelling by car;
- 5 man-days processing replies;
- 5 man-days report writing, and presentation of results.

The total costs are currently (1990) estimated in the order of magnitude of Dfl 60 000 (US $30 000), including expenses and labour time. Thus, only a fraction of the costs of the ongoing research (market sampling and research surveys) were sufficient to fill in a serious defect in the assessment and management of the fisheries.

References

Dekker, W. (1987) Preliminary assessment of the IJsselmeer eel fishery based on length frequency samples. ICES C.M. 1987/M: 22 (mimeo, 14 pp).
Willemsen, J. (1977) Population dynamics of Percids in Lake IJssel and some smaller lakes in the Netherlands. *J. Fish. Res. Bd. Canada* **34**: 1710−1719.

Chapter 23
The applicability of catch per unit effort (CPUE) statistics in fisheries management in Lake Oulujärvi, Northern Finland

PEKKA HYVÄRINEN and KALERVO SALOJÄRVI *Finnish Game and Fisheries Research Institute, P.O.Box 202, 00151 Helsinki, Finland*

With a view to standardizing and rationalizing the collection of CPUE data, a study was made of the sources of variability (effect of fishing season, place, wind conditions, fisherman and the saturation of gill nets). The data were collected from Lake Oulujärvi, Northern Finland, between 1974−87. The material consists of 155144 lifted gill nets. The gill nets were divided into four different groups according to their mesh sizes. The CPUEs of vendace, whitefish, pike and burbot were studied. The frequency distributions of CPUE data were normalized by transformation $(1/(1 + CPUE))$. Gill net saturation was observed in winter fishing. Significant differences were observed between years, months and fishing areas. Wind affected the vendace catches off open shores in October, but not those of whitefish. CPUE can be a useful index of the size of a given fish stock, if the sources of variability are minimized through standardization of data collection. A decreasing trend in CPUE does not always indicate overexploitation, if recruitment depends on population size. There are indications that in some cases CPUE can increase as a function of increasing fishing effort, as in vendace fishing in Finland. This means that factors affecting the success of recruitment should be known when the CPUE index is used for making decisions in fisheries management.

23.1   Introduction

According to the Finnish Fisheries Act, the fishing rights and the responsibility for fisheries management belong to the owners of water areas. Water areas (lakes and rivers) in general are jointly owned by the landowners in the associated village. Fisheries management under joint ownership is organized at meetings of the owners (Fisheries Associations), who decide on a policy and elect an executive committee to realize the policy. Associations of the owners for a given water area, e.g. a central lake and the rivers and lakes flowing into it, form a larger management unit, known as a Fisheries Area. The Fisheries Areas also include representatives from clubs for sports fishermen and the unions of professional fishermen.

The Fisheries Areas and Fisheries Associations of the water owners manage most of the Finnish fisheries and, according to the Fisheries Act, these should be run on a sustainable basis.

There are thousands of fisheries management units in Finland and the persons involved in fisheries management are ordinary fishermen, without any formal education in fisheries management or fish biology. Therefore they need simple, rapid, cheap and efficient methods of estimating the state of the most important fish stocks. The morphoedaphic index (Ryder, 1982) was developed for this purpose, but there are indications that it does not work in Finnish conditions (Myllymaa & Ylitolonen, 1977; Lindström & Ranta, 1988; Ranta & Lindström, 1989).

There are in fact very few methods suitable for use by ordinary fisheries managers in estimating the state of fish stocks at the local level in Finland. The fishing effort (number of licences sold) is generally known, but the total catch and catch by species is unknown. Therefore the CPUE cannot be obtained from fisheries statistics. The CPUE can be calculated, however, from the data collected by a given group of fishermen (book-keeping fishing). Generally, the CPUE is assumed to be proportional to the average density of a fish stock (Gulland, 1983). The most serious problems connected with the fishing effort and CPUE are their great variability and the difficulty of measuring and standardizing them.

The purpose of this paper is to study the sources of variation of the CPUE obtained from fishing records kept in Lake Oulujärvi. The sources of variation studied are 'saturation' of gill nets, and the effects of the fishing area, season, fisherman, target species and wind conditions. Only gill net fishing was considered. The final objective was to draw up recommendations for ordinary fisheries managers on how to standardize the collection of CPUE data and how to apply CPUE to local decision-making in the management of fisheries.

23.2   Study area

Lake Oulujärvi is the central lake in the Oulujoki water system, which drains into the Gulf of Bothnia (Fig. 23.1). It is one of the largest lakes in Finland and its main physical characteristics are presented in Table 23.1.

Lake Oulujärvi is an exceptional Finnish lake in many respects besides its size. The number of private Fisheries Associations is only 14 and large parts of the lake (especially the open pelagic areas) are state-owned (areas 1 and 10, see Fig. 23.1). The whole lake belongs to one Fisheries Area (the Oulujärvi Fisheries Area). The Oulujärvi fisheries have been studied for nearly 20 years (Salojärvi *et al.*, 1981, 1985; Hyvärinen, 1989), which is not the case with most Finnish lakes. The prerequisites for efficient and successful fisheries management are thus good.

The number of fishermen has increased since the beginning of the 1970s, and currently stands at *c.* 8000. Many types of gear are used in Lake Oulujärvi, such as gill nets of different mesh sizes, seines, trawls, fyke nets, wire traps, hooks on long lines, and rod fishing by spinning and lures (summer angling and ice fishing).

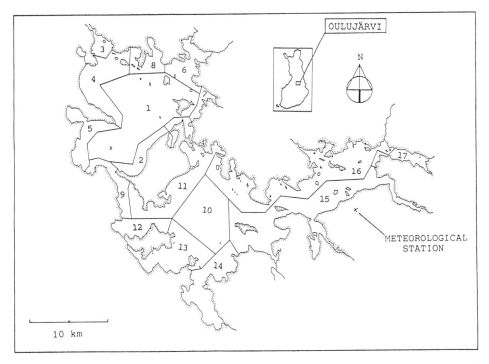

Fig. 23.1 Map of the study area. The CPUEs of the species in the numbered areas are compared using the analysis of variance (ANOVA). Borders also delimit the areas of the Fisheries Associations

Table 23.1 The main physical characteristics of Lake Oulujärvi (National Board of Waters, Finland 1977)

Physical characteristics	
Drainage basin	$19\,506\,\text{km}^2$
Mean water surface (natural shore line)	$928\,\text{km}^2$
Minimum water surface (regulated)	$778\,\text{km}^2$
Maximum water surface (regulated)	$944\,\text{km}^2$
Amplitude of water level regulation	$2.7\,\text{m}$
Maximum depth	$36\,\text{m}$
Mean depth	$7.6\,\text{m}$
Mean elevation above sea level	$121\,\text{m}$
Mean outflow	$216\,\text{m}^3\,\text{sec}^{-1}$

Gill nets are the most important fishing gear and their number has increased from less than 10000 in 1973 to *c.* 15000 at present. Marked changes in the gear have taken place during the last 20 years. In the early 1970s, seines were the most important gear used by professional fishermen. At the end of the 1970s and the beginning of the 1980s, mainly gill nets were used (Arvola, 1989). Professional fyke net fishing for whitefish was introduced in 1984 and trawlers entered the fishery in 1987.

Standardized fishing effort has been estimated with sample surveys since 1984. The effort is calculated for the different gears by multiplying the number of fishing days by the mean number of gears used (Salojärvi *et al.*, 1990). There are indications, however, that for the purpose of estimating the CPUE index, it is very difficult to get reliable estimates of effort by this method. In regulating the fishing effort the Fisheries Associations and Fisheries Area use 'gear units'. The basic unit is a gill net (length 30 m and height 3 m). A seine is equalled to 30 units and a fyke net to 10 units. The total number of gear units used at present is estimated to be 41 000 and the total allowable number of gear units is 90 000. This means that no active regulation of fishing effort has taken place, because the lake is thought to be underexploited.

During the last 15 years the total annual catch from Lake Oulujärvi has varied between 350 and over 600 tonnes (Fig. 23.2). Over 90% of the total fish catch is composed of six species: whitefish, vendace, perch (*Perca fluviatilis* L.), pike (*Esox lucius* L.), burbot (*Lota lota* L.) and roach (*Rutilus rutilus* (L.)). The catch of stocked brown trout (*Salmo trutta* L.) is now *c*. 15 tonnes. The catch variation is mainly due to the strong fluctuation of the vendace stock. Marked changes in the species composition have also occurred since the early 1970s. Fish stocking has considerably increased the whitefish and brown trout catches from the lake. More than half of the total catch (*c*. 60%) is taken with gill nets and these are the most important gear in subsistence and recreational fishing. Most of the professional catch (*c*. 90%) is taken with trawls and fyke nets.

23.3   Material and methods

Book-keeping data (daily records of catches grouped by species and gear) were collected from fishermen between 1974–87, inclusive. These data comprise the number and type of gears lifted each fishing day, the catch in kilograms for each species and the fishing area. The nets were classed by mesh size as follows: under 20 mm, 27–33 mm, 34–40 mm, over 40 mm (bar length).

The mean CPUE index for a given species and net class in each sample was calculated from the following equation:

$$\text{CPUE} = \Sigma \ (Y/n)/N \tag{1}$$

where CPUE is the mean catch per unit effort for a given species and net class, Y is the catch in weight of a given species in one lift, n the number of nets lifted and N the number of lifts.

The saturation of gill nets was studied, because in winter fishing the interval between lifts varies from one day to two weeks and this can affect the catchability of the nets. A graphical study of the mesh size >40 mm, the net class used most in winter, was made. To achieve this the CPUE for each year, when nets were lifted at different intervals, were compared.

To identify how net fishing in Oulujärvi was directed towards different species by the season and mesh size, it was assumed that if the catch of a given species

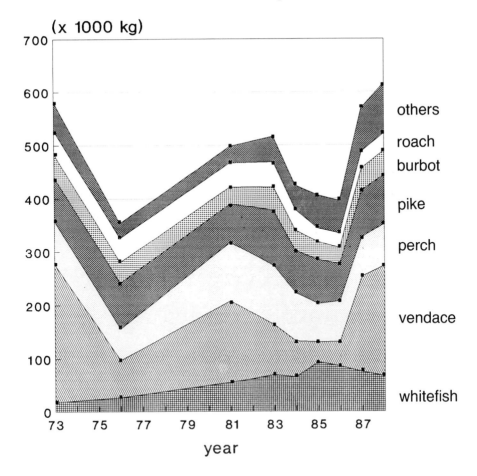

Fig. 23.2 Total catch in Lake Oulujärvi grouped by species in the years 1973−88 (Salojärvi *et al.*, 1990)

was more than 50% of the total catch in one lift, fishing was directed to that species. If the catch of any one species was not more than 50% of the total catch in one lift, it was assumed that the fishing target was mixed species.

The differences in CPUE between years, months and areas were studied using ANOVA tests. Similarly, ANOVA analysis was used to test for differences in the CPUE of two fishermen fishing in the same area (area 11, see Fig. 23.1) and the effects of wind direction on CPUE in three fishing areas (areas 5, 11 and 16, see Fig. 23.1).

Data on wind direction were available from the meteorological station of the airport of Kajaani (Fig. 23.1). The wind observations were made every third hour and the measurements were averages of ten-minute periods. For statistical analysis the direction of the wind was divided into four sectors. If at least four values for a day were placed in the same sector, this was accepted as the prevailing direction of the wind. If eight values for a day were divided equally between two sectors, the wind was considered to be variable. In all other cases the wind was also treated as variable. In the statistical tests, the CPUE for a given day was compared with the wind of the day when the nets were set.

CPUE data have been shown to be negatively distributed with a binomial distribution highly skewed to the right (e.g. Bannerot & Austin, 1983; Virapat, 1986). To use parametric tests different kinds of transformations are needed (e.g. logarithmic, square root or reciprocal) (Sokal & Rohlf, 1981; Ranta *et al.*, 1989). The frequency distribution of the book-keeping data was studied by tests of normality and by examining the correlation of the standard error to the square of the sample means. The statistical calculations were made with a VAX computer and the statistical software of SAS (SAS Institute Inc., 1985).

23.4   Results

The CPUE of gill nets (mesh size >40 mm) lifted at varying intervals indicates that saturation occurs (Fig. 23.3); thus the CPUE could actually be higher if the nets are lifted at shorter intervals. This leads to the conclusion that the catch can be increased in winter fishing by shortening the intervals between lifts. The number of fishing days is unimportant in winter fishing; only the number of lifts is significant. In winter mainly pike and burbot are caught and gill net saturation should be considered if the CPUE of these species is standardized.

In Finnish lakes different kinds of nets are used to catch different fish species. Small mesh nets ($< = 20$ mm) are used for vendace and the best fishing season is autumn associated with spawning in Lake Oulujärvi (Fig. 23.4(a)). Gill nets with mesh sizes of $27-33$ mm and $34-40$ mm are used to catch mixed species, but in October they are used mainly for whitefish fishing (Figs. 23.4(b) and (c)) on the spawning grounds around the lake. Gill nets of mesh sizes over 40 mm are used to catch pike and burbot (Fig. 23.4(d)), but the best fishing seasons for these two species differ. The best season for pike is in spring, because in Lake Oulujärvi pike spawn in May, and the best season for burbot is in winter, from December to April; burbot spawn in February. In standardizing the CPUE the target species of gill net fishing should therefore be considered.

Analysis of the CPUE data showed that the frequency distributions of CPUE were highly skewed and were close to negative binomial distributions. A positive linear correlation was also found between the standard error and the square of the sample means of the CPUE for gill nets. In such a case reciprocal transformation ($1/x$) is recommended (Sokal & Rohlf, 1981; Ranta *et al.*, 1989), but because of zero catches the transformation $1/(1 + x)$ was used.

Analysis of variance (ANOVA) tests showed significant differences ($P < 0.05$) in CPUE between years, months and areas. The results were similar with all the

all species, mesh > 40 mm

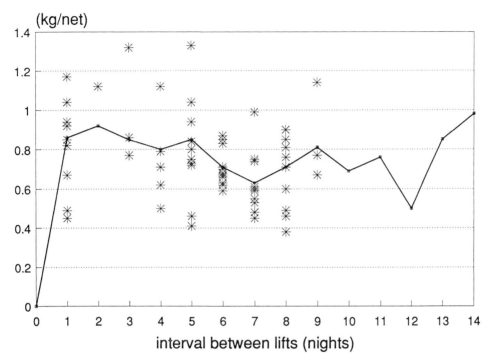

Fig. 23.3 Mean catch per net (mesh size >40 mm) for all species in each year (data for December, January–April) when nets were lifted at different intervals. Only cases for which there were at least ten observations are presented. The line indicates the mean catch per net for all data in December, January–April

species studied (vendace, whitefish, pike and burbot). There were also significant differences ($P < 0.05$) between fishermen fishing in the same area (area 11, see Fig. 23.1) in the CPUE of vendace and burbot, but not in the CPUE of whitefish or pike (Table 23.2). The reason for the significant differences between fishermen in the CPUE of vendace lay in the size of the nets used for catching these fish. The significant differences in the CPUE of burbot were due to the fact that one of the fishermen used trammel nets. Another reason for the difference in the catches was evidently the choice of fishing places. For example, Area 11 is rather large (see Fig. 23.1).

Wind direction seemed to affect the vendace catches in September in area 11 and in October in areas 5 and 11, but not in area 16 (Figs. 23.1 and 23.5(a)–(c)). Areas 5 and 11 are open to southern and eastern winds and area 16 is relatively sheltered, which could explain the differences in the results (Table 23.3). The results also agree with observations made by local fishermen. In the present data the whitefish CPUE did not seem to be influenced by the wind (Table 23.3).

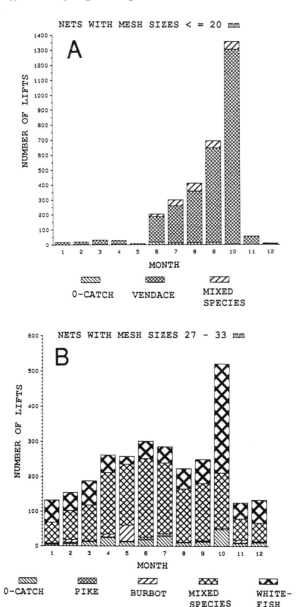

Fig. 23.4 (a)–(d) Fishing for a given species with different mesh sizes in different fishing seasons

The CPUE indices for the species studied reveal wide variation between months during the study period (1974–87) (Figs. 23.6(a)–(e)). Part of this variation can be explained by the low number of lifted nets. For all the species the standard error of the sample means is large until the number of lifts is at least 100 (Figs. 23.7(a)–(d)).

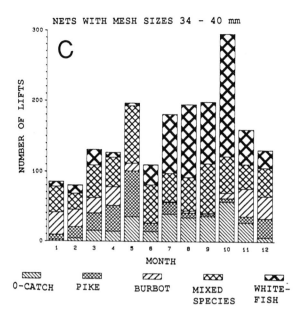

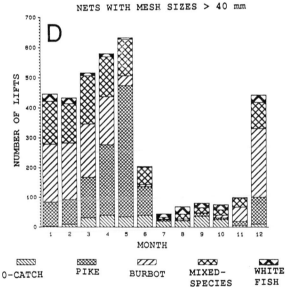

Fig. 23.4 Continued

Pike catches have been relatively stable since 1974, though a slight increase can be seen in the 1980s. A similar trend is seen in the CPUE curve for the winter months (January, February, March and April) (Fig. 23.6(d)). By contrast, the CPUE curve for May is at a much higher level and shows a decreasing trend from 1974 to the beginning of the 1980s, followed by a marked increase up to 1985. Since 1985 the CPUE of pike has again decreased, but is still at a very high level.

Table 23.2 Values of F (from analysis of variance) for transformed data (1/(1 + CPUE)) of different species grouped by year, month, area and fisherman. The tested samples were chosen separately for each species. Under each F-value are shown the samples tested for the years, months and areas. Tests for two fishermen's catches were done for data from area 11. Nets for vendace were of mesh size <=20 mm, for whitefish 27–40 mm, for pike >40 mm and for burbot >40 mm. **P < 0.05

Species	Years	Months	Areas	Fishermen
Vendace	48.39**	48.82**	44.84**	19.33**
Samples	74, 78–87	6–10	5, 11, 14, 16	
Whitefish	24.04**	49.22**	34.54**	0.01
Samples	74, 78, 81–86	5–10	5, 6, !1, 14, 15, 16	
Pike	12.01**	50.12**	7.67**	2.41
Samples	79–87	12, 1–5	5, 6, 9, 11, 14, 15, 16	
Burbot	6.18**	115.37**	20.22**	17.75**
Samples	79–87	12, 1–5	5, 6, 9, 11, 14, 15, 16	

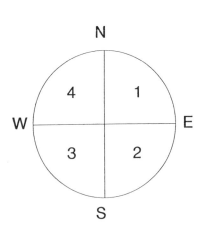

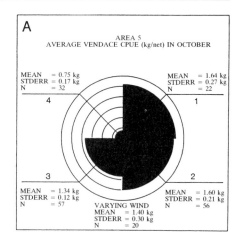

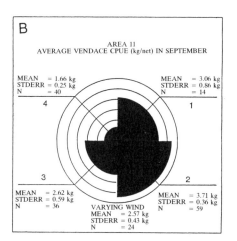

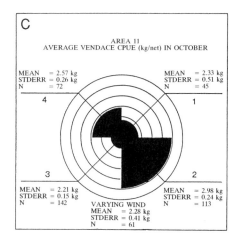

Fig. 23.5 (a)–(c) Monthly mean catch rates (kg/net) for vendace in areas 5 and 11 in relation to wind direction. The cases in which the analysis of variance (ANOVA) showed differences significant at the 95% confidence level are highlighted. (N = number of lifts, STDERR = standard error)

Table 23.3 Values of F (from analysis of variance) for transformed data $(1/(1 + \text{CPUE}))$ of vendace and whitefish tested for influence of wind direction (categorized as sectors 1 to 4 and variable) in three months and areas. Nets for vendace of mesh size $<=20$ mm and for whitefish of mesh sizes $27-40$ mm.

| Area | V | | | W | | |
	August	September	October	August	September	October
Area 5	1.75	1.45	5.03**	1.18	1.90	1.97
Area 11	1.43	5.80**	4.69**	0.97	0.94	0.36
Area 16	0.58	1.25	0.84	1.38	0.78	0.55

Burbot catches have varied by 10 tonnes during the study period (1973−88) (*c*. 30−40 tonnes) and the CPUE has varied even more (December) (Fig. 23.6(e)). There are indications that the CPUE of burbot is inversely correlated to the CPUE of pike.

The CPUE of vendace shows clear fluctuation (Fig. 23.6(a)). The CPUEs for September and October, in particular, follow the known fluctuation in the vendace stock in Lake Oulujärvi. The catch of vendace was high at the end of the 1970s. Although no catch statistics are available, this was indicated by serious problems in marketing the professional vendace catch at that time.

Whitefish catches increased considerably from 1973−85 and since then the catch has slowly decreased (Figs. 23.6(b) and 23.6(c)). The CPUE from whitefish was calculated for two net classes (mesh size 27−33 mm and 34−40 mm). The CPUE trends for different months (1974−87) are contradictory and the variation in CPUE was very high at the end of the 1970s and the beginning of the 1980s (Figs. 23.6(b) and 23.6(c)). There are many reasons for this, such as deficiencies in the CPUE data and the low number of lifted nets (Fig. 23.7(a)). The main cause, however, is stocking with peled fingerlings, which was initiated in the mid-1970s.

The peled whitefish behaves differently from the endemic whitefish forms. Schooling behaviour is more typical of this whitefish as it lives in shallower and more sheltered areas than the local forms. Both of these factors increase the variability of the CPUE.

The biomass of the whitefish stocks showed a considerable increase in the 1980s (Fig. 23.8), which had possibly already begun at the end of the 1970s. The biomass of the peled whitefish is unknown and therefore the total whitefish biomass in 1977−82 is also unknown. The whitefish biomass and CPUE for gill net fishing could be compared only after 1983 and the comparison indicates that there is a positive relationship between these two variables, though no statistical significance was found (Figs. 23.9(a) and 23.9(b)).

23.5 Discussion

Kennedy (1951) showed that the fishing effort exerted in Great Slave Lake by a net cleared of fish after a certain interval was not generally directly comparable

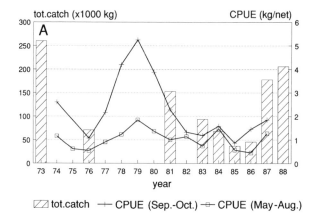

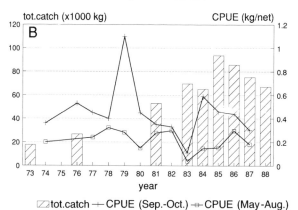

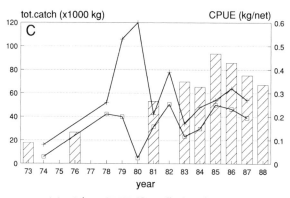

Fig. 23.6 (a)–(e) Average pike, burbot, whitefish and vendace CPUE (kg/net) from book-keeping data compared with total annual catch of the respective species. The vendace CPUE was calculated for a net mesh < = 20 mm, the pike and burbot CPUE for >40 mm and the whitefish CPUE for two mesh size classes, 27–83 mm and 34–40 mm, separately.

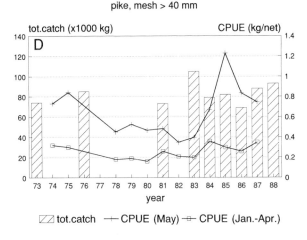

Fig. 23.6 Continued

with the fishing effort exerted by a similar net that is cleared of fish after a different interval. He showed that the greater the catch per net that can be made when nets are cleared daily, the smaller will be the relative increase in the catch per net when they are cleared every two days. He also showed that it is possible to 'saturate' nets, after which they will catch no additional fish. The idea of the saturation of gill nets has also been suggested by Van Oosten (1935) (ref. Kennedy, 1951) and Baranov (1948). Meth (1970) (ref. Hamley, 1975) suggested that saturation depends on the twine material: nylon nets can be expected to be saturated sooner than the less efficient cotton nets, and the longer the nets are in the water, the smaller will be the advantage of nylon over cotton.

In the ice-free season in Finland, nets are usually lifted once a day. Consequently the material collected during summer is homogeneous and standardization of the effort is not complicated by saturation of the nets. During winter, the situation is

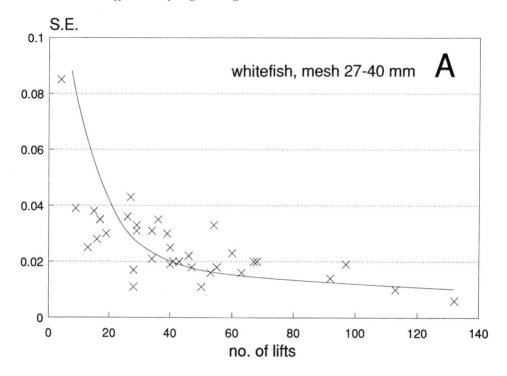

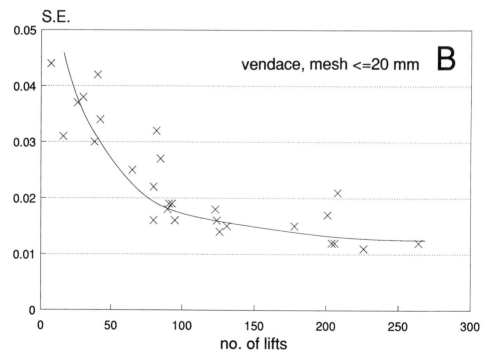

Fig. 23.7 (a)−(d) Standard errors of means (from transformed data) shown in Fig. 23.6 compared with number of observations in each sample. (Line drawn by hand)

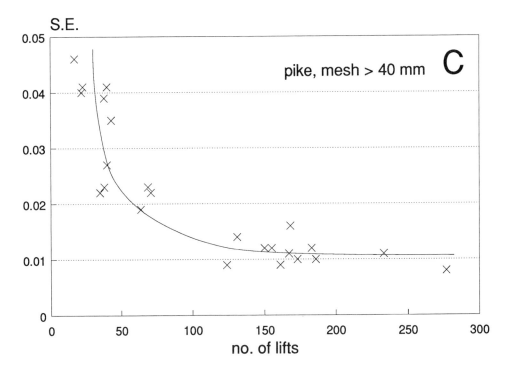

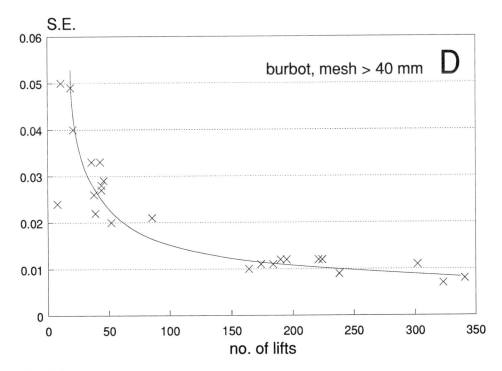

Fig. 23.7 Continued

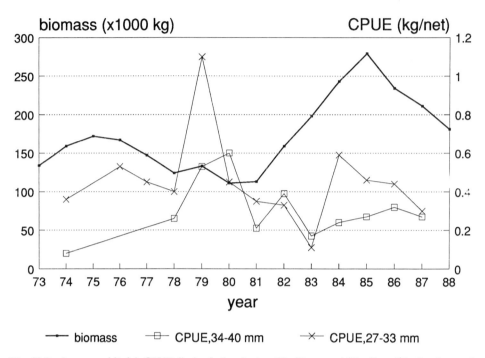

Fig. 23.8 Average whitefish CPUE (kg/net), (mesh sizes 27–33 mm and 34–40 mm) in October and September and the biomass of whitefish stock (> = 3 years old) calculated by virtual population analysis (Salojärvi *et al.*, 1990)

different, because the interval between lifts varies greatly between fishermen. The present data were not good for studying the saturation effect. There were few observations relating to nets lifted at different intervals, but under otherwise similar conditions, e.g. the same year, month and area, and all the known sources of variation could not be taken into consideration at the same time. However, the present results concur with those of Kennedy (1951).

Collins (1987) studied the increased catchability of the deep monofilament nylon gill net and its expression in a simulated fishery. His simultaneous catch comparison between the two gears showed that deep nets were 1.7 times more efficient (but varied seasonally) for whitefish. The increase in efficiency exceeded that expected from the increase in area of the deeper nets. In this study the height and length of the nets were not strictly standardized. The nets used by the fishermen were *c.* 2 m deep and 30 m long.

Every fishing gear can catch a large variety of species, and many different species occur on most fishing grounds. Very few fisheries are based solely on single species. In practice the interpretation of catch and effort data concerning one species has to take into account the effects on the fisherman's tactics and strategy of possible catches of other species. Gear saturation effects are also more

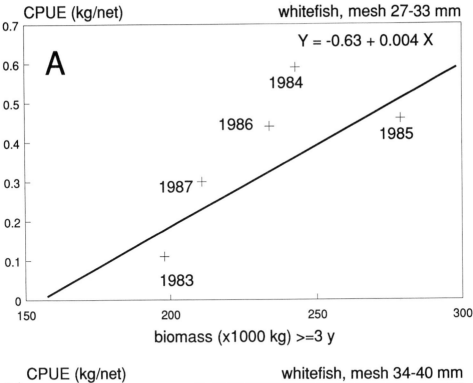

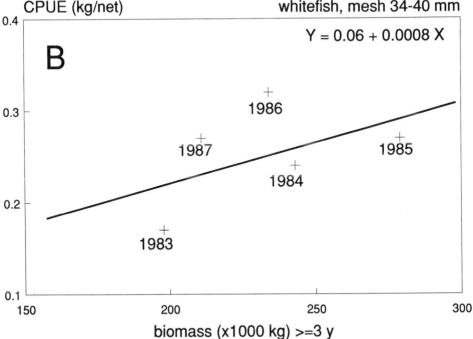

Fig. 23.9(a)−(b)   The relationship between whitefish biomass ($> = 3$ years old) and the CPUE (mesh sizes 27−33 mm and 34−40 mm) of gill net fishing in the years 1983−87. Regression analysis showed no statistical significance

complicated when a mixed-species catch is concerned. Because of saturation, it is possible that the CPUE for a given species will decrease when the stock size of other species is increasing, even if the stock size of the studied species is stable. The results from Lake Oulujärvi indicate that the best season for collecting book-keeping data on gill net fishing is the principal fishing season for the species being studied (usually the spawning season).

Eberhardt and Gilbert (1975) give instructions for estimating the number of test fishings necessary in various situations. The sample size is evaluated by the ratio of the means and coefficient of variation and normal or log normal distribution of the data is assumed. The data of Lake Oulujärvi showed a close to negative binomial distribution with a large number of zero catches. Thus transformation by $(1/1 + CPUE)$ was required to normalize the data. The standard errors of the normalized data were of a high level until the sample size was at least 100.

There are indications in many fisheries around the world that weather conditions have a marked influence on the catches (e.g. Harden Jones & Scholes, 1976; Taggart & Legget, 1987; Rose & Legget, 1988). According to the local fishermen, the wind direction in Lake Oulujärvi affects the vendace catches and has some influence on the whitefish catches. They have also suggested that the effect of the wind is different in different parts of the lake. The results of this study at least partly confirm these observations.

In Lake Oulujärvi there are three fishing methods which require more detailed analysis. The data provided on trawling comprise the total catches of different species and the number of fishing hours in every month; therefore these results are sufficient for analyses. Fyke net and seine fishing are also important, due to the fairly large vendace and whitefish catch, but the fishermen using fyke nets and seine nets are not obliged to report their catch. There is thus a need to collect statistics from all fishermen using such gears. Even when these three fishing methods are treated separately, there is still rather wide variation, for example in the total catch. To decrease the variation, the remainder of the material can be divided between households selling their catch and households fishing for recreation or subsistence.

As CPUE is, at least in theory, proportional to the average density of a fish stock, it can serve as an index of the size of the fish stock (Gulland, 1983). The results obtained for whitefish after 1983 are in agreement with this conclusion. It is generally believed that there is an inverse relationship between stock size and fishing effort and that CPUE is inversely related to fishing effort. These assumptions may not hold, however, for all fish species. If the recruitment of a species depends on the size of the spawning stock, the CPUE of that species can increase with the fishing effort. There are indications from Finland (e.g. Huusko, 1990) that the CPUE of vendace increased, when the fishing effort was increased. Gear selectivity is another problem, which may result in an inverse relation between the CPUE of a given species and the stock density, due to density-dependent growth. In this case, however, the CPUE is positively related to the size of the catchable stock.

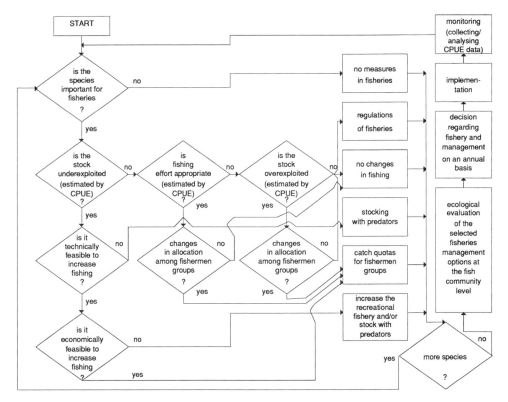

Fig. 23.10 Decision scheme for Fisheries Areas. The CPUE data are collected and analysed in the monitoring box and used in evaluating the state of a given stock in the under-/overexploitation or appropriate fishing boxes

23.6   Application to fisheries management

To regulate fisheries properly, the status of the important fish stocks should be known. Evaluation of the state of the fish stocks is mostly based on the fisheries managers' own experience and on interviews with other fishermen. Instead of this circumstantial evidence, the CPUE by species and by gears could serve as a useful and objective index. The material needed to calculate the CPUE can be collected rapidly (from two to four weeks) and cheaply (material collected from normal fishing), and the CPUE can be calculated using simple mathematics. Problems are posed, however, by the variability of the CPUE and the difficulty in interpreting the results.

If the collection of the CPUE data is standardized, the variance can be kept reasonably low, and the CPUE of a given species can be considered to be an index of the catchable stock size. For standardization, the CPUE data for a given species should be collected during the best fishing season with standardized gear and fishing techniques and using the same fishing places from year to year. In

Lake Oulujärvi the best fishing season for vendace is September-October, for whitefish October, for pike May and for burbot December. If gill nets are used for data collection, the number of observations (lifts) should be at least 100 in each sample. It is preferable to collect CPUE data from unselective gear, such as seine, trawl and fyke nets. The information obtained from the routine fishing records could usefully be supplemented by taking fish stock samples.

The interpretation of the CPUE data can be a difficult problem. The CPUE can reasonably be assumed to be positively related to the stock size, or at least to the catchable stock size. However, the CPUE of predator species may be inversely related and the CPUE of dense pelagic planktivorous small-sized species positively related to the fishing effort. More ecological research is needed, for example on the density-dependent mechanisms of population regulation.

It may be that the CPUE of gill nets is not a very accurate index of stock size from the statistical point of view, but there are very few practical alternatives. Moreover the CPUE has other uses. It is an index of fishing profitability and can be used to determine the prices of fishing licences, though the demand for licences may be more suitable for this purpose. It can also be used to compare the fishing efficiency of different gears.

Fisheries management operations are generally considered species by species. The fishing effort (number of licences sold) is the regulator and the opinions and feelings of fisheries managers are the indicator of the effect of fishing on the fish stocks. It is recommended that, instead of this circumstantial evidence, the catch per unit effort (CPUE) could be used as the indicator of the state of the fish stocks. As shown in the decision scheme presented in Fig. 23.10, the CPUE index could be used together with other information to determine whether the fish stock is underexploited, overexploited or properly fished.

References

Arvola, I. (1989) Kalavesien käyttö- ja hoitosuunnitelma. Oulujärven kalastusalue. Osa I nykytila. Kainuun kalatalouspiiri. 25 p.

Bannerot, S.P. and Austin, C.B. (1983) Using frequency distributions of catch per unit of effort to measure fish stock abundance. *Trans. Am. Fish. Soc.* **112**: 608−617.

Baranov, F.I. (1948) *Theory and assessment of fishing gear.* Pischepromizdat, Moscow. (Ch. 7 Theory of fishing with gill nets translated from Russian by Ont. Dep. Lands For., Maple, Ont., 45 p.)

Collins, J.J. (1987) Increased catchability of the deep monofilament nylon gillnet and its expression in a simulated fishery. *Can. J. Fish. Aquat. Sci.* **44** (Suppl 2): 129−135.

Eberhardt, C.C. and Gilbert, R.O. (1975) Biostatistical aspects. In *Environmental Impact Monitoring of Nuclear Power Plants.* Source Book of Monitoring Methods. National Environmental Studies Project. Atomic Industrial Forum 2. pp. 783−918.

Gulland, J.A. (1983) *Fish stock assessment: a manual of basic methods.* FAO/Wiley series on food and agriculture. 1. Chichester: John Wiley & Sons, 223 p.

Hamley, J.M. (1975) Review of gillnet selectivity. *J. Fish. Res. Board. Can.* **32**: 1943−1969.

Harden Jones, F.R. and Scholes, P. (1976) Wind and the catch of Lowestoft trawlers. *J. Cons. int. Explor. Mer,* **39**: 53−69.

Huusko, A. (1990) Kuusinkijoen vesistöalueen kalatalousselvitys. Manuscript.

Hyvärinen, P. (1989) Yksikkösaaliin vaihtelu ja siihen vaikuttavat tekijät Oulujärvellä. M. Sc. Thesis. Univ. Helsinki. 71 p.

Kennedy, W.A. (1951) The relationship of fishing effort by gillnets to the interval between lifts. *J. Fish. Res. Board. Can.* **8**: 264−274.

Lindström, K. and Ranta, E. (1988) Is the relationship between the morphoedaphic index and fish yield in Finnish lakes a statistical artefact? *Aqua Fennica* **18,2**: 205−209.

Meth, F. (1970) Saturation in gill nets. M. Sc. Thesis. Univ. Toronto. Toronto. Ont. 39 p.

Myllymaa, U. and A. Ylitolonen (1977) Kuusamon vesistötutkimus vuonna 1977. *Vesihallitus. Tiedotus* **191**: 1−164.

Ranta, E. and Lindström, K. (1989) Prediction of lake-specific fish yield. *Fisheries Research* **8**: 113−128.

Ranta, E., Rita, H. and Kouki, J. (1989) *Biometria*. Tilastotiedettä ekologeille. Helsinki. Yliopistopaino. 569 s.

Rose, G.A. and Legget, W.C. (1988) Atmosphere-ocean coupling and Atlantic cod migrations: effects of wind-forced variations in sea temperatures and currents on nearshore distributions and catch rates of Gadus morphua. *Can. J. Fish. Aquat. Sci.* **45**: 1234−1243.

Ryder, R.A. (1982) The morphoedaphic index − use, abuse, and fundamental concepts. *Trans. Am. Fish. Soc.* **111**: 154−164.

Salojärvi, K., Auvinen, H. and Ikonen, E. (1981) Oulujoen vesistön kalatalouden hoitosuunnitelma. Helsinki. RKTL, kalantutkimusosasto. Monistettuja julkaisuja 1. 277 s.

Salojärvi, K., Partanen, H., Auvinen, H., Jurvelius, J., Jäntti-Huhtanen, N. and Rajakallio, R. (1985) Oulujärven kalatalouden kehittämissuunnitelma. Osa I: Nykytila. Helsinki. RKTL, Kalantutkimusosasto. s. 1−273.

Salojärvi, K., Moilanen, P. and Hyvärinen, P. (1990) Oulujärven siian kalastus, siikojen ekologia, istutustoiminnan tulokset ja ekologiset vaikutukset. Manuscript.

SAS Institute Inc. (1985) *SAS User's Guide: Statistics, Version 5 Edition.* 956 p.

Sokal, R.R. and Rohlf, F.J. (1981) *Biometry.* San Francisco: Freeman, 2nd ed.

Taggart, C.T. and Legget, W.C. (1987) Wind forced hydrodynamics and their interactions with larval fish and plankton abundance: a time-series analysis of physical-biological data. *Can. J. Fish. Aquat. Sci.* **44**: 438−451.

Van Oosten, J. (1935) Logically justified deductions concerning the Great Lakes fisheries exploded by scientific research. *Trans. Am. Fish. Soc.* **65**: 71−75.

Vesihallitus (1977) Oulujoen vesistön vesien käytön p.kokonaissuunnitelma. *Osa I. Tiedotus* **125**: 1−102.

Virapat, C. (1986) Use of catch per unit of effort in fish stock assessment of Kiantajärvi lake. M. Sc. Thesis. Helsinki University. 91 + 7 p.kokonaissuunnitelma. Osa I. Tiedotus 125: 1−102.

Chapter 24
A logbook scheme for monitoring fish stocks; an example from the UK bass (*Dicentrarchus labrax* L.) fishery

G.D. PICKETT and M.G. PAWSON *Ministry of Agriculture, Fisheries and Food, Fisheries Laboratory, Lowestoft, Suffolk, NR33 OHT, UK*

Annual catch and fishing effort in the recreational and commercial bass (*Dicentrarchus labrax* L.) fisheries around the coasts of England and Wales have been estimated by a combination of census and sampling. As with many fisheries in large inland lakes, this is a multi-species fishery prosecuted by full-time, part-time and casual commercial fishermen and for sport. It is a seasonal and opportunistic fishery and though individual landings are generally small they have a high unit value, and there is considerable competition between commercial fishermen and sport anglers. As a consequence, catch reporting is subject to strong bias and a high proportion of landings goes undetected by routine monitoring.

The paper describes a scheme for sampling catch and effort by logbooks which employs a three-dimensional stratification of fishing effort, and compares catch estimates with those obtained by collection at markets and merchants statistics and by random questionnaire and postal surveys. It is argued that the logbook technique is particularly applicable to fisheries in which there is a diversity of effort types, and can be used to provide several independent indices of the size of the exploited population which, with a knowledge of relative catchability and the overall distribution and quantity of the various fishing effort categories, can be assessed.

24.1   Introduction

The European sea bass (*Dicentrarchus labrax* L.) is regarded as a prime sport fish by anglers in England and Wales and also as a valuable commercial resource. The commercial fishery has two main components: offshore mid-water trawling for adult fish in winter and early spring, mainly by French vessels, but with increasing Scottish, Danish and English interest; and netting and lining inshore along the coast and in estuaries and creeks by smaller boats in the warmer months. The recreational fishery is conducted from inshore boats and from the shore. This paper is concerned only with assessment of the inshore fishery, which is similar in many ways to those in large inland lakes.

Since the early 1970s, commercial interest in sea bass has risen and concern that the stock was in danger of being overexploited has resulted in increased research being devoted both to the fish and its fishery. In 1982 the Directorate of Fisheries Research at Lowestoft (DFR) was given the task of providing information on catch and effort in both the commercial and recreational sectors as a basis for advice to ministers on the management of this fishery. From the outset, it was clear that official government statistics collected by MAFF Sea Fisheries Inspectorate (SFI) contained little reliable catch and effort data for bass. Separate landings figures for bass have been available since 1972, but these have been based only on information obtained through markets and merchants at the major ports. They ignore the majority of landings made by the inshore fishery. Some trawlers' catches are probably accurately recorded in the figures shown in Table 24.1, but these may, at best, only be used to indicate trends.

This paper describes a method of assessing an extensive fishery which involves small boat and shore-based effort, operating seasonally and opportunistically with a wide variety of catching methods.

24.2   Methods

24.2.1   *Assessment strategy*

Before attempting any assessment, fishery managers need a clear view of the use to which catch and effort data will be put. Historically, catch data for many English and Welsh fisheries have been collected with no clear management objective in mind. They therefore do not necessarily provide a reliable foundation upon which to base management decisions. In the present study, data were collected to evaluate the state of the bass stock and its response to exploitation, and to help predict the effects that management measures would have on the fishery.

The method of data collection is also determined by the timescale of the study; for the UK bass fishery a work programme of around five years was envisaged. In the event, useful catch and effort statistics have been collected for the boat fishery for bass from 1985 onwards. Comprehensive studies on large or widely distributed

Table 24.1   Annual UK bass landings in tonnes, official SFI statistics

Year	Weight	Year	Weight
1972	20.7	1980	118.0
1973	28.2	1981	130.9
1974	52.4	1982	134.0
1975	79.3	1983	234.1
1976	101.7	1984	138.1
1977	104.9	1985	105.8
1978	115.3	1986	103.0
1979	103.7	1987	125.1
		1988	176.9

fisheries can be labour intensive, and the expense of setting them up, which might take one or two years, tends to overshadow annual running costs. Given the objectives of the bass investigation, it was considered necessary to establish a time-series of catch at age or size group data in order to determine annual catch in the fishery, stock mortality, exploitation patterns and yield dynamics, and to provide stock abundance indices. Although it was not a priority for this work, cohort analysis could be carried out using these statistics, provided there was adequate biological sampling of the catch for size (length and weight) and age distributions.

Before deciding on the most suitable assessment method, some knowledge of the UK bass fishery, its characteristics, seasonality and geographical extent were needed. Considerable qualitative data were obtained from a review of the coastal fisheries of England and Wales carried out in 1981−82 (Pawson & Benford, 1983). A more up-to-date summary is given in Pawson and Pickett (1987). The recreational fisheries are scattered along the English and Welsh coasts from Yorkshire in the north-east to Cumbria in the north-west, whilst the main commercial fishery extends from the Thames Estuary to South Wales. Effort is often directed solely at bass in season; June−August in the north, all year round in south-west England and generally May−October in the intermediate locations, although much of the commercial catch is taken as a bycatch in trawls and gill nets. Over 20 different fishing gears are used for taking bass. These can be grouped into seven main types of gear which catch in different ways and with varying efficiencies.

Around 2000 boats, mainly under 10 m, are involved in the commercial inshore fishery, the majority being used on a part-time basis. There are around 50 specialized angling charter boats and a few inshore trawlers. The number of anglers who fish for bass at some time is thought to be in the region of 0.5 million, with over 800 boats being used in the recreational fishery.

The bass catch tends to bypass the main port markets and auctions, and the main sales outlet is to small merchants, much of the catch going for export. Although some rod and line-caught bass are sold or traded and sport anglers may return fish to the water, much of the recreational catch is taken for home consumption. The very high monetary value obtained for all sizes of bass has encouraged a vigorous 'free' market.

24.2.2   *Chosen option*

The census of catch as carried out by SFI was considered inappropriate for fisheries as widespread, fragmented and diverse as that for bass. Its use was therefore restricted to the few ports where coverage of the local bass fishery was considered adequate. It was beyond the scope and budget of this project to evaluate the catch and effort of shore anglers in the UK, though in 1986 MAFF commissioned an independent economic evaluation of both this and the commercial sector of the bass fishery, which has provided relevant data (Dunn *et al.*, 1989).

This present study therefore was directed at estimating catch and effort in the inshore boat fisheries.

Whilst abandoning the concept of total catch census, it was considered that a census of fleet size and structure was both possible and necessary. Pawson and Benford's (1983) survey included information on a much larger number of ports and landing places (300) than are covered by SFI statistics (90). It provided a knowledge of the number of vessels that might be used to catch bass, thus giving a measure of potential effort. This could be stratified into categories on the assumption that it had characteristic levels and patterns of exploitation by region, boat size or fishing power and gear, (Table 24.2). Effort was designated as being part-time or full-time, with the latter classification applying only where bass are the main target species in the appropriate season.

To obtain effort frequency (days fishing for bass) and estimate total actual effort and catch, individual boats were sampled within the various effort census strata for daily catch and effort data. Total catch for each stratum was derived by raising each sample (catch per fishing boat year) by total estimated effort in that stratum. The chosen sampling tool was a fisherman's logbook, in which catches were recorded for each day fished during the local bass fishing season. Due to the diversity of gears and variations in their efficiency, effort designations were simplified by using boat days as a standard effort unit. By this means and by stratifying effort by fishing power (full- or part-time and number of crew) the complications of attempting to standardize units of effort between various gears has been avoided.

Effort census

The number of boats working on bass in various effort categories (strata) were recorded at each port against a prescribed list covering five main geographical regions: the ICES fishing area divisions abutting the English and Welsh coasts, (Fig. 24.1). Effort data for boats working in the current year were obtained by site visits of Department of Fisheries Research staff and by follow-up correspondence with local fisheries officers.

Catch sampling

Although a sampling level of one logbook per effort stratum [5 (regions) × 7 (gears) × 6 (fishing powers) = 210] would meet the statistical requirement of covering about 10% of the fleet, our resource limitations made this sampling level impossible. The strategy adopted was therefore one of stratified random sampling with no attempt to satisfy mathematical criteria, other than to achieve a similar minimum level of cover in each of the five regions. It was not always necessary to select vessels by main gear type, since many commercial vessels use several methods within one season and thus data on catches by various gears were often

Table 24.2 Distribution and numbers for region 107d, 1988 showing F = full-time; P = part-time; 7 = trawl; 41 = drift net; 50 = gill net; 52 = trammel net; 71 = long lines; 72 = angling; 73 = hand lines.

Port	Code	Charter/casual angling			Single handed							2+ handed						
		71	72	73	7	41	50	52	71	72	73	7	41	50	52	71	72	73
Dungeness	623P	—	2	—	—	—	—	2	—	—	—	—	—	—	—	—	—	—
Rye	626P	—	—	—	—	—	2	1	—	—	—	—	—	—	—	—	—	—
	F	—	—	—	—	—	—	—	—	—	—	—	—	—	—	—	—	—
Hastings	628P	—	—	—	—	—	1	2	—	—	—	—	—	—	—	—	—	—
Bexhill	630P	—	5	—	—	—	2	2	—	5	—	—	—	2	10	—	—	—
Eastbourne/Langney	632P	—	10	—	—	—	5	5	—	2	2	—	—	2	—	1	20	—
Newhaven	635P	—	10	—	—	—	4	2	—	—	—	10	—	2	1	1	2	—
	F	—	—	—	—	—	2	1	—	—	—	—	—	2	2	—	—	—
Brighton	638P	—	5	—	—	—	—	—	—	—	—	—	—	—	—	—	—	—
Shoreham	639P	—	20	—	—	—	—	10	2	—	—	4	—	5	5	3	—	—
	F	—	—	—	—	—	5	—	—	—	—	—	—	—	2	—	—	—
Worthing	640P	—	—	—	—	1	—	2	1	—	—	—	—	2	20	2	—	—
	F	—	—	—	—	1	—	2	—	5	—	—	—	2	4	—	—	—
Littlehampton	645P	—	15	—	—	—	2	3	—	—	—	—	—	7	6	2	5	—
	F	—	—	—	—	—	7	3	—	—	—	—	—	2	—	2	—	—
Bognor	646P	—	5	—	—	—	2	3	2	—	—	—	—	2	2	2	4	—
Selsey	647P	—	—	—	—	—	5	2	0	—	—	—	—	1	—	2	2	—
	F	—	—	—	—	—	6	5	1	—	—	—	—	6	—	—	6	—
Emsworth	651P	—	10	—	—	—	6	—	—	—	—	—	—	—	—	—	—	—
	F	—	—	—	—	—	2	—	—	—	—	—	—	—	—	—	—	—
Hayling Island	653P	—	10	—	1	—	—	—	10	—	—	8	—	3	—	—	10	—
	F	—	—	—	—	—	—	—	1	—	—	—	—	—	—	—	—	—
Portsmouth	654P	—	40	—	6	—	2	2	1	2	—	6	3	3	4	6	4	—
	F	—	—	—	—	—	—	—	—	—	—	—	2	4	4	6	—	—
Southampton	655P	—	16	—	—	—	2	—	—	—	—	2	2	3	—	1	2	—
	F	—	—	—	—	—	—	—	—	—	—	—	—	1	—	—	—	—
Lymington	661P	—	25	—	—	—	—	—	—	—	—	—	—	2	—	3	—	—
Poole	662P	—	40	—	4	—	20	5	1	5	—	10	—	8	8	2	5	—
	F	—	12	—	—	—	6	—	—	—	—	—	3	—	—	—	5	—
Swanage	663P	—	3	—	—	4	—	—	—	6	—	—	—	—	—	—	2	—
Christchurch	664P	—	10	—	—	8	—	—	—	6	—	—	—	—	—	2	5	—
	F	—	—	—	—	2	—	—	—	1	—	—	3	—	—	—	—	—

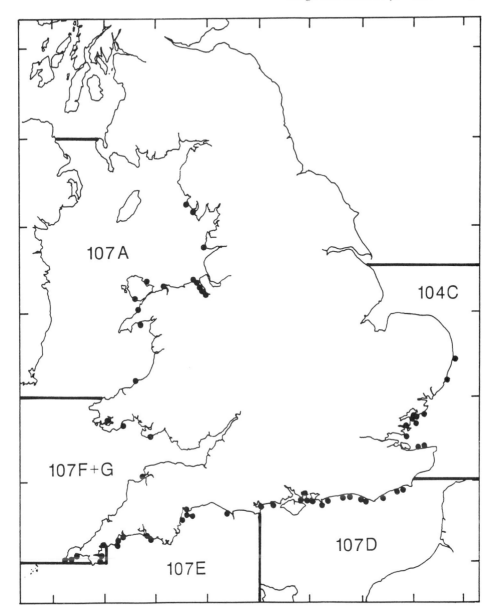

Fig. 24.1 The distribution of logbooks in the bass fishery during 1987, and the ICES sub-divisions used for regional catch assessments shown as MAFF sub-division codes

available from a single logbook. Excluding gear categories, one sample of each of the six fishing power categories per region (i.e. $6 \times 5 = 30$ logbooks) was considered the absolute minimum requirement to produce an assessment of national catch, and ten samples in each ICES (International Council for the Exploration of the Seas) division are needed to produce regional catch estimates. On a yearly basis, sampling frequencies were in the range 1.6–3.3% of the fishing fleet, which has made it necessary to devise a system for weighting sampled catches in order to substitute for unsampled strata and to help correct any bias in sample distributions.

Catch of individual boats

The details required for each logbook were boat type (fishing power), full- or part-time designation, port of landing and main fishing method, with which the vessel was allocated a position in the effort census matrix. The information requested from each fisherman was: day/date, grounds fished, gear used, weight and numbers of bass (and mullet, *Chelon labrosus*, *Liza ramada* and *Liza aurata*, which are frequently taken in the same fishery) caught in three weight ranges, and numbers and total weight of other species caught. A simple weekly form (Fig. 24.2) was compiled into books of 25–50 sheets, depending on the length of the local fishing season.

On return of these books, daily data were extracted and input to a menu-driven computer program written specifically to compile and analyse the logbook and effort census data (Pickett, 1990). With this it is a simple matter to calculate total catch (weekly, monthly or seasonally) or catch per unit of effort (weight or number of bass per boat day or per fishing trip) for individual boats.

24.2.3   Calculation of total catch

The total catch for the bass fishery can be estimated from a weighted mean of all daily catches, raised by total effort in boat days. Monthly mean catch weights per

SKIPPER CODE:				WEEK:				PORT-CODE	For Fish Lab use	
Date	Grounds Fished	Gear Used	Species	Enter numbers of fish in size range			Total Weight (lb)	Other species (numbers and wt of each type)		
				under 1½(lb)	1½– 6(lb)	over 6(lb)			rect	gear
SUN			BASS							
			MULLET							
MON			BASS							
			MULLET							
TUE			BASS							
			MULLET							
WED			BASS							
			MULLET							
THUR			BASS							
			MULLET							
FRI			BASS							
			MULLET							
SAT			BASS							
			MULLET							

Fig. 24.2 A bass logbook weekly data sheet

boat are derived from the logbook database and allocated to strata using the matrix shown in Table 24.2. The corresponding effort values, stored on the effort census file, are used to raise the mean catch by the total number of boats in each stratum. Mean catches are weighted according to the distribution of samples over the effort census, by multiplying sampled catch with the appropriate fishing power ratio. These were originally set arbitrarily according to the number of fisherman-shares individual boats represented, but are now derived from observed catch rates in the various effort strata. Weighting values are allocated to each stratum where significant differences in mean catch occur between part-time and full-time boats and between the various fishing powers, but are only used when a particular stratum within one region is not sampled. In effect, this results in the addition of estimated monthly catches in each region for each gear group. As the effort census values are recorded by port, it is possible to estimate monthly and annual catches at each port, even though only around 40% of the ports are sampled by logbooks.

24.2.4 *Calculation of total effort and CPUE*

Total fishing effort each year is calculated by raising the effort recorded in logbooks (i.e. number of days spent fishing for bass) by the stratified effort census, (i.e. number of boats by region, fishing power, gear and full/part-time designation), and is given in boat days. Three forms of catch per effort can be derived:

(1) Catch per effort of particular individual boats, which are directly comparable between years and can be used as local stock density indices.
(2) Mean catch per effort (unraised) of the sampled fleet, split by gear type if required. Where the same boats are used from year to year these provide a stock index, but the values may be biased in relation to the whole fleet.
(3) Estimated mean catch per effort of the total fleet (weighted and split by gear type), which is usually lower than the unweighted mean because sampling is biased towards full-time boats. By separating regional and gear values, the relative catching power of different gears in different areas can be assessed.

24.3 Results

24.3.1 *Logbook distribution and returns*

Initially problems were encountered in finding fishermen willing to participate in the scheme, and considerable time was spent visiting ports to distribute logbooks. The minimum requirement of 50 (ten in each of five regions) completed logbooks was achieved in most years, with a maximum reached in 1986, when 66 books were returned. Return rates have averaged 50% of the books given out, and although this improved over the first few years of the scheme, it was subsequently necessary to recruit new logbook holders where a stratum was found to be poorly sampled or where individual fishermen dropped out of the scheme.

24.3.2   *Catch and effort*

Estimates of catch and effort obtained from bass logbooks and effort census for the years 1985−88 are shown in Table 24.3 by gear-type and region. These data do not include SFI data (mainly trawl landings) which are added to the estimated catch each year. Between 1875 and 2039 vessels were used in the fishery, and estimated effort has risen from 98 000 to 121 000 boat days each year. The annual catch appears to have remained stable apart from in 1987, when particularly high catch rates by gill nets increased landings considerably. The decline in rod and line and hand-line catches, which prompted this study, has been seen to continue.

24.3.3   *Catch per unit of effort*

The results provide no evidence of any consistent trends in the abundance of bass over the years 1985−88, CPUE being much more variable between gears within the same region than with the same gear between regions. This highlights the impact that changes in the fishing power of catching gears has on catch rates.

24.3.4   *Catch at age*

Although it is outside the scope of this paper to describe the system used to obtain estimates of bass catches in numbers of each age group, these are a principal component of the assessment. Their quality depends to a large extent on the adequacy of biological sampling of catches (for length and age) throughout the fishery, but they have been sufficiently robust to enable the year-classes from 1976 onwards to be monitored as they progress through the fishery, and characteristic exploitation patterns have been observed for the various catching gears in each region.

24.4   **Discussion**

A system for obtaining catch and effort data in a relatively large and dispersed fishery, which is not well assessed by the established statistics collection system, has been described. The main shortcoming of the fishermen's voluntary logbook sampling method is the low level of sampling attained and the difficulty in obtaining a complete effort census each year. Because individual effort strata have seldom been sampled by more than one logbook, it is not possible to determine errors due to between-boat variability. We are confident, however, that these logbook data represent actual catches, since there is little incentive to misreport and catch per effort trends in the various effort strata are consistent from year to year.

Table 24.3 Estimated bass catch and effort in inshore boat fisheries in England and Wales derived from DFR logbooks and effort census

		1985		1986		1987		1988	
	Region	Catch (kg)	Boat days	Catch (kg)	Boat days	Catch (kg)	Boat days	Catch (kg)	Boat days
Otter	104C	5 884	1 260	2 088	624	4 386	1 389	6 552	819
trawl	107D	34 092	7 309	8 977	2 610	9 452	2 951	20 254	2 415
	107E	12 752	2 728	1 355	638	2 075	886	4 668	580
	107F + G	832	73	1 428	287	1 694	506	5 574	680
	107A	–	–	–	–	–	–	–	–
	Total	53 560	11 370	13 848	4 159	17 607	5 732	37 048	4 494
Drift	104C	15 916	285	4 722	596	1 577	276	3 933	757
net	107D	590	70	7 050	265	15 201	1 712	9 514	1 644
	107E	1 323	270	10 731	2 124	54 297	2 167	3 819	719
	107F + G	528	104	8 774	595	45 843	2 178	7 119	282
	107A	1 238	137	2 193	311	2 117	845	5 669	280
	Total	19 595	866	33 470	3 891	119 035	7 178	30 054	3 682
Gill	104C	7 609	2 566	19 622	2 591	52 207	2 428	40 902	2 231
net	107D	24 426	11 286	32 745	7 769	93 141	6 405	74 534	4 594
	107E	22 458	6 583	43 209	7 480	102 155	8 041	45 612	6 726
	107F + G	7 056	2 268	20 839	3 583	28 716	5 070	6 448	6 314
	107A	16 100	8 799	24 669	11 167	17 544	6 599	28 399	1 9021
	Total	77 649	31 502	141 084	32 590	293 763	28 543	195 895	38 886
Trammel	104C	1 937	444	2 107	1 032	3 674	827	10 088	2 602
net	107D	9 948	2 355	10 346	4 139	14 302	4 302	26 068	7 424
	107E	7 080	1 172	4 788	1 518	8 125	2 734	4 986	2 058
	107F + G	14	7	331	414	–	–	1 428	347
	107A	66	131	5 137	5 786	2 705	1 132	499	390
	Total	19 045	4 109	22 709	12 889	28 806	8 995	41 988	13 902
Long	104C	292	87	10 786	1 182	1 180	313	2 292	249
lines	107D	253	23	45 258	3 489	10 801	1 408	15 593	1 386
	107E	–	–	1 846	269	294	16	127	20
	107F + G	–	–	–	–	445	56	1157	124
	107A	–	–	2 718	243	860	80	1 525	154
	Total	545	110	60 608	5 183	13 580	1 873	20 694	1 933
Angling/	104C	33 995	2 553	19 621	3 244	21 971	5 240	24 318	5 990
headlines	107D	110 268	16 647	73 407	14 288	40 245	17 150	56 000	16 225
	107E	188 801	19 350	51 253	10 447	50 963	11 515	48 150	11 355
	107F + G	35 894	4 503	69 106	7 853	85 550	10 989	41 440	12 026
	107A	39 816	7 176	112 840	11 713	88 417	9 785	16 020	9 505
	Total	408 774	50 229	326 227	47 545	287 146	54 679	185 928	55 011
Total	104C	65 633		58 946		84 995		88 085	
catch	107D	180 310		181 464		183 142		201 963	
(kg)	107E	231 619		111 225		217 909		107 362	
	107F + G	44 368		98 754		162 248		63 166	
	107A	57 220		147 557		111 643		52 003	
	Total	579 168		597 946		759 937		512 079	

Table 24.4 Estimated bass CPUE (kg per boat day) and effort in inshore boat fisheries in England and Wales (weighted means)

		1985		1986		1987		1988	
	Region	Boat days	× (kg) CPUE	Boat days	× (kg) CPUE	Boat days	× (kg) CPUE	Boat days	× (kg) CPUE
Otter	104C	1 260	4.67	624	3.35	1 389	3.16	819	8.00
trawl	107D	7 309	4.66	2 610	3.44	2 951	3.20	2 415	8.39
	107E	2 728	4.67	638	2.12	886	2.34	580	8.05
	107F + G	73	11.40	287	4.98	506	3.35	680	8.20
	107A	–	–	–	–	–	–	–	–
	Mean		4.70		3.30		3.10		8.20
Drift	104C	285	*55.85	596	7.92	276	5.71	757	5.20
net	107D	270	4.90	2 124	5.05	2 167	7.01	1 644	5.79
	107E	104	5.08	295	14.75	2 178	24.93	719	5.31
	107F + G	70	8.43	265	26.60	1 712	26.78	282	25.24
	107A	137	9.04	311	7.05	845	2.51	280	20.25
	Mean		22.60		8.60		16.60		8.20
Gill	104C	2 566	2.97	2 591	7.57	2 428	21.50	2 231	18.33
net	107D	11 286	2.16	7 769	4.21	6 405	14.54	4 594	16.22
	107E	6 583	3.41	7 480	5.78	8 041	12.70	6 726	6.78
	107F + G	2 268	3.11	3 583	5.81	5 070	5.66	6 314	1.02
	107A	8 799	1.83	11 167	2.21	6 599	2.66	19 021	1.49
	Mean		2.50		4.30		10.30		5.00
Trammel	104C	444	4.36	1 032	2.04	827	4.44	2 602	3.88
net	107D	2 355	4.22	4 139	2.50	4 302	3.32	7 424	3.51
	107E	1 172	6.02	1 518	3.15	2 734	2.97	2 058	2.42
	107F + G	7	2.00	414	0.80	–	–	347	4.12
	107A	131	0.50	5 786	0.89	1 132	2.39	390	0.78
	Mean		4.60		1.80		3.20		3.00
Long	104C	87	3.36	1 182	9.13	313	3.77	249	9.20
lines	107D	23	11.00	3 489	12.97	1 408	7.67	1 386	11.25
	107E	–	–	269	6.86	16	18.38	20	6.35
	107F + G	–	–	–	–	56	7.95	124	9.33
	107A	–	–	243	11.19	80	10.75	154	9.90
	Mean		5.00		11.70		7.30		10.70
Angling/	104C	2 553	13.32	3244	6.05	5 240	4.19	5 990	4.06
handlines	107D	16 647	5.55	14 288	5.14	17 150	2.35	16 225	3.45
	107E	19 350	9.76	10 447	4.91	11 515	4.43	11 355	4.24
	107F + G	4 503	7.97	7 853	8.80	10 989	7.79	12 026	3.45
	107A	7 176	5.55	11 713	9.63	9 785	9.04	9 505	1.69
	Mean		8.10		6.90		5.30		3.40

* This figure is thought to be artificially high due to an underestimate of effort for drift nets – many drift netters being recorded as gill-netters.

This study has shown that the commercial and recreational boat fisheries for bass can be assessed in this way. Total annual catch appears to be relatively stable, and fluctuations between regions may be due as much to the varying accessibility of bass to the inshore fisheries as to errors in our estimates. Rough weather limits the fishing activity of small boats but does not usually affect all regions similarly, and annual changes in the weather pattern also affect year class strength and the distribution of adult bass.

The main sources of variance from the fish population viewpoint are the overall abundance of the stock, the seasonal movements of the fish and their distribution in the fishery, and for those fleets that target recruiting 3−5 year olds, the relative strength of successive year classes. A tendency to increase effort on the more vulnerable younger fish has been noted. There has been a marked consistency in these catch-at-age trends during this study, which suggests that the system's output is satisfactory for our purposes.

Complaints by some anglers and professional hand-liners, that their catch rates of bass have declined, seem justified by the CPUE figures given in Table 24.4, though this trend was not observed so clearly in any other part of the fishery. Bearing in mind the weighting procedure, which will tend to smooth data between adjacent sampled and unsampled strata (but not between gears), a main source of variance is the distribution of logbooks through the fishery. This is particularly significant if only one, unrepresentative, vessel is used as a sample for a regional catching power stratum. CPUE figures for shore anglers have been derived in an economic assessment of the bass fishery using postal questionnaires for 1987 and 1988 (Dunn *et al.*, 1989). Despite low individual catch rates, the high overall level of effort leads to an estimated total shore angling catch in the range 660−694 t for 1987, only 100 t less than the total boat catch of 760 t estimated by DFR. Estimates of the average annual earnings in the commercial bass fishery are remarkably similar between the logbook system and Dunn's econometric evaluation.

A warning must be highlighted against the dangers of basing indices of stock abundance on catch per effort data derived in the manner described. The common effort units (boat days) used do not necessarily reflect proportional catching powers of the gears or the manner in which they are used. The prime aim is to estimate catch for biologically sampled gear groupings and thus to arrive at numbers at age caught. Abundance indices can then be calculated (for each year class or age group) by dividing numbers at age by total estimated effort.

In conclusion, it was found that the logbook system is an enduring and relatively inexpensive method of catch and effort data collection. It could be applied to moderate or large fisheries in salt or fresh water in which there is little or no prospect of total catch census. Where it is a condition of fishing licences to provide catch data, usually as part of a management scheme, this often results in misleading reporting, and an independent sampling logbook scheme may help to reveal and resolve any discrepancies.

References

Dunn, M., Potten, S., Radford, A. and Whitmarsh, D. (1989) An Economic Appraisal of the Fishery for Bass (*Dicentrarchus labrax* L.) in England and Wales. A report to the Ministry of Agriculture, Fisheries and Food. Vol. 1, 217 pp. Marine Resources Research Unit, Portsmouth Polytechnic.

Pawson, M.G. and Benford, Teresa, E. (1983) The Coastal Fisheries of England and Wales, Part 1: A review of their status in 1981. MAFF DFR Internal Rep. No. 9. Lowestoft. 54 pp.

Pawson, M.G. and Pickett, G.D. (1987) The bass (*Dicentrarchus labrax* L.) and management of its fishery in England and Wales. MAFF DFR Lab Leaflet No 59. Lowestoft 38 pp.

Pickett, G.D. 1990 (in press) Assessment of the UK bass fishery using a logbook based catch record system. MAFF DFR Tech. Report. Lowestoft. 27 pp.

Chapter 25
Local catch per unit effort as an index of global fish abundance

DAVID B. SAMPSON *Centre for Marine Resource Economics, Portsmouth Polytechnic, Milton, Southsea, Hampshire, PO4 8JF, UK*

Fisheries scientists often use catch per unit effort (CPUE) as an index of fish abundance. They implicitly assume that catch rates are proportional to the number of fish local to the fishing operations, that CPUE is proportional to the size of the fish population. But it may be that only a part of the population is within range of the fishing gear. How does CPUE relate to abundance when some fish are inaccessible? This is examined by considering the properties of two simple models of the catch process.

In the first model fishing occurs within a harvest region on a local population which is replenished from an unexploited source. It is shown that CPUE is a reasonable measure of abundance for the entire population provided there are sufficient rates of movement between the exploited and unexploited portions. In the second model fish migrate one-way through the harvest region. In this case CPUE is not a good measure of global abundance; the ratio of CPUE to abundance is not independent of the duration of fishing.

25.1   Introduction

Many of the conventional analyses of fisheries science are based on the catch equation:

$$C(t) = N(0) \ F \ \{1 - \exp \ [-(F + M)t]\}/(F + M)$$

where $C(t)$ is the number of fish caught during time interval $(0, t)$, $N(0)$ is the number of fish in the population at time $t = 0$, F is the instantaneous rate of fishing mortality, and M is the instantaneous rate of natural mortality. Forms of this equation occur in yield-per-recruit analysis (Beverton & Holt, 1957), in virtual population analysis (Gulland, 1965), and in various estimation techniques based on mark-recapture experiments (Seber, 1982). The catch equation also provides the basis for the use of catch per unit effort (CPUE) as a measure of fish abundance.

If the units of fishing gear act independently and without interference, then the instantaneous mortality due to fishing is just the sum of the fishing mortalities generated by the individual units of gear. If each unit produces equal fishing intensity, or if gear types with differing intensities are measured relative to some standard, then the instantaneous fishing mortality is $F = f q$, where f is the number of standardized units of gear and q, the catchability coefficient, is the instantaneous fishing mortality created by one standard measure of fishing gear. Given these conditions the relationship between population size and accumulated catch and accumulated fishing effort ($f\,t$) can be expressed as:

$$\text{CPUE} = C(t)/(f\,t) = q\,\overline{N}(t)$$

where

$$\overline{N}(t) = N(0)\,\{1 - \exp[-(F + M)t]\}/[(F + M)t]$$

is the time-averaged population size. CPUE is proportional to abundance and the catchability coefficient is the constant of proportionality. These relationships are developed and discussed in many of the standard references of fisheries science (e.g. Beverton & Holt, 1957; Ricker, 1975; Gulland, 1983).

Implicit in the catch equation is the assumption that all fish in the population are equally susceptible to fishing. However, at any given instant it is likely that some portion of the fish population will be at such a distance that the fish in this portion are not liable to capture. The standard catch equation describes the relationship between the observed catches and the unobserved local abundance; the model on which the equation is based does not consider the relationship between the local abundance and the fish that are outside the range of the fishing operations.

In some fisheries, particularly those in which the fishing gear is mobile, one could argue that the fishing operations randomly sample from the universe of regions of local abundance and that the sample average abundance is therefore an unbiased estimate of the true average, which in turn is proportional to the total size of the fish population. But such an approach is problematic. Alternatively, one can analyse the relationship between local catches and global abundance by means of models in which catches are taken from a local population and the local population is replenished from an external source.

This paper examines the properties which emerge from two opposing mechanisms whereby the replenishment might occur. In both cases a population of fish is distributed initially in different geographic regions and harvesting occurs in only one region. No new recruits join the unexploited portion; the global population declines steadily due to losses from natural and fishing mortality. In the first model, there is random interchange between the harvested and unharvested portions of the population. This type of model might be appropriate for a lake fishery or one on the open ocean. In the second model, there is a directed flow of fish past the fishing operations. This type of model might be appropriate for a fishery on a migratory species on a river or along a coastline.

25.2   Model 1: random movement between regions

Consider the following deterministic model (Fig. 25.1). There is some geographic region H within which fish are harvested. No fishing occurs outside this region. There are a total of f fishing operations which operate independently and without interference and each of them removes fish at a rate of q per fish; the instantaneous rate of fishing mortality is $F = f q$. Fish immigrate independently into region H at an instantaneous rate of I and emigrate independently at an instantaneous rate of E. Natural mortality occurs in both regions at an instantaneous rate of M.

Let $N_H(t)$ denote the number of living fish interior to H at time t, let $N_O(t)$ denote the number of fish outside of H, and let $C(t)$ denote the cumulative catch during the interval $(0, t)$. The behaviour of N_O, N_H, and C through time is governed by the following system of differential equations:

$$\dot{N}_H = I\, N_O(t) - (E + M + F)\, N_H(t)$$

$$\dot{N}_O = -(I + M)\, N_O(t) + E\, N_H(t)$$

$$\dot{C} \;\; = F\, N_H(t)$$

The notation '$\dot{y}$' symbolizes the derivative of the dependent variable y with respect to the independent variable time t. The solution to the system can be derived by eliminating N_O from the first two equations; this results in a second-order differential equation in N_H:

$$[D^2 + (2\,M + I + E + F)\, D + M^2 + M\,(I + E + F) + I\,F]\, N_H = 0,$$

where 'D' denotes the differential operator d/dt. The solutions to the differential equations for N_O and C can be obtained directly from the solution for N_H.

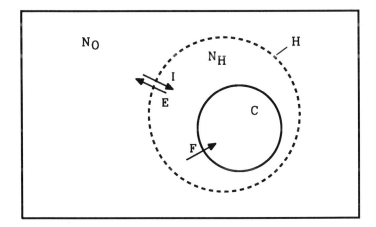

Fig. 25.1 Schematic diagram of a catch process $C(t)$ which harvests fish within a harvest region H from a local population $N_H(t)$ which is replenished from an outside population $N_O(t)$ which is unexploited.

The system of equations has the following general solution:

$$N_H(t) = k_1 \exp (r_1 t) + k_2 \exp (r_2 t)$$

$$N_O(t) = (k_1/I) (r_1 + E + M + F) \exp (r_1 t) + (k_2/I) (r_2 + E + M + F) \exp (r_2 t)$$

$$C(t) = C(0) + (F k_1/r_1) [\exp (r_1 t) - 1] + (F k_2/r_2) [\exp (r_2 t) - 1]$$

where

$$r_1 = -M - \{I + E + F - [(I + E + F)^2 - 4 I F]^{\frac{1}{2}}\}/2 < 0$$

$$r_2 = -M - \{I + E + F + [(I + E + F)^2 - 4 I F]^{\frac{1}{2}}\}/2 < r_1$$

The terms r_1 and r_2 are the roots of the auxiliary equation:

$$D^2 + (2 M + I + E + F) D + M^2 + M (I + E + F) + I F = 0$$

The terms k_1, k_2, and $C(0)$ are the arbitrary constants which arise from integrating the differential equations.

If one assumes that the net rate of migration between the two populations N_O and N_H is zero at time $t = 0$, then $I N_O(0) = E N_H(0)$ and the arbitrary constants k_1 and k_2 take the values

$$k_1 = - N_H(0) (r_2 + M + F)/(r_1 - r_2)$$

and

$$k_2 = N_H(0) (r_1 + M + F)/(r_1 - r_2)$$

If one begins counting the catch at time zero, then $C(0) = 0$.

The catch from the harvest region accumulates as a simple catch process and the CPUE is proportional to the time-averaged abundance of the fish within the harvest region; the ratio of CPUE to N_H is q. However, the ratio of CPUE to time-averaged abundance for the entire fish population is:

$$\left\{ \frac{C(t)}{f t} \right\} \bigg/ \left\{ \frac{1}{t} \int_0^t [N_O (u) + N_H (u)] \, du \right\} = q' (t)$$

$$q' = \frac{q I \{k_1 r_2 [\exp (r_1 t) - 1] + k_2 r_1 [\exp (r_2 t) - 1]\}}{k_1 r_2 [\exp (r_1 t) - 1] [r_1 + W] + k_2 r_1 [\exp (r_2 t) - 1] [r_2 + W]}$$

where

$$W = I + E + M + F$$

The derivative of $q'(t)$ with respect to t is less than zero for all non-negative values of t. The function $q'(t)$ decreases steadily with increasing t from:

$$Q_0 = \lim_{t \to 0} q' (t) = \frac{q I N_H(0)}{k_1 (r_1 + W) + k_2 (r_2 + W)} = \frac{q I}{I + E}$$

at time zero to

$$Q_i = \lim_{t \to \infty} q'(t) = \frac{q \ I \ [k_1 \ r_2 + k_2 \ r_1]}{k_1 \ r_2 \ (r_1 + W) + k_2 \ r_1 \ (r_2 + W)}$$

at infinite time.

The expression for the limiting value Q_i has a complicated form but the following provides a good approximation:

$$Q_i \doteq \tilde{Q}_i = \frac{q \ I}{1 + E + a \ F} \qquad \text{where } a = \frac{E}{1 + E}$$

The term $a = E/(1 + E)$ is equivalent to the fraction of the total population that is initially inaccessible to harvesting:

$$E/(1 + E) = N_O(0)/[N_O(0) + N_H(0)]$$

From an examination of the relative error of the approximation (Fig. 25.2) one can see that the absolute value of the relative error increases with F and M and decreases with I and E. For reasonably small values for F and M, say $F < 2$ and $M < 0.2$, the magnitude of the relative error is not too large. The approximation is always slightly smaller than the true Q_i; the approximation exaggerates the difference between Q_0 and Q_i.

If CPUE is to be used as a reliable index of abundance, then the ratio of CPUE to average global abundance should not vary with either the intensity or the duration of fishing. With the model developed in this section, the ratio of CPUE to global abundance decreases through time from Q_0 to Q_i. However, the expressions for Q_0 and $\tilde{Q}_i$ differ only in the presence of the term $(a \ F)$ in the denominator of $\tilde{Q}_i$. Therefore, unless the instantaneous rate of fishing mortality is large relative to the rates of immigration and emigration, the ratio of catch per unit effort to time-averaged global abundance is approximately constant for all values of t. Unless fishing mortality is large or the fish are relatively immobile, the local CPUE provides a reasonable measure of the total fish abundance regardless of the amount of time spent fishing.

In section 10 of Beverton and Holt (1957) the authors present an analysis of a fishery in which there are movements of fish between geographic regions and in which fishing effort is not uniformly distributed with respect to fish density. Their approach is similar to the one used above. They derive an equation (10.31) which relates total fish abundance and catch per unit effort. Their index of abundance is the sum over all regions of the CPUE values in each region. They conclude that 'if part of the area containing fish is not fished, the density in it must nevertheless be measured directly or estimated and included in (10.31)'. On small temporal or spatial scales one will almost certainly not have complete coverage over the entire geographic range of the fish population.

The analysis in this paper shows that CPUE can be a valid index of abundance even though the fishing operations exploit only a limited portion of the area

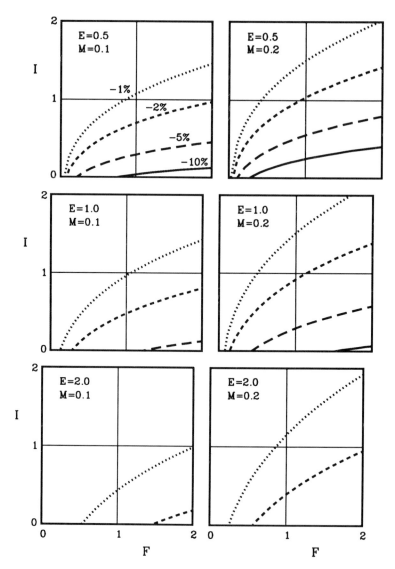

Fig. 25.2 Relative error of an approximation for the limiting value as time goes to infinity of the ratio of catch per unit effort to total time-averaged abundance. In each panel the lines represent contours of relative error, $(Q_i - \hat{Q}_i)/Q_i$

occupied by the fish population. However, the constant of proportionality between CPUE and the time-averaged global abundance is $q\, I/(I + E)$ rather than just q, the catchability coefficient. The term $I/(I + E)$ is the probability that a given individual fish is within the region local to the fishing operations.

25.3   Model 2: directed migration between regions

Consider now an alternative model of a fishery (Fig. 25.3). Again, there is some geographic region within which there is harvesting and outside of which there is

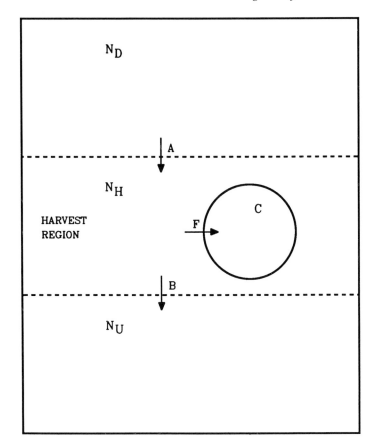

Fig. 25.3 Schematic diagram of a catch process $C(t)$ which removes fish from a harvest region through which fish migrate from a downstream population $N_D(t)$ to an upstream population $N_U(t)$

none. But in this case the fish migrate in one direction through the harvest region. Fish immigrate into the harvest region from a downstream population $N_D(t)$ at an instantaneous rate of A and emigrate from the harvest region into an upstream population $N_U(t)$ at an instantaneous rate of B. Both A and B are greater than zero. As in the previous model $N_H(t)$ denotes the number of living fish within the harvest region at time t, $C(t)$ denotes the cumulative catch during $(0, t)$, natural mortality occurs in all regions at an instantaneous rate of M, and there are a total of f fishing operations which operate independently and without interference and each removes fish at an instantaneous rate of q.

The behaviour of $N_D(t)$, $N_H(t)$, $N_U(t)$, and $C(t)$ are determined by the following system of differential equations:

$$\dot{N}_D = -(A + M)N_D(t)$$

$$\dot{N}_H = A\,N_D(t) - (B + M + F)N_H(t)$$

$$\dot{N}_U = B\,N_H(t) - M\,N_U(t)$$

$$\dot{C} = F\,N_H(t)$$

The system can be solved in a sequential manner beginning with the equation for N_D:

$$N_D(t) = N_D(0)\, e^{-(A+M)\,t}$$

$$N_H(t) = k\, e^{-(A+M)\,t} + [N_H(0) - k]\, e^{-(B+M+F)\,t}$$

$$N_U(t) = e^{-M\,t}\{N_U(0) + B\, k/A\, [1 - e^{-A\,t}] + B\,[N_H(0) - k]/(B+F)\,[1 - e^{-(B+F)\,t}]\}$$

$$C(t) = C(0) + F\,\{k/(A+M)\,[1 - e^{-(A+M)\,t}] + [N_H(0) - k]/(B+M+F)\,[1 - e^{-(B+M+F)\,t}]\}$$

where $k = A\, N_D(0)/(-A + B + F)$. For simplicity it has been assumed that $A \neq B + F$.

If the entire population of fish is initially downstream from the harvest region, then $N_H(0)$ and $N_U(0)$ are both equal to zero and the following equation describes the evolution of the time-averaged global abundance:

$$\bar{P}\,(t) = (1/t)\int_0^t \{N_D(x) + N_H(x) + N_U(x)\}\, dx$$

$$= \frac{N_D\,(0)}{t}\left\{\frac{B}{B+F}\,\frac{1 - e^{-M\,t}}{M} + \frac{F}{-A+B+F}\left[\frac{1 - e^{-(A+M)\,t}}{A+M}\right.\right.$$

$$\left.\left. - \frac{A}{B+F}\,\frac{1 - e^{-(B+M+F)\,t}}{B+M+F}\right]\right\}$$

Average global abundance declines steadily with time from $N_D(0)$ and eventually decays to zero. Some fish are lost to fishing mortality while they pass through the harvest region; those that avoid the fishing operations succumb to natural mortality.

If one begins counting the catch at time zero, then $C(0) = 0$ and the following equation describes the evolution of catch per unit effort:

$$\text{CPUE}\,(t) = \frac{q\,k}{t}\left\{\frac{1 - e^{-(A+M)\,t}}{A+M} - \frac{1 - e^{-(B+M+F)\,t}}{B+M+F}\right\}$$

The behaviour of this function is illustrated in the upper panel of Fig. 25.4. Initially there are no fish in the harvest region and the CPUE is zero. As fish begin entering the harvest region, the CPUE increases. As the fish continue on their migration, the CPUE reaches a maximum. Eventually all the surviving fish join the upstream portion $N_U(t)$ and the CPUE returns to zero. The lower panel of Fig. 25.4 shows the evolution of the time-averaged global abundance.

In this system the ratio of CPUE to time-averaged abundance does not remain constant through time; it varies in a manner similar to the CPUE. Initially the ratio is zero but with the passage of time it increases to a maximum and then returns to zero. For a migrating fish population the relationship between CPUE and global abundance is not independent of the duration of fishing and therefore CPUE is not a good measure of abundance.

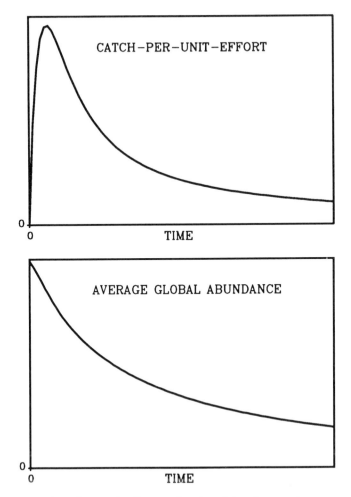

Fig. 25.4 Upper panel: catch per unit effort as a function of time for a catch process operating on a migrating fish population. Lower panel: time-averaged global abundance

25.4   Discussion

In the sections above it has been shown that the relationship between catch per unit effort and fish abundance depends in a fundamental manner on the mechanisms of fish movement and redistribution. But the models are simplifications of the real world; in practice, additional complications are likely to arise. Paloheimo and Dickie (1964) investigate the consequences of fish occurring in aggregations and they show that this can distort the simple linear relationship between CPUE and abundance. Rothschild (1977) discusses the problems of gear saturation and competition; these can upset the simple linear relationship between fishing mortality and fishing effort.

A further problem arises because fishing is a stochastic rather than a deterministic process (Sampson, 1988). However, stochasticity in the catch process does not invalidate the general conclusions of this paper. Provided that the individual fish in the different geographic regions behave independently, the solutions to the deterministic systems considered in this paper also describe the time trajectories for the corresponding expected values in stochastic versions of the models. This result is a general feature of stochastic compartmental models (Matis & Hartley, 1971).

References

Beverton, R.J.H. and Holt, S.J. (1957) *On the Dynamics of Exploited Fish Populations*. Fishery Investigations Series II, Volume XIX. London: Her Majesty's Stationery Office. 533 p.

Gulland, J.A. (1965) Estimation of mortality rates. Annex to *Arctic Fisheries Working Group Report*. ICES CM Gadoid Fish Committee. 9 p. (mimeo).

Gulland, J.A. (1983) *Fish Stock Assessment: A Manual of Basic Methods*. New York: John Wiley & Sons. 223 p.

Matis, J.H. and Hartley, H.O. (1971) Stochastic compartmental analysis: model and least squares estimation from time series data. *Biometrics* **27**: 77–102.

Paloheimo, J.E. and Dickie, L.M. (1964) Abundance and fishing success. *Rapp. Cons. int. Explor. Mer* **155**: 152–163.

Ricker, W.E. (1975) *Computation and Interpretation of Biological Statistics of Fish Populations*. Bull. Fish. Res. Board Can. 191. Ottawa: Fisheries Research Board of Canada. 382 p.

Rothschild, B.J. (1977) Fishing effort. In *Fish Population Dynamics*, (Ed. by J.A. Gulland). New York: John Wiley & Sons. pp. 96–115.

Sampson, D.B. (1988) Fish capture as a stochastic process. *J. Cons. int. Explor. Mer* **45**: 39–60.

Seber, G.A.F. (1982) *The Estimation of Animal Abundance and Related Parameters*. New York: Macmillan. 654 p.

Chapter 26
A new approach to the analysis of age structure of fish stocks using surveys for various water basins and behavioural patterns of fish concentrations

Y.T. SECHIN *All-Union Research Institute of Pond Fisheries (AURIPF), Rybnoye, Dmitrov Region, Moscow Province 141 821 USSR*
W.I. BANDURA *State Research Institute of Lake and River Fisheries (GosNIORH), Volgograd Department, Pugachiov str. 1, Volgograd 400 001 USSR*
S.W. SHIBAYEV *State Research Institute of Lake and River Fisheries (GosNIORH) Gorkij Department, Nizhnie − Wolzhskaya quay 7/8, Gorkij 603 001 USSR*
V.V. BLINOV *All-Union Research Institute of Marine Fisheries and Oceanography (VNIRO) 17 a Verkhne-Krasnoselskaja, 107 140 Moscow USSR*

The evaluation of absolute abundance of fishes which inhabit water basins of various types usually has a low degree of reliability even when modern instrumental means are applied for ichthyological surveys. Samples in the sites where the bottom is littered with snags, stones, bushes, sunk logs should be taken using gill nets of known catchabilities and selectivities rather than with any active type of gear. Gill net catches are known to depend on the activity of the fish and the densities of fish shoals which are, in their turn, influenced by the season of year, weather, water temperature and transparency, depth for net setting and other factors. Reliable fish abundance indices provide the basis for the proposed method for estimating the age structure of the stock and general mortality rates (Z) of the different ages. The method proposed deals with the analysis of age structure of the population and ratios of particular age group numbers. The procedure uses data from pairs of years in successive computations to provide values of Z. Dimensioning of the Z values is fulfilled from the condition of synchronous changes in neighbouring ages in two adjacent years. Input data for the proposed method can be obtained from gill net and trawl catches if they are representative and cover a wide range of ages of fish.

An experiment on Lake Il'mien' where simultaneous fishing by gill nets and trawls was carried out in 1989 is briefly described. Interpretation of the experimental results is given in terms of densities of fish concentrations and fish activity, both of which have an influence upon the success of the fishery. Difficulties with the calibration of gill nets over trawl catches are emphasized.

26.1   Introduction

Most methods of forecasting fish catch are largely based on estimates of age group abundances. However, estimates of absolute fish abundance appear, in some cases, to exhibit a lot of uncertainty and cannot be relied upon. This implies that catch projections have low reliability. Absence of accurate quantitative population characteristics precludes development of population dynamics models which are ultimately aimed at estimating optimum exploitation parameters.

In the USSR, these negative features of research on inland freshwater basins can be removed by determining the dynamics of changes in population age structure rather than the absolute fish abundance.

In the present paper a method is described for estimating mortality rate coefficients in fish populations and serves as the background for choosing relevant types of fish gears for surveys which study the population structure of different fish species in the basin.

26.2   Estimating mortality rate coefficients using abundance indices for particular age groups of the stock

The method proposed (Shibayev, 1987) is based upon well-known fisheries concepts and deals with well-known notions.

Suppose that abundance of some age group t in year x is $N_{x,\,t}$. Baranov's (1918) equation predicts that for the same year class a year later $(x+1)$ the number of fish has reduced to $N_{x+1,\,t+1}$:

$$N_{x+1,\,t+1} + N_{x,\,t} \times \exp\left(-Z_t\right) \tag{1}$$

where Z_t is the instantaneous mortality rate of fish at age t.

For the next age group $(t+1)$ in the adjacent year $x+1$ a similar equation is valid:

$$N_{x+1,\,t+2} + N_{x,\,t+1} \times \exp\left(-Z_{t+1}\right) \tag{2}$$

On the basis of equations (1) and (2), the corresponding coefficients of mortality (Z) can be defined as:

$$Z_t = -\ln\left(N_{x+1,\,t+1}/N_{x,\,t}\right) \tag{3}$$

$$Z_{t+1} = -\ln\left(N_{x+1,\,t+2}/N_{x,\,t+1}\right) \tag{4}$$

A difference between the coefficients of mortality is designated between the two consecutive groups as:

$$\Delta Z_t = Z_{t+1} - Z_t \tag{5}$$

Substituting equations (3) and (4) into expression (5) yields the expression:

$$\Delta Z_t = -\ln\left(N_{x+1,\,t+2}/N_{x+1,\,t+1}\right) + \ln\left(N_{x,\,t+1}/N_{x,\,t}\right) \tag{6}$$

or

$$\Delta Z_t = Z'_{x+1,\ t+1} - Z'_{x,\ t} \tag{7}$$

As can be seen from (7), the Z values are determined by the ratio of numbers in two consecutive age groups for the same year. To calculate the Z it is sufficient to employ data on population age structure reflected in the indices $n_{x,\ t}$, where n represents the proportion of each group in the catch, instead of the corresponding absolute abundances (N).

When applying the expression (7) to the age structure indices it is possible to obtain age specific changes in mortality rates, i.e. to construct the relationship for the mortality rate of the population $Z(t)$ for the particular conditions in the basin. The resulting curve $Z(t)$ reflects the changes in mortality rate but generally fails to provide absolute values for these coefficients. The only possible way to identify the absolute value of Z, which has not received clear recognition previously, is to find an equal pair of consecutive ages for two adjacent years, i.e. such arrays of Z for which the equality

$$Z_{x,\ t} = Z_{x+1,\ t} \tag{8}$$

is valid for whatever base age group. Existence of such a value (or values) are not proven here, and the problem is open to resolution. It is only emphasized that such a situation is likely to occur for some range of stability in the population parameters and will be reflected in a stable structure of Z for some range of ages where Z_t tends to zero, and true value of Z is thus found :

$$Z_t = Z'_{x,\ t} = Z'_{x+1,\ t} \tag{9}$$

The coefficients of mortality Z for other age groups are easily computed because values of ΔZ_t for other pairs of years are already known.

26.3   Examples of the implementation of the proposed method

Consider some age-specific dynamics of mortality rates for a hypothetical population, the parameters of which are given in Table 26.1. The occurrence of a high abundant year class (at age 4; see Table 26.1) is anticipated for the array of numbers, and natural mortality of fish is assumed to be an exponential decline.

The equality of the conditional coefficients Z' (obtained on the basis of the abundance indices) takes place at age 6 (Table 26.1) and thus it is possible to accept the Z values as being absolute values for this age group. Other coefficients of mortality are calculated according to equation (9). The resulting curve for total mortality rate corresponds to the initial conditions and is described by the usual curve.

As a second example, the age specific mortality dynamics for the sterlet (*Acipenser ruthenus* L.) population which inhabits the Cheboksar Reservoir is examined. Materials concerning the population age structure for the present

Table 26.1 Computation of total mortality coefficients on the base of abundance indices for age groups of the stock (artificial data)

Age of fish	Abundance $N_{x, t}$	$N_{x+1't}$	Abundance index % $n_{z, t}$	$n_{x+1, t}$	Conditional coefficients $Z'_{x, t}$	$Z'_{x+1, t}$	ΔZ_t	Instantaneous mortality rate
1	100	100	36.5	39.5	0.51	0.69	−0.29	0.69
2	60	50	21.9	19.8	0.69	0.22	0.00	0.40
3	30	40	10.9	15.8	−0.51	0.69	0.11	0.40
4	50	20	18.2	7.9	0.92	−0.40	0.40	0.51
5	20	30	7.3	11.9	0.91	1.32	−0.22	0.91
6	8	8	2.9	3.2	0.69	0.69	0.70	0.69
7	4	4	1.5	1.6	0.69	1.39	−	1.39
8	2	1	0.8	0.3	−	−	−	−

computations were gathered from experimental catches during the period 1983–86 (Table 26.2).

Sterlet is a species with a fairly long life span. Individuals reach 28 years of age in the Cheboksar Reservoir, though few individuals older than 11–12 years appear to a significant extent in trawl catches. Fish between ages 2 and 4 dominate in the catches representing 13% and 21% of the catch respectively. Individual sterlet from older ages represent between 2% and 7% of the catches. This percentage is only weakly dependent on the age of the fish.

Sterlet in this basin are regarded as a species to be protected, and annual catches do not exceed 5 tonnes, the total stock size being assessed as 400 tonnes. Consequently, the dynamics of the sterlet population age structure and the rate of mortality of the fish are both predominantly the result of natural causes. Bearing in mind these circumstances and accounting for the long life span of this species, its mortality coefficients would be expected to be low.

Table 26.2 Age structure of *Acipenser ruthenus* L. in Cheboksar Reservoir, %

Age of Fish	Years 1983	1984	1985	1986
1	15.63	11.97	7.01	5.95
2	21.53	18.61	15.02	14.18
3	21.13	19.82	21.13	21.17
4	13.70	14.17	16.05	15.97
5	7.33	8.51	9.72	9.77
6	5.69	6.91	7.92	8.06
7	3.95	4.97	5.71	5.95
8	2.75	3.85	4.48	4.79
9	2.42	3.31	3.84	4.16
10	2.55	3.28	3.78	4.05
11	1.90	2.61	3.03	3.31
12	1.41	1.99	2.32	2.61

In this stock, values for the conditional coefficients (Z'), which approximately correspond to the equality described in equation (8), have been accepted as the initial points for subsequent computations. In most cases the approximate equivalence of $Z'_{x,\,t}$ and $Z'_{x+1,\,t}$ is valid for age group 5 (see Table 26.3).

From the results (Table 26.4) it is possible to conclude that the sterlet mortality dynamics over the period of study have a lot of similarities from year to year. This situation is evidently due to the stability of the population and a sparse level of fishing. Annual mortality reaches a maximum at age 4, followed by a gradual decrease over the age range 4–8, and then a slight increase as the fish get older (Table 26.4). These features are in agreement with general views of biologists about age specific changes in natural mortality rates among fish populations (Severtsov, 1941; Tiurin, 1962; Nikol'skij, 1974).

Table 26.3 Conditional coefficients of total mortality (Z') for the population of *Acipenser ruthenus* L. in Cheboksar Reservoir, %

Age of fish	Years			
	1983	1984	1985	1986
1	−37.70	−55.39	−114.38	−138.16
2	1.84	−6.53	−40.68	−49.32
3	35.16	28.51	24.03	24.56
4	46.50	39.93	39.46	38.82
5	22.34	18.81	18.52	17.51
6	30.57	28.05	27.90	26.15
7	30.52	22.51	21.57	19.54
8	11.96	14.21	14.23	13.14
9	−5.31	0.94	1.57	2.63
10	25.20	20.40	19.93	18.24
11	26.15	23.78	23.33	21.07

Table 26.4 Computed coefficients of total mortality rate for the population of *Acipenser ruthenus* L. in Cheboksar Reservoir, %

Age of Fish	Years		
	1983	1984	1985
1	0	0	0
2	19.91	0.63	0
3	41.67	29.13	25.27
4	45.96	39.99	39.82
5	18.00	18.60	18.00
6	24.03	27.71	25.67
7	15.21	21.19	17.06
8	0	12.78	8.15
9	0	0	0
10	10.96	19.12	13.39
11	9.26	22.09	14.63

Similar results were obtained for another long-lived species — sheatfish (*Silurus glanis* L.) an inhabitant of Tsymliansk Reservoir (Bandura *et al.*, 1987). In this case it became clear that the occurrence of one or several year classes of high abundance had no influence on the accuracy of the assessment of the total mortality for the sheatfish population. Other methods applied in this case experienced much greater difficulties.

The degree of accuracy of the present method depends evidently on the available input data. Underestimation of some age groups due to various causes might lead to errors in the estimates of the abundance indices. In the sterlet population such uncertainties probably account for negative values for the mortality coefficients, Z, at ages 8 and 9. Consequently, in order to obtain a reliable pattern of age structure for the population it is necessary to conduct surveys using fishing gears of different types. This would ensure that any aggregations of fish are sampled.

26.4   Choice of fishing gears for surveying abundance and age structure of fish population

A method for straight accounting of the abundance of fish resources in some basins provides fairly good results only if all or most of the bottom area of the basin is accessible to fishing gears of the active type. Unfortunately, most lakes and reservoirs in the USSR have considerable bottom area which is covered with stones, sunken logs, and bushes that make it impossible to fish with those gears. Only former river-bed sites in a number of reservoirs are suitable for trawling.

The problem of assessing the fish density, the species mix and length structure on sites with inhibitive bottoms can be solved if gill nets with known catchabilities and selectivities are used. Gill net characteristics can be obtained from a calibration procedure provided that the densities and age structure of all the species of interest are known from trawl and seine net catches.

It is well known that gill net catches depend considerably on density and behaviour of fish concentrations that, in turn, are determined by factors such as year, season, weather, water temperature, transparency and net setting depth. If the gill net samples are taken when these conditions are accounted for any calibration of the gear will be unnecessary.

If the assortment of net mesh sizes is designed to capture a particular target species, then the length composition of the catches will approach that for the stock of the target species. For some species, the combined selectivity assigned to the net assortment should account for the variability in fish body exterior, girth length, mesh size, and the degree of compression exerted onto a fish body by the mesh filament (Sechin, 1969).

Gill net surveys as a rule have prolonged durations compared to trawl and seine net surveys. In order to minimize the complexity of the survey, more effort is made in favour of gears of the active types so long as the gill net catches remain representative. Comparisons of the catches taken by gears of all types over the same sites are performed on the basis of relative (R) abundance indices and/or absolute (A) abundance estimates for the particular species. Consequently, a research worker sometimes has two (R and/or A) calibration age series for particular species or sites of the basins. With these procedures all characteristics of the catches are adjusted to volumetric or real quantities in order to allow comparisons or to estimate site totals.

26.5   Comparison of gill net and trawl catches

An experiment was undertaken on Lake Il'mien' to obtain catches at the same time and site by different fishing gears. In late August 1989 two gill-net gangs, with various mesh sizes between 30–100 mm, were set in the central part of lake Il'mien' (area 1000 km², silt-covered bottom, average depth 3.5 m, water temperature 14°C). One gang was set north–south and the other west–east. Each gang consisted of 10 nets which were 3 m high with a hanging ratio of 0.5. Nets were examined once a day. In the vicinity of both gangs (within the distance 100–300 m) trawl hauls (one or two a day) were carried out in various directions. Altogether seven 30-minute trawl hauls were made during five days with a standard 18-m otter-trawl. The latter had a codend mesh size of 30 mm and vertical opening of 3 m when hauled at a speed of 4.5 km h⁻¹. If those parameters remain constant during the fishing operations the area swept would be approximately 30 ha. Water depth for the trawl hauls and gill-net gangs was 4 m. Gale weather lasted during the whole experimental period. One net gang was set for two days, and the other for three days. Four fish species formed the bulk of the catches in both types of gears: *Abramis brama, Abramis ballerus, Stizostedion lucioperca, Pelecus cultratus*.

The length composition of the catches (Table 26.5) was not representative of the populations. The low catches in number reflect either low densities of fish or lack of movement. Trawls are known to capture fish if the fish have moderate swimming speed and rushing reactions; a moving trawl overcomes those fish activities. Trawl catches reflect the densities of fish rather than other particular features of the species provided that the reactions of individual fish are within moderate limits.

Catches of fixed (or relatively fixed) gill nets are affected by the densities of the fish concentrations, the collective movement of fish in some predominant direction (migrations, rheophilic movement), and also the random movements of individual fish. Behavioural patterns among fishes vary greatly depending on their physiological status. In this case collective movements were absent and although individuals of *Abramis brama* and *Pelecus cultratus* have similar swimming activities the density of *Abramis* was much higher than that for *Pelecus cultratus*.

On the basis of this the trawl catch data can be redesignated $n_t(l)$ in Table 26.5 as $n_t(l) = \gamma(l)$ where γ is proportional to the density of fish (if $\gamma(l) = n_t(l)/V_{tr}$, where V_{tr} is water volume swept during the survey trawl). The factors which determine catches are either due to specific promotion by existing fish densities in the hauled site, or due to individual fish activity. However, when this is high

Table 26.5 Length composition of catches taken by bottom trawl and set gill net, individuals in Lake Il'Mien', 1989

Fish species	Fishing gear	Length of fish, cm																									
		6	7	8	9	10	11	12	13	14	15	16	17	18	19	20	21	22	23	24	25	26	27	28	29	30	31
Bream, *Abramis brama*	Trawl	5	1	2	4	14	44	229	207	249	240	230	337	266	282	358	216	178	164	170	153	78	61	66	52	41	36
	Net	—	—	—	—	—	—	—	—	—	—	—	—	1	—	1	1	—	—	1	2	—	—	—	1	1	1
Siniets, *Abramis ballerus*	Trawl	1	—	2	—	—	—	1	—	2	11	3	14	56	89	125	119	106	82	68	34	9	3	1	1	—	—
	Net	—	—	—	—	—	—	—	—	—	—	—	—	5	21	30	17	29	28	9	7	2	1	—	—	—	1
Zander, *Stizost. luciop.*	Trawl	—	—	—	—	—	—	—	—	—	—	—	—	—	—	—	—	—	—	—	—	—	—	—	1	1	—
	Net	—	—	—	—	—	—	—	—	—	—	—	—	—	—	—	—	—	—	—	—	—	—	1	—	2	4
Chekhon', *Pelecus cultratus*	Trawl	—	—	—	—	—	—	—	—	—	—	—	—	—	—	—	—	—	—	—	—	3	2	2	—	—	1
	Net	—	—	—	—	—	—	—	—	—	—	—	—	—	—	—	—	—	—	—	—	3	5	3	3	3	13

Table 26.5 (continued)

Fish species	Fishing gear	Length of fish, cm																			Total
		32	33	34	35	36	37	38	39	40	41	42	43	44	45	46	47	48	49	50	
Bream, *Abramis brama*	Trawl	31	26	21	39	29	10	99	5	–	–	–	–	–	–	–	–	–	–	–	3941
	Net	–	–	1	–	–	–	–	–	–	–	–	–	–	–	–	–	–	–	–	11
Siniets, *Abramis ballerus*	Trawl	–	–	–	–	–	–	–	–	–	–	–	–	–	–	–	–	–	–	–	724
	Net	–	–	–	–	–	–	–	–	–	–	–	–	–	–	–	–	–	–	–	151
Zander, *Stizost lucioperca*	Trawl	1	1	1	1	–	1	1	–	–	1	–	4	4	2	4	7	–	2	3	34
	Net	1	1	1	1	2	1	–	–	1	–	3	1	3	3	2	1	–	–	–	28
Checkhon', *Pelecus cultratus*	Trawl	1	3	–	1	1	–	2	1	–	–	–	–	–	–	–	–	–	–	–	17
	Net	10	4	5	3	3	6	2	4	1	2	–	–	–	–	–	–	–	–	–	70

density of fish and low individual activity, gill net catches are likely to be low (e.g. for bream as judged by the data in Table 26.5). Therefore, after considering all these factors, it should be recognized that the dominant influence on the net catches is the fish swimming activity. The gill net data from Table 26.5 is thus redesignated as $n_c(l) = A(l)$ and can be treated as a measure of swimming activity in the site where gill nets are set (complicated process, e.g. the reactions of fish to the net are ignored).

$\gamma(l)$ and $A(l)$ are functions of the same argument, and eliminating l from both functions a phase diagram can be drawn (Fig. 26.1). The latter provides an approximation for both factors which govern the occurrence of catches in the trawl and the gill nets. Fig. 26.1 shows that trawl catches indicate high densities of bream concentrations whilst scarce net catches underline that the swimming activity of bream at the same time was low. Concentrations of zander and chekhon' (*Pelecus cultratus*) appeared to be very diluted but with higher swimming activity compared with bream.

The data most suitable for the calibration of gill net catches over trawl catches appeared to be those for *Abramis ballerus*. In the phase diagram (Fig. 26.1) the data for this species is elliptical. However, there are indications of clustering of these data. The upper parts of the distributions $n_c(l)$ and $n_t(l)$, in the vicinity of the largest lengths (in the narrow range of *Abramis ballerus* lengths $l = 19-23$ cm), form the upper cluster (it is surrounded by the dashed line in Fig. 26.1). This signifies that those fish were present in dense concentrations and that they move actively. Individuals of *Abramis ballerus* of the lengths $19 > l > 23$ cm were present in lower densities and were not so active. Differences between those clusters are significant and are likely to be due to some ecological or physiological cause.

In addition to the phase diagram, a calibration line of relative length composition $\gamma_{n/tr}$ of gill net catches to trawl catches, i.e. $\gamma_{n/tr} = n_n(l)/n_t(l)$ (Fig. 26.2) can be drawn for *Abramis ballerus*. The multi-modal relative distribution given in Fig. 26.2 can be explained by insufficient statistics for joint catches due to the brevity of the experimental programme. Moreover, uncertainties in the $\gamma_{n/tr}$ values are evidently at least as high or higher than those both in the numerator and denominator. A possible average calibration line is drawn in Fig. 26.2. It shows that the variability of $\gamma_{n/tr}$ is high, and to determine it reliably larger volumes of data need to be gathered, probably ten times more than the present study. One of the main conclusions is that a short experiment with a small number of different fishing gears often fails to provide stable calibration relationships for the desired type of gears, and that a strategy for such an experiment must account for specific fish behaviour.

If the summed trawl and gill net catches for the whole experiment are used and treated conditionally as being related to the unit of fishing effort (more accurately combined effort), the density of fish concentrations computed through the gill net catches would differ from those obtained from the trawl survey data: about 400 times lower for bream (*Abramis brama*), five times lower for *Abramis ballerus*, similar for *Stizostedion lucioperca* and overestimated by four times for *Pellecus cultratus*.

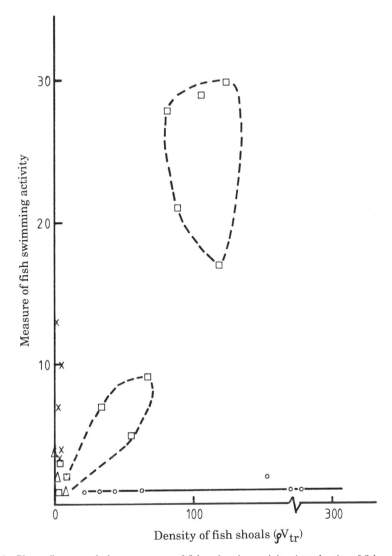

Fig. 26.1 Phase diagram relating a measure of fish swimming activity *A* to density of fish concentrations, scaled as ρ *V$_{tr}$* (individuals). The data are based on the results obtained in short experimental fishing trials by bottom trawls and gill nets at Lake Il'mien' in 1989.
(○ bream; □ chekhon'; △ siniets; × zander)

Thus, it can be inferred from the above material that estimating densities of fish on the basis of gill net catches is a fairly complex problem. Nevertheless, research should be concentrated in this direction and the strategy should be adjusted to particular basins, seasons of year and status of gears (i.e. their appropriate parameters), and features of the fish populations. This might provide the base on which to gain the necessary generalized characteristics and relationships which are the most relevant to the system 'population − environment − fishery' (PEF) in the basin.

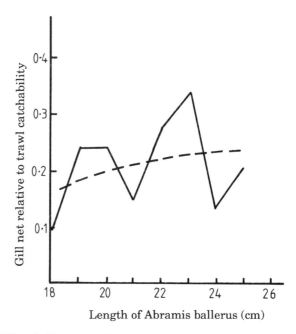

Length of Abramis ballerus (cm)

Fig. 26.2 Catchability of gill nets relative to that of a bottom trawl, $\gamma_{n/t}$, with regard to siniets between 18–25 cm in length

26.6   Discussion

The following are the main points behind the proposed method for analysing age structure of the stock and estimating total mortality rates at age, $Z(t)$.

(1)   The quantity Z has a logarithmic nature. Z is determined as the logarithm of the ratio of the final and initial abundances of the same year class in relation to a particular time interval (most frequently, a yearly interval). In other words, Z is a function of the so called 'logarithmic difference' of the stock abundance which can be conceived as reflecting the dynamics of the so called 'perturbed layer' of the year class abundance or of the stock as a whole.

(2)   There is a likelihood of a similar pattern of mortality of fishes in consecutive age groups for two adjacent years (especially when the requirements for the accuracy of estimations are not very strong) if the ecosystem conditions which promote stability in the population are not markedly deviated. Based on this perception the occurrence of situations which are distinguished by the condition that $Z = 0$ in expression (7) seems to be very realistic. For some conditions sustained in the system 'population − environment' (*PE*) such a situation ought to take place for all adjacent pairs of years.

(3)   Finally, it would be useful to know a wide set of states of the system *PE* for which the age structure dynamics among the fish population could reasonably be studied with the use of the method proposed here.

With reference to Fig. 26.1, the problem of comparing catches taken by active and passive fishing gears remains up to now unresolved because of the great, and even fundamental, differences in the processes underlying the catchabilities of those gears. However, trawl catchabilities relative to different fish species have been better investigated, and so, trawls are recommended as a standard gear for future estimations of densities of fish and age structure of stocks.

Gill net catches are affected by a far greater set of factors and more by the 'fine' nature and structure of the behavioural features which exist among fishes. The process of capturing fish in calm water is especially complicated, and many parameters and processes are involved in determining the reaction of a fish as it approaches the net.

Thus, in order to improve estimates of absolute fish abundances and age structure of a stock on the basis of the method proposed and using data from gill net and trawl catches, investigations should be carried out which aim to determine gill net catchability and to elaborate the models for the processes governing fish capture by gill nets in the particular *PE* system.

References

Bandura, W.I., Dronov, W.G., Shibayev, S.W. (1987) Parameters for optimal fishing of sheat-fish stock in Tsymliansk Reservoir. *Sbornik nauchnykh trudov GosNIORH*, **270**: 23–39 (In Russian).

Baranov, F.I. (1918) K woprosu o biologicheskikh osnowanijakh rybnogo khoziajstwa. *Izwiestija otdiela rybolowstwa i nauchno-promyslowykh issliedowanij* **1**: 84–128 (In Russian).

Nikol'sky, G.W. (1974) *Theory of fish stock dynamics.* Moscow: Pishchiewaja promyshliennost', 447 p (In Russian).

Sechin, You.T. (1969) Optimum assortment of nets for reservoirs. *Trudy Saratovskogo otdelienija GosNIORH* **9**: 8–63 (In Russian).

Severtsov, S.A. (1941) *Population dynamics and adaptive evolution of animals.* Moscow: Izd. AN USSR, 315 p (In Russian).

Shibayev, S.W. (1987) A method of investigation of mortality dynamics in fish at age within unstable populations. *Shornik nauchnykh trudov GosNIORH* **270**: 87–93 (In Russian).

Tiurin, P.W. (1962) Faktor jestiestwiennoj smiertnosti i jego znachienije pri regulirowanii rybolowstwa. *Worposy ikhtiologii* **2**: 3 (24), 403–427 (In Russian).

Chapter 27
Forecasting fish catches in USSR fresh waters

Y.T. SECHIN *All-Union Research Institute of Pond Fisheries (AURIPF), Rybnoye, Dmitrov Region, Moscow Province, 141 821 USSR*

About 20 regional Fisheries Research Institutes and a number of academic institutes take part in the development of long- and short-term forecasting of fish catches in USSR inland freshwater basins. Scientific institutions apply different methods for assessing fish abundance and for catch forecasting, mainly because of peculiarities in the various basins and in relation to the available technical equipment. The branch-oriented methodology recommends research workers to use standard fishing gears and direct quantitative evaluation of fish abundance for most large freshwater basins based on trawl and seine net surveys. This provides data on absolute fish abundance in the basin for computation of allowable catches for one to three years ahead. Fish abundance indices, forecasts of water level and the share of the catch between different gear types in the fishery are used in the forecasting procedures. Fishery statistics as a whole and disaggregated to particular fishing gears are predominantly used for catch forecasting in the medium and small lakes and reservoirs. The relevant information from 60 basins with relatively long histories of exploitation has been collected since 1985 to give a database for the State Cadaster of fishes. Fish stock size and catch level for main fish species, reproductive success, hydrochemical state of the basin, abundance of forage organisms and other data are incorporated in an automated system for forecasting fish catches.

27.1 Introduction

In the USSR there is an abundance of fresh waters for actual and potential fisheries. The total length of all rivers is about 5×10^{-5} km; the overall area of lakes is about 2×10^{-5} km^2 and that of reservoirs 7×10^{-4} km^2. Many of the larger waters are continuously exploited for their fishery resources and are thus regularly investigated. Others, particularly the small (<10 km^2) and middle ($10-100$ km^2) sized lakes holding less valuable ichthyofauna, are only fished periodically, perhaps once in $2-3$ years.

A great many of the lakes and rivers, especially in the west and east Siberia, are not exploited because the fisheries have no economic value. According to statistics of the State Inland Fishery Boards, total fish catch in the USSR freshwater basins in

1988 made up about 3×10^5 tonnes; 8×10^4 tonnes in lakes, 1.5×10^5 tonnes in rivers and 7×10^4 tonnes in reservoirs. Moreover, according to assessments of the Fish Resource Protection Board in 1988 recreation fishermen harvested up to 1.2×10^5 tonnes. The bulk of commercial fish catches consists of: 20% *Abramis brama*, 8% *Coregonus spp.*, 6% *Rutilus rutilus caspiens*, 5% *Acipenseridae*, 5% *Stizostedion lucioperca*, 3.7% *Esox lucius*, 3.3% Osmeridae, 3.3% *Silurus glanius*, 3.1% *Cyprinus carpio*, 42.6% ordinary fish species (*Rutilus rutilus*, *Perca fluviatilis*, *Abramis ballerus* and others).

All fish species exhibit considerable fluctuations in abundance, with abundant year classes sometimes ten times greater than poor ones. These features of inland basins require monitoring of stock status of the main fish species, and with information on environmental conditions, provide a basis on which to develop scientifically substantiated catch forecasts.

Twenty fishery Institutes, four Research Institutes of the Academy of Science and several Higher Education Institutes and Universities take part in the investigation of fish resources and elaboration of short- and long-term catch forecasts for the USSR freshwater basins.

The head organization in the USSR Inland Fishery Branch dealing with fish catch forecasting is the All-Union Research Institute of Pond Fisheries (AURIPF). In 1985 the Council of Directors of Fishery Institutes formed the Methodic Council for fish catch forecasting, which includes leading specialists in the field from all relevant institutes. Representatives from the USSR Ministry of Fisheries, industrial enterprises and Fish Resource Protection Boards are invited to participate in activities of the Methodic Council. The main purpose and task of the Methodic Council as prescribed is to increase the quality of catch projections, to reduce their errors, so that:

(1) The fish catch forecasting programme is developed annually, and made more accurate.
(2) New methods for fish catch forecasting are discussed and evaluated.
(3) Final results of all institutes forecasting activities are discussed, analysed and evaluated.
(4) Activities of the institutes are coordinated in the facets of mathematical modelling, implementation and application of computer programs.
(5) Catch projections are highly substantiated to ensure a stable resource base for fishery which would be able to provide maximum allowable catches.

The problem with developing short-term catch forecasts (for $1-3$ years ahead) can be split into two main steps:

(1) Collection and primary treatment of the ichthyological material gathered.
(2) Computation of total and possible allowable catches.

The present paper concerns the first step in this work.

27.2   Methods for assessment of fish abundance

Otter trawls, pair trawls, beach and boat seines, set nets, drift and towed nets, operated from small research and fishing vessels with engine power 40–150 hp, and auxiliary boats are used for quantitative stock assessments of the main exploited fish species. For most basins, however, only a few gears are recommended for collection of material on species abundance, for example, where sites are inaccessible for active gears (trawls). Table 27.1 shows the recommended gears for surveys in different seasons of the year.

The most representative material on abundances of fishes in rivers is gathered during the spring when migrating fish are captured by beach seine nets and drift nets. Spring surveys, however, are labour intensive because of great increases in the basin area due to spring floods.

Autumn and summer are the most convenient seasons for surveys in lakes and reservoirs when all types of fishing gears (active and passive) can be used successfully. Trawls and boat seine gill nets are used most effectively during late autumn when fish become more passive.

The quality and quantity of fish samples are dependent on various morphometric peculiarities of the basins, their species composition, differences in fishing gear available for research workers and fishermen, and the number of research workers. The branch-orientated methodological instructions, however, recommend specific designs of set gill nets and trawls.

Set gill nets should be single-walled in design with a hanging ratio of 0.5 and a working height of 2 m; catches being adjusted to a net length of 25 m, i.e. the standard net for research work has an area of $50\,m^2$ and is defined as the unit of fishing effort when it is used for one day. The working height of the net should not surpass the water depth, thus necessitating the need to have gill nets of different heights. Their catches are adjusted to the standard unit area of $50\,m^2$. The relative catchability of a single-walled gill net with regard to different species of fish and length can be computed from the formula provided by Sechin (1969).

Table 27.1 Summary of gears preferred for sampling fisheries at different times of the year

Fishing gears		Seasons of year			
		Spring	Summer	Autumn	Winter
Nets	set	+	+	−	−
	drift	+	+	+	−
	towed	+	+	+	−
Trawls	otter	+	+	+	−
	pair	+	+	+	−
Seine	boat	−	+	+	+
nets	beach	+	+	−	−

The arrangement of a set of gill nets which allows all length groups of fish of the same species to be captured with equivalent efficiency is of the form:

$$\bar{a}_{n+1} = \gamma\,\bar{a}_n + \sigma \tag{1}$$

where $\bar{a}_n$ is the smallest gill net mesh size, mm, $\bar{a}_{n+1}$ the compared mesh size, mm and γ and σ are coefficients which are assumed constant and species specific.

If the length composition of the fish stock being studied and research catches for two successive years plus commercial catches during the survey are known the absolute abundance of the fish stock can be assessed. The accuracy of such computations depends on the volume and quality of field material gathered. Gill net gangs are set, as a rule, according to the sites in a grid chosen to provide fishing in all parts of the basin with equal intensity at different depths. The number of gill nets set in the basin should allow complete sampling of the catches, for length-weight measurements, age, etc. Gill netting is carried out during the same period every year, with the same number of settings of net gangs and fishing times.

Assessment of absolute fish abundance based on gill net catches necessitates that catches have to be split into age and length groups. When disaggregating catches into different length-age groups the relationship

$$l_{t+1} = A + B \times l_t, \tag{2}$$

is applied, where l_t is the fish length at age t and l_{t+1} the fish length at age $t+1$. Thus, the first length-age group would comprise fishes having lengths between l_1 and l_2, the second group between l_2 and l_3, and so on. The coefficient of annual natural mortality at age (M) should also be known. Values of M have been assessed by Tiurin (1972), Gulin (1971), Shibayev (1985), Zykov (1988), Sechin (1988) and others.

The principal scheme for computation of absolute fish abundance is as follows. In the i-th year of observation the number of fishes at age t in gill net catches would be N_i, and in year $(i+1)$ the number of fishes at age $(t+1)$ would be N_{i+1}, so that

$$N_i - N_{i+1} = N_F + N_M, \tag{3}$$

where N_F is the number of fishes in the annual catch and N_M the number of fish having died through natural causes.

Relating equation (3) to N_i:

$$1 - (N_{i+1}/N_i) = (N_F + N_M)\,N_i = K_{(F+M)} \tag{4}$$

If the value $(N_M/N_i) = K_M$ is known the value of

$$K_F = (N_F/N_i) = K_{(F+M)} = K_M \tag{5}$$

can be obtained, and N_i and successively N_{i+1} are found.

27.3   Determination of fish abundance in the basin on the basis of trawl catches

Trawls are considered to be the most easily operated and reliable means of surveying the basins. Bottom and mid-water trawls are used in a single or pair trawling scheme depending on degree of roughness of the bed of the basin and on species-specific fish behaviour.

The 25-metre trawl of GosNIORH design, equipped with rectangular doors of 180×90 cm, is used if surveys are carried out by ships with an engine power of 150 hp. Ships with an engine power 90 hp haul operate a 180-metre trawl of GosNIORH design which uses doors of 160×80 cm. Theoretical and experimental substantiation of the selectivity and absolute catchabilities of these trawls and the optimum speed and duration of hauling have been made by a number of investigators between 1969 and 1983 (Romanenko & Sechin, 1969; Sechin & Kuznietsov, 1973; Judanova, 1974; Sechin, 1980, 1983). Duration of trawling, at $4-5 \, \text{km} \, \text{h}^{-1}$, varies between 5 and 60 minutes depending on fish productivity in the basin and type of the trawl.

The area fished by the otter-trawl is assumed to be the product of the distance of hauling and the distance between the trawl doors. In pair trawling this quantity is determined by trawling distance and the horizontal opening of the trawl wings.

Calculation of the absolute fish abundance in the basin is determined by:

$$N = (S \times Y \times 10^6)/(l \times v \times t \times n \times K_t), \tag{6}$$

where S is the area of the basin, Y the trawl catch during the survey, l the distance between the trawl doors during hauling, (for pair trawling − the horizontal opening of the trawl), v the trawling speed, t the duration of trawling, (the unit of fishing effort), n the number of trawls during the survey and K_t the coefficient of absolute catchability of the trawl relative to the given fish species.

The following values for K_t relative to various fish species are recommended for use in computations: *Abramis brama* − 0.6; *Stizostedion lucioperca* − 0.4; *Abramis ballerus* − 0.5; *Perca fluviatilis* L. − 0.4; *Rutilus rutilus* − 0.4; *Lucioperca volgensis* − 0.3; *Pelecus cultratus* − 0.5; *Silurus glanis* − 0.6. These coefficients have been experimentally determined.

27.4   Assessment of stock abundance from seine net catches

Data on commercial catches of both beach and boat seine nets are widely applied to quantitative evaluations of fish densities in the basins. Various seine nets, differing in design, size and webbing properties are used, depending on the peculiarities of the particular basin (bottom character, depth, streams, and so on), of the target species and pulling characteristics of seine winches. In this respect it is not possible to recommend any standard seine net for research work in all basins. Moreover, seine netting needs a number of personnel and special machinery, thus research bodies generally ask teams of fishermen with their own seine net equipment to carry out ichthyological surveys. Field material from seine net catches must include samples that are representative of species and size composition.

The area swept by the seine net is usually that swept by the webbing part of the net.

Computation of the absolute stock abundance is determined by:

$$N = (S' \times Y \times 10^6)/(S_s \times K_s) \tag{7}$$

where S' is the area of the basin from which commercial catches are taken, S_s the area swept by the seine net during the survey, Y the seine net survey catch and K_s the absolute catchability coefficient. Values of K_s are determined in relation to the peculiarities of the basin in question, seine net design and fishing operations, and by applying fish tagging techniques. Values of K_s for commercial seine nets vary from 0.27 to 0.73, according to the fish species inhabiting a particular basin in a particular fish community (Novozhylov, 1969; Treshchev, 1983; Denisov, 1978).

27.5   Fish catch forecasting

On the basis of information on fish stock abundance one can compute a Total Allowable Catch (TAC), i.e. the part of the stock which can be taken by the fishery whilst preserving the reproductive potential of the stock. When there is some doubt over stock abundance or accessibility and the state of fishery development, TAC assessment in some cases is tentatively reduced to the value called Possible Allowable Catch (PAC). In some exploited basins PACs are computed on the base of the actual commercial catches with corrections for different biological characteristics such as abundance of recruiting year classes, relative level of CPUE, growth rates, age-length composition of catches, natural mortality of fish, changes in environmental factors. Many biological characteristics reveal poor correlations with catches and stock size, and PAC values are such that they contain a lot of subjective features. When recruitment to the stock is relatively constant and fishery statistics are properly kept, PAC computation for one to three years ahead using regression analysis lead to fairly good short-term forecasts. Such an approach contains some drawbacks and does not detect important higher levels of exploitation of the stock. Experience on fishery management at a number of basins has shown that this forecasting and management strategy has sometimes resulted in an implied fish stock abundance decline with fall in catches or lower commercial catch rates suggesting corresponding underexploitation of the resource in the basin.

To assist in the forecasting procedure, the USSR Ministry of Fisheries, its system of research institutes and the Fish Stock Protection Board, make efforts to update the State Cadaster of fishes, commercial invertebrates and sea mammals. The information collected includes:

- species composition of catches;
- absolute abundance and biomass at age for main species targeted by commercial fisheries;
- estimates of annual age specific mortality due to fishing and natural causes;

- biomass of forage organisms for particular fish species;
- evaluation of fish stock status and possible new fishery resources;
- measures to improve reproduction in the stock.

These data are used in the estimation of short-term fish catch forecasting. It should be emphasized that TACs and PACs are computed with the intention of preserving the abundance and biomass of parent stock. Data on sex ratios and fecundity at different ages, maturity ogives and assessed length composition of the exploited stocks, will be included in the Cadaster information from 1990.

References

Denisov, L.I. (1978) *Rybolowstwo na wodokhranilishchakh*. Moscow: Pishchiewaja promyshliennost', 286 p. (In Russian).

Gulin, V.V. (1971) Theoretical principles and practical elaboration of methods for estimation of total fishing and natural mortalities of fish in inland water basins. *Izviestija GosNIORH* **73**: 33–75. (In Russian).

Judanova, N.M. (1974) On swimming ability of bream. *Rybnoje khoziajstwo* **9**: 35–36. (In Russian).

Methodological instructions to assessment of fish abundance in freshwater basins (1986). Elaborated by Dr You.T. Sechin, Moscow: AURIPF, 50 p. (In Russian).

Novozhylov, E.P. (1969) Investigation of catchability of river beach seine nets. *Trudy Saratovskogo otdielienija GosNIORH* **9**: 63–72. (In Russian).

Romanienko, W.I., Sechin, You.T. (1969) Wlijanije skorosti tralienija na izbiratiel'nyje swojstwa eliektrotrala. *Trudy Saratowskogo otdielienija GosNIORH* **9**: 85–98. (In Russian).

Sechin, You.T. (1969) Optimum assortment of nets for reservoirs. *Trudy Saratovskogo otdielienija GosNIORH* **9**: 8–63. (In Russian).

Sechin, You.T. (1980) Development of trawl fishery at reservoirs. *Sbornik nauchnykh trudov GosNIORH* **151**, 8–87. (In Russian).

Sechin, You.T. (1983) Obosnowanije osnownykh kharacteristik tralowoj sistiemy dlia otsenki zapasov ryb na wnutriennikh wodojomakh. *Sbornik nauchnykh trudov GosNIORH* **198**: 134–161. (In Russian).

Sechin, You.T. (1988) Otsenka parametrov krwoj jestiestwiennoj smiertnosti ryb w oblawliwajemoj popopuliatsii (na primierie leshcha oziera Il'mien'). *Sbornik nauchnykh trudov VNIIPRH* **54**: 128–133. (In Russian).

Sechin, You.T. and Kutuzov, A.T. (1973) Rigging of the footrope of bottom trawl. *Rybnoje khoziajstwo* **11**: 27–28. (In Russian).

Sechin, You.T. and Karagoishijev, K.K. (1983) Methods for Determination of Absolute Catchability Coefficients for Bottom Trawl. *Sbornik nauchnykh trudov GosNIORH* **198**: 162–1988. (In Russian).

Tiurin, P.W. (1972) 'Normal' curves of survival and natural mortality rates in fish as the theoretical basis for fishing regulation. *Izviestija GosNIORH* **71**: 71–128. (In Russian).

Treshchev, A.I. (1983) *Intensity of fishery*. Moscow: 'Ljogkaja i pishchewaja promyshliennost', 236 p. (In Russian).

Zykov, L.A. (1986) Metod otsenki koeffitsientow jestiestwiennoj smiertnosti differientsirowannykh po wozrastu ryb. *Sbornik nauchnykh trudov GosNIORH* **243**: 14–21. (In Russian).

Chapter 28
Characteristics in the annual variation of yield from professional fisheries in freshwater bodies of the temperate and the tropical zones

A.D. BUIJSE, W.L.T. VAN DENSEN and M.A.M. MACHIELS

Department of Fish Culture and Fisheries, Agricultural University Wageningen, P.O. Box 338, 6700 AH Wageningen, The Netherlands

Long-term statistics were used to analyse the inter-annual variation in the yield per species (n = 51) from 39 freshwater bodies. Fisheries per species and location were categorized according to characteristics in the inter-annual variation in the yield. These characteristics were defined with respect to the size of the inter-annual variation, both absolute and relative to the long-term variation around the mean yield, and the long-term trend in the catch levels. Intensive exploitation will increase the inter-annual variation in the yield because of a higher sensitivity of the yield to fluctuations in year-class strength. The uncertainty can only partially be diminished by lowering the fishing effort. Fisheries managers will always have to account for uncertainty in the annual yield from a fishery. Categorizing species and situations by characterizing the inter-annual variation in the yield might help the fisheries manager in embedding his management advice in the context of inevitable uncertainty generated by the environment.

28.1   Introduction

The variable yield of a fishery has direct consequences for the economic status of the fishermen. Steadily increasing catches often stimulates investment which in turn can lead to increased fishing effort. Steadily falling catches are responded to more directly by increasing fishing effort and thus making a downward trend in total yield less dramatic than the actual decrease in stock size. In the case of an extremely spasmodic variation in annual yield the fishermen tend to target other species. Specialists have more difficulty in coping with natural variability than generalists (Smith & McKelvey, 1986). Fish traders and processors are also sensitive to this variability. They can react on a variable availability of fish by, for instance, storing the fish, or by diversifying the fish species they buy and sell. However, little effort has been directed at estimating the true costs of stock variability to the industry as a whole. Coupling biological models of the fish stock, which describes its variability, with socio-economic models might help reveal suitable management strategies (Silvert, 1982).

Reasons for the variability in the yield are fluctuations in recruitment, growth rate, survival and, more indirectly, alterations in mesh size and fishing effort. By modelling the population and the fishery the sensitivity of the annual yield for these factors can be evaluated (Jacobson & Taylor, 1985).

In this chapter the variation in the annual yield of a number of freshwater fisheries is characterized and a categorization of the fisheries is proposed. Variation is considered both on a long-term and short-term basis with respect to single species and individual water body. Thus, no data series with taxonomical or geographical groupings were used. Finally the possible use of the categorization in fisheries management is discussed.

28.2   Materials and methods

Time series of annual catches were taken from the literature, mostly by extraction of data presented in graphical form.

28.2.1   *Indexing variation*

The variation experienced by the fishermen on a long-term basis is expressed by the coefficient of variation (CV) in the annual yield. In the short-term the fishermen relate their catches (Y_i) in a given year to that in the previous year (Y_{i-1}). In a long-term data series, the mean of the absolute differences in the annual catch between successive years can be related to the mean catch according to:

$$U_a = \frac{|dY|}{Y} = 100 \times \frac{\text{mean } |Y_i - Y_{i-1}|}{Y} \ (\%)$$

where U_a is an index of absolute variation, and $Y =$ the mean long-term catch. Since the fishermen generally experience the short-term variation in their catch as a percentile change, the relative variation (U_r) was calculated from the mean of the absolute difference of log-transformed catches (r) as:

$$r = \sum_{i=2}^{n} |\log_{10}(Y_i/Y_{i-1})|/n - 1$$

where n is the length of the time series (years) and:

$$U_r = 100 \times 2 \times ((1 - 1/10^r)/(1 + 1/10^r)) \ (\%)$$

If $U_r > U_a$ variation is inversely related to the yield but if $U_r < U_a$ relative variation is directly related to the yield. Part of the variation in the annual yield might come from a long-term downward or upward trend in catches. To investigate whether catches follow trends, 1st (linear trends) and 2nd order polynomials were fitted to the data using time (years) as the independent variable and catch as the dependent variable. The estimated coefficients of variation are named respectively, $CV0$, $CV1$ and $CV2$. The number indicates the order of polynomial fitted to the data.

28.2.2 Characteristic types of variation in the fishery

A theoretical time series of data are shown in Fig. 28.1. The first situation illustrates an extremely stable fishery with small short- and long-term variation. The second fishery is characterized by both a high CV and a high U_a and is highly unstable. In the third fishery, although there is a large long-term variation the fishermen only experience minor differences from one year to the next and U_a is the same as in the first fishery.

28.3 Results

In all, 121 cases were selected for examination (Table 28.1). The length of the time series varied between 8–88 years, of which approximately 50% were *c.* 20–30 years long (Fig. 28.2). The coefficients of variation, assuming no trends, were mainly between 40–80% (Fig. 28.3). In Fig. 28.4 the CV of 1st and 2nd order polynomials are plotted against a lower order fit. Points on the bisectrix mark situations where no long-term trend exists. Points markedly below this line have a significant trend (see also Table 28.1). The frequency distribution for CVs had a mode at 50% for the 1st and 40% for the 2nd order polynomial.

The frequency distribution for U_a was positively skewed with the mode around 30–40%, suggesting that, in general, fishermen experience a variation of *c.* one third of their catch from one year to the next (Fig. 28.5). The mean U_a and U_r were 34% and 39%. U_r was significantly larger (paired t-test: n = 118, t = −6.33, p < 0.001) than U_a indicating that variation was inversely related to yield, resulting in a higher uncertainty when catches are low.

Examples of characteristic time series were selected from Table 28.1. The Arctic charr, *Salvelinus alpinus*, from Lake Vättern (Grimas *et al.*, 1972) showed a very stable yield over a *c.* 50-year period (Fig. 28.6), but both the blue pike, *Stizostedion vitreum glaucum*, from Lake Erie (Nepszy, 1977) (Fig. 28.7), and the lake whitefish, *Coregonus clupeaformis*, from Lake Michigan (Taylor *et al.*, 1987) (Fig. 28.8) exhibited a large variation. Although the CV of the lake whitefish catches was high compared with blue pike, the inter-annual differences were small. The latter combination is only able to exist without a long-term trend when some periodicity occurs. This periodicity encompasses time intervals of about ten years or more of steadily rising or declining catches.

An example of a long-term upward trend is the Nile perch, *Lates niloticus*, fishery in Lake Victoria (CIFA, 1988) (Fig. 28.9). In the period after its introduction an estimate of the long-term sustainable yield was attempted (Ssentongo & Welcomme, 1985), but the catches have surpassed this level by far (Ligtvoet & Mkumbo, 1991). Although U_a is low, the uncertainty is high, because the moment of levelling off and the downfall thereafter have not yet been experienced. It is

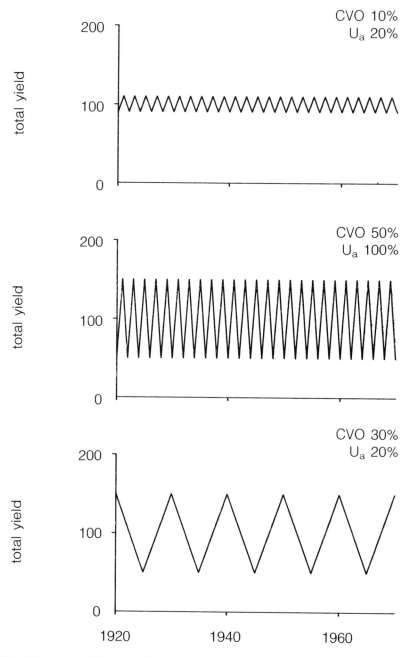

Fig. 28.1 Three types of long-term data series on annual yields without a long-term trend and with characteristic combinations of *CV0* and U_a: (a) low *CV0*, low U_a; (b) high *CV0*, high U_a; (c) high *CV0*, low U_a. See text for further explanation

Table 28.1 Cases sorted by species and water body, including the length of the time series in years (*n*), the mean annual yield, the coefficient of variation for 0-, 1- and 2-nd order polynomials (%), the indices (%) for absolute (*U_a*) and relative (*U_r*) short-term variation and the index (*dY/Y*) for long-term trends (%). See text for further explanation.

Species	Water body	Part of the water body	Country	n	Start	End	Yield	Unit	CV0	CV1	CV2	Ua	Ur	dY/Y	Source
Alburnus alburnus	Mikti Prespa	Greek	Greece	14	1973	1986	87.38	metric tons	86	65	67	41	55	-3	Crivelli 1990
Alosa pseudoharengus	Michigan		USA	24	1960	1983	11770.01	metric tons	53	47	29	25	27	3	Hanson 1987
Anguilla anguilla	Corrib		Ireland	8	1979	1986	25.04	metric tons	31	29	27	31	29	-5	Moriarty 1990
Anguilla anguilla	Corrib		Ireland	8	1979	1986	3.04	metric tons	71	62	68	81	101	32	Moriarty 1990
Anguilla anguilla	IJssel		The Netherlands	29	1959	1987	547.21	metric tons	21	21	20	16	16	-1	Densen et al. 1990
Anguilla anguilla	Lough Neagh		Ireland	15	1972	1986	195.93	metric tons	32	33	21	19	19	0	Moriarty 1990
Anguilla anguilla	Lough Neagh		Ireland	15	1972	1986	605.80	metric tons	17	15	15	12	12	1	Moriarty 1990
Anguilla anguilla	Shannon		Ireland	19	1968	1986	28.94	metric tons	45	30	29	38	42	4	Moriarty 1990
Bagrus docmac	Victoria	Kenya	Kenya	18	1968	1985	1149.91	metric tons	57	59	58	45	52	-5	CIFA 1988
Chondrostoma sp.	Mikti Prespa	Greek	Greece	14	1973	1986	18.99	metric tons	61	29	30	30	31	-15	Crivelli 1990
Cirrhinus jullieni	Ubolratana		Thailand	15	1969	1983	423.10	metric tons	58	42	43	58	69	13	Bhukaswan 1985
Clarias gariepinus	Victoria	Kenya	Kenya	18	1968	1985	1862.43	metric tons	37	36	28	29	30	-4	CIFA 1988
Coregonus acronius	Constance	upper basin	Switz., Germ. and Aust.	23	1946	1968	6.27	metric tons	66	64	47	37	46	1	Nümann 1972
Coregonus artedii	Superior	Thunder Bay	Canada	32	1948	1979	257.00	metric tons	60	52	53	37	40	-1	Jacobson et al. 1987
Coregonus artedii	Superior	Canada	Canada	29	1956	1984	926.17	metric tons	27	27	26	26	27	2	MacCallum and Selgeby 1987
Coregonus artedii	Superior	USA	USA	13	1971	1983	171.24	metric tons	96	48	27	35	47	-25	MacCallum and Selgeby 1987
Coregonus artedii	Superior	Black Bay	Canada	29	1951	1979	446.77	metric tons	53	42	30	35	49	3	Jacobson et al. 1987
Coregonus clupeaformis	Michigan		USA	83	1900	1982	950.00	metric tons	70	70	65	22	28	3	Taylor et al.
Coregonus clupeaformis	Michigan		USA	12	1973	1984	1980.16	metric tons	32	19	20	24	22	5	Eck and Wells 1987
Coregonus clupeaformis	Superior		USA/Canada	25	1960	1984	547.42	metric tons	54	23	15	12	12	8	MacCallum and Selgeby 1987
Coregonus fera	Constance	upper basin	Switz., Germ. and Aust.	23	1946	1968	13.04	metric tons	45	43	40	35	37	-5	Nümann 1972
Coregonus hoyi	Huron	USA	USA	18	1950	1967	0.41	kg ha^{-1}	102	92	72	36	44	1	Brown et al. 1987
Coregonus hoyi	Huron	Canada	Canada	34	1950	1983	0.35	kg ha^{-1}	115	114	108	33	36	-3	Brown et al. 1987
Coregonus hoyi	Michigan		USA	34	1950	1983	1.51	kg ha^{-1}	58	33	31	19	31	-3	Brown et al. 1987
Coregonus lavaretus	Constance		Switz., Germ. and Aust.	43	1945	1987	424.26	metric tons	56	56	56	45	48	-1	Löffler 1990
Coregonus macrophthalmus	Constance	upper basin	Switz., Germ. and Aust.	23	1946	1968	26.06	metric tons	76	40	40	30	38	8	Nümann 1972
Coregonus palea	Geneva		Switz. and France	36	1950	1985	94.06	metric tons	54	47	46	42	43	-1	Gerdeaux 1990
Coregonus sp.	Hallwil		Switzerland	20	1967	1986	7.51	kg ha^{-1}	123	72	59	49	53	22	Müller 1990
Coregonus sp.	Sarnen		Switzerland	20	1967	1986	16.06	kg ha^{-1}	31	31	31	34	37	1	Müller 1990
Coregonus sp.	Superior		USA/Canada	23	1960	1982	793.56	metric tons	35	35	31	28	32	-4	MacCallum and Selgeby 1987
Coregonus sp.	Thun		Switzerland	37	1951	1986	17.57	kg ha^{-1}	87	86	78	40	40	1	Müller 1990
Coregonus sp.	Vänern	Unknown	Sweden	32	1942	1974	5.53	metric tons	67	52	55	35	35	7	Svärdson 1976
Coregonus wartmanni	Constance	Upper Basin	Switz., Germ. and Aust.	23	1946	1968	343.87	metric tons	65	66	55	47	55	-2	Nümann 1972
Corica goniognathus	Ubolratana		Thailand	10	1974	1983	242.38	metric tons	45	47	50	42	56	7	Bhukaswan 1985
Cyprinus auratus	Alaotra		Madagascar	13	1955	1967	138.46	metric tons	81	27	17	22	31	-18	Moreau et al. 1988
Cyprinus carpio	Alaotra		Madagascar	21	1955	1975	508.22	metric tons	80	70	45	23	22	-13	Moreau et al. 1988

Table 28.1 Continued

Species	Water body	Part of the water body	Country	n	Start	End	Yield	Unit	CV0	CV1	CV2	Ua	Ur	dY/Y	Source
Cyprinus carpio	Mikti Prespa	Greek	Greece	14	1973	1986	6.06	metric tons	116	98	72	54	57	−34	Crivelli 1990
Esox lucius	Geneva		Switz. and France	36	1950	1985	11.03	metric tons	54	38	34	26	28	−3	Gerdeaux 1990
Esox lucius	Hjälmaren		Sweden	10	1966	1975	19.11	metric tons	31	19	21	16	19	−5	Rundberg 1977
Esox lucius	Mälaren		Sweden	12	1964	1975	36.02	metric tons	37	25	21	21	21	−11	Rundberg 1977
Lates niloticus	Victoria	Kenya	Kenya	18	1968	1985	33 134.89	metric tons	74	41	17	17	15	13	CIFA 1988
Lates stappersi	Tanganyika	Burundi	Burundi	28	1956	1983	1175.84	metric tons	83	61	62	46	58	5	Roest 1988
Leuciscus cephalus	Mikti Prespa	Greek	Greece	14	1973	1986	10.58	metric tons	49	51	52	44	49	3	Crivelli 1990
Lota lota	Archipelago sea		Finland	24	1963	1986	92.00	metric tons	47	46	33	29	30	2	Lehtonen and Hudd 1990
Lota lota	Quark		Finland	24	1963	1986	73.13	metric tons	71	63	64	53	48	−5	Lehtonen and Hudd 1990
Mirogrex terraesanctae	Kinneret		Israel	41	1936	1976	595.04	metric tons	62	20	20	16	19	5	Ben-Tuvia 1978
Morulius chrysophekadion	Ubolratana		Thailand	15	1969	1983	85.97	metric tons	60	46	44	41	47	−9	Bhukaswan 1985
Osmerus mordax	Erie		USA/Canada	23	1953	1975	4782.61	metric tons	54	33	29	26	28	7	Nepszy 1977
Osmerus mordax	Superior		USA/Canada	23	1960	1982	782.61	metric tons	56	56	50	45	42	−2	MacCallum and Selgeby 1987
Osteochilus hasselti	Ubolratana		Thailand	15	1969	1983	115.19	metric tons	61	40	41	48	47	−14	Bhukaswan 1985
Pelecus cultratus	Balaton		Hungary	72	1902	1973	114.68	metric tons	62	62	60	42	42	−1	Biro 1977
Perca flavescens	Clair	Ontario	Canada	21	1947	1961	4.38	metric tons	74	61	60	55	53	6	Johnston 1977
Perca flavescens	Erie		USA/Canada	72	1915	1987	4796.99	metric tons	71	62	60	33	35	1	Nepszy 1977, Hatch et al. 1990
Perca flavescens	Michigan	WM-5, 6	USA	22	1954	1975	214.36	metric tons	116	103	97	56	56	−3	Wells 1977
Perca flavescens	Michigan	Illinois	USA	22	1954	1975	85.95	metric tons	71	71	71	59	66	−1	Wells 1977
Perca flavescens	Michigan	Wisconsin waters of Green Bay	USA	45	1936	1982	404.05	metric tons	65	52	53	35	41	−1	Milliman et al. 1987
Perca flavescens	Michigan	Green Bay	USA	22	1954	1975	258.45	metric tons	84	72	70	39	51	−2	Wells 1977
Perca flavescens	Michigan	WM-3, 4	USA	18	1954	1975	53.89	metric tons	93	64	66	44	56	−4	Wells 1977
Perca fluviatilis	Red		USA	46	1930	1975	102.68	metric tons	54	55	48	35	43	0	Smith Jr. 1977
Perca fluviatilis	Constance		Switz., Germ. and Aust.	43	1945	1987	343.28	metric tons	79	67	65	49	51	2	Löffler 1990
Perca fluviatilis	Geneva		Switz. and France	36	1950	1985	501.26	metric tons	66	67	56	42	46	1	Gerdeaux 1990
Perca fluviatilis	Hallwil		Switzerland	20	1967	1986	0.50	kg ha^{-1}	130	120	81	68	49	14	Müller 1990
Perca fluviatilis	Hjälmaren		Sweden	10	1966	1975	29.76	metric tons	31	15	18	19	19	−8	Rundberg 1977
Perca fluviatilis	IJssel		The Netherlands	41	1947	1987	1.73	kg ha^{-1}	76	48	47	27	30	3	Densen et al. 1990
Perca fluviatilis	Mälaren		Sweden	12	1964	1975	9.26	metric tons	38	30	19	15	16	−9	Rundberg 1977
Perca fluviatilis	Sarnen		Switzerland	20	1967	1986	1.49	kg ha^{-1}	71	72	74	79	64	−1	Müller 1990
Perca fluviatilis	Vänern		Sweden	32	1942	1974	5.49	metric tons	42	32	32	26	27	−3	Svärdson 1976
Protopterus aethiopicus	Victoria	Kenya	Kenya	18	1968	1985	1007.84	metric tons	86	40	38	32	49	−15	CIFA 1988
Punctioplites proctozysron	Ubolratana		Thailand	14	1970	1983	207.26	metric tons	78	81	55	57	55	−0	Bhukaswan 1985
Punctius spp.	Ubolratana		Thailand	14	1970	1983	407.22	metric tons	26	19	16	20	20	−4	Bhukaswan 1985
Rastrineobola argentea	Victoria	Kenya	Kenya	18	1968	1985	7518.83	metric tons	96	39	25	21	26	20	CIFA 1988
Rutilus rutilus	Geneva		Switz. and France	36	1950	1985	164.64	metric tons	38	31	32	33	35	2	Gerdeaux 1990
Rutilus rutilus caspius	River Volga		USSR	28	1932	1959	25 967.26	metric tons	40	41	37	29	31	1	Poddubnyi 1979

Species	Water body	Region	Country	n	Start	End	Yield	Units						Trend	Reference
Salmo salar	River Rhine	Dutch part	The Netherlands	82	1863	1945	31 520.00	number	78	58	51	22	26	−2	Groot 1989 a
Salmo salvelinus	Constance	upper basin	Switz., Germ. and Aust.	22	1946	1968	0.54	metric tons	104	104	88	53	71	−6	Numann 1972
Salmo trutta	Constance	upper basin	Switz., Germ. and Aust.	23	1946	1968	10.36	metric tons	38	34	30	21		−3	Numann 1972
Salmo trutta	Loutaure		Sweden	21	1946	1966	0.30	metric tons	53	27	24	22	22	−7	Svärdson 1976
Salmo trutta	Skalsvattnet		Sweden	25	1949	1973	0.06	metric tons	40	41	36	36	38	1	Svärdson 1976
Salmo trutta	Geneva		Switz. and France	36	1950	1985	14.33	metric tons	40	34	30	27	28	4	Gerdeaux 1990
Salmo trutta lacustris	River Rhine		The Netherlands	35	1952	1986	0.94	metric tons	101	102	101	54	59	1	Groot 1989 b
Salmo trutta trutta	River Rhine		The Netherlands	49	1902	1951	973.76	number	85	85	81	52	52	2	Groot 1989 b
Salmo trutta trutta	Zug		Switzerland	75	1900	1974	41 743.52	number	77	32	32	24	32	−3	Ruhle 1977
Salvelinus alpinus	Geneva		Switz. and France	36	1950	1985	10.34	metric tons	71	54	52	30	33	−2	Gerdeaux 1990
Salvelinus alpinus	Loutaure		Sweden	21	1946	1966	0.43	metric tons	66	28	28	28	40	10	Svärdson 1976
Salvelinus alpinus	Skalsvattnet		Sweden	25	1949	1973	0.30	metric tons	50	34	35	32	29	3	Svärdson 1976
Salvelinus alpinus	Tunhovdfjord		Norway	20	1961	1980	2.70	kg ha^{-1}	25	23	23	19	19	0	Aass 1990
Salvelinus alpinus	Vättern		Sweden	41	1917	1957	56.68	metric tons	12	12	9	4	4	0	Grimas *et al.* 1972
Salvelinus namaycush siscowet	Superior	USA	USA	13	1971	1983	61.10	metric tons	58	35	33	43	68	10	MacCallum and Selgeby 1987
Sarotherodon macrochir	Alaotra		Madagascar	18	1958	1975	1563.64	metric tons	47	48	27	16		4	Moreau *et al.* 1988
Stizostedion canadense	Clair	Ontario	Canada	15	1947	1961	2.04	metric tons	81	74	75	61	52	0	Johnston 1977
Stizostedion canadense	Erie	Western	USA	41	1914	1954	827.87	metric tons	83	45	41	34	40	−6	Rawson and Scholl 1978
Stizostedion canadense	Winnipeg	South Basin	Canada	25	1945	1974	120.15	metric tons	49	49	46	39	44	10	Schlick 1978
Stizostedion canadense	Winnipeg	South Basin	Canada	15	1958	1974	45.28	metric tons	78	78	75	65	90	−9	Schlick 1978
Stizostedion lucioperca	Balaton		Hungary	65	1902	1973	105.05	metric tons	42	40	37	18	20	2	Biro 1977
Stizostedion lucioperca	Balaton		Hungary	26	1960	1985	1.80	kg ha^{-1}	35	27	24	20	21	−3	Biro 1990
Stizostedion lucioperca	Firth of Szczecin		Poland	28	1948	1975	340.08	metric tons	31	30	27	25	25	1	Nagiec 1977
Stizostedion lucioperca	Firth of vistula		Poland	28	1948	1975	178.97	metric tons	35	29	20	18	20	−1	Nagiec 1977
Stizostedion lucioperca	Hjälmaren		Sweden	10	1966	1975	156.93	metric tons	37	29	13	18	19	−5	Rundberg 1977
Stizostedion lucioperca	IJssel		The Netherlands	41	1947	1987	2.25	kg ha^{-1}	95	93	93	65	63	−4	Densen *et al.* 1990
Stizostedion lucioperca	Mälaren		Sweden	12	1964	1975	144.75	metric tons	29	30	20	20	21	−2	Rundberg 1977
Stizostedion lucioperca	?		Poland	26	1948	1975	166.24	metric tons	21	15	15	14	14	2	Nagiec 1977
Stizostedion vitreum	Erie	Western	USA/Canada	73	1915	1987	1411.65	metric tons	99	99	92	27	34	2	Schneider and Leach 1977, Hatch *et al.* 1990

Table 28.1 Continued

Species	Water body	Part of the water body	Country	n	Start	End	Yield	Unit	CV0	CV1	CV2	Ua	Ur	dY/Y	Source
Stizostedion vitreum	Erie	New York	USA	29	1950	1978	28.11	metric tons	62	49	37	28	38	4	Wolfert 1981
Stizostedion vitreum	Erie	Ohio and Michigan	USA	16	1955	1970	710.48	metric tons	132	76	39	32	65	−24	Busch et al. 1975
Stizostedion vitreum	Erie		USA/Canada	54	1915	1968	1618.52	metric tons	92	88	79	28	32	−1	Schneider and Leach 1977
Stizostedion vitreum	Huron	Georgian Bay	Canada	47	1924	1970	37.23	metric tons	45	43	44	30	34	−1	Schneider and Leach 1977
Stizostedion vitreum	Huron	Northwestern	USA	40	1892	1970	11.25	metric tons	70	57	58	50	62	−2	Schneider and Leach 1977
Stizostedion vitreum	Huron	Southern	USA/Canada	40	1930	1969	133.75	metric tons	25	25	25	19	19	1	Schneider and Leach 1977
Stizostedion vitreum	Huron	North Channel	Canada	52	1924	1975	25.71	metric tons	83	51	39	24	29	−4	Schneider and Leach 1977
Stizostedion vitreum	of the Woods	Ontario	Canada	79	1895	1973	0.83	kg ha^{-1}	49	40	34	25		1	Schupp and Macins 1977
Stizostedion vitreum	of the Woods	Minnesota	USA	82	1889	1973	1.21	kg ha^{-1}	63	63	47	28	33	1	Schupp and Macins 1977
Stizostedion vitreum	Ontario	Ontario	Canada	53	1918	1970	32.57	metric tons	79	79	79	28	33	−0	Schneider and Leach 1977
Stizostedion vitreum	Ontario	New York	USA	49	1918	1966	4.27	metric tons	126	96	89	48	45	−2	Schneider and Leach 1977
Stizostedion vitreum	Red	New York	USA	46	1930	1975	278.93	metric tons	37	35	29	20	25	−1	Smith Jr. 1977
Stizostedion vitreum	Shoal		Canada	20	1956	1975	2.41	kg ha^{-1}	57	54	48	38	43	0	Schupp and Macins 1977
Stizostedion vitreum	St. Clair and connecting waters		Canada	78	1893	1970	66.57	metric tons	97	91	43	23	27	−4	Schneider and Leach 1977
Stizostedion vitreum	Superior	Ontario	Canada	88	1881	1968	48.23	metric tons	61	45	43	35	41	0	Schneider and Leach 1977
Stizostedion vitreum	Winnipeg	South Basin	Canada	15	1958	1974	26.85	metric tons	120	100	59	61	95	−7	Schlick 1978
Stizostedion vitreum	Winnipeg	South Basin	Canada	25	1945	1974	146.57	metric tons	85	84	59	31	40	2	Schlick 1978
Stizostedion vitreum glaucum	Erie		USA/Canada	46	1915	1960	5504.51	metric tons	55	56	53	41	48	−4	Nepszy 1977
Tilapia rendalli	Alaotra		Madagascar	21	1955	1975	494.37	metric tons	78	75	77	37	26	3	Moreau et al. 1988

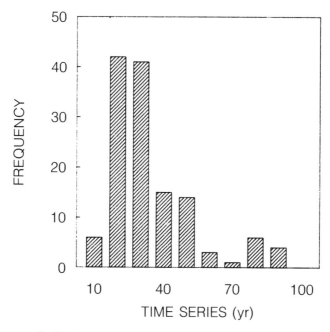

Fig. 28.2 Frequency distribution of the length of the time series on yield statistics of 51 species from 39 freshwater bodies investigated

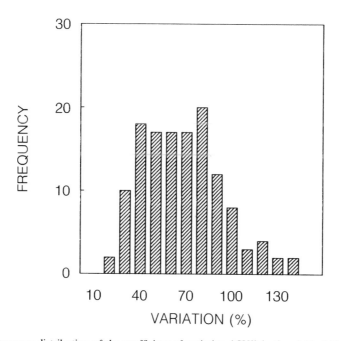

Fig. 28.3 Frequency distribution of the coefficient of variation (*CV0*) in the yield of 51 species from 39 freshwater bodies

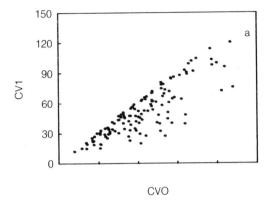

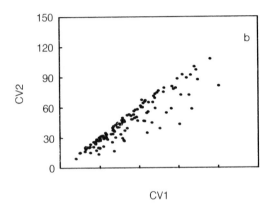

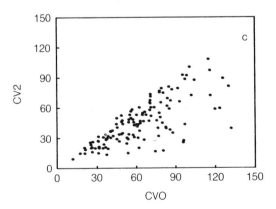

Fig. 28.4 Plots of the coefficient of variation (*CV*) in the catch of successive fitting procedures with polynomials of increasing order − *CV0*: without a trend; *CV1*: with a linear trend; *CV2*: with a second order polynomial

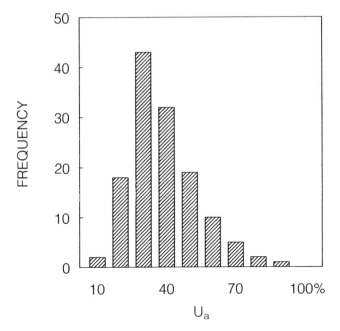

Fig. 28.5 Frequency distribution of the index for short-term variation in the yield (U_a) of 51 species from 39 freshwater bodies. See text for further explanation.

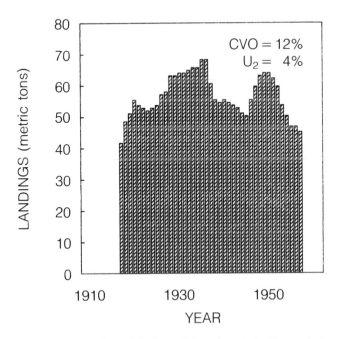

Fig. 28.6 Annual yield of Arctic charr, *Salvelinus alpinus*, from Lake Vättern in Sweden, 1917–57 (Grimas *et al.*, 1972)

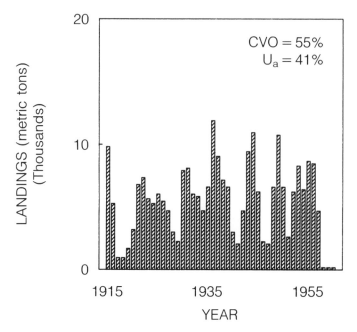

Fig. 28.7 Annual yield of blue pike, *Stizostedion vitreum glaucum*, from Lake Erie, 1915–60 (Nepszy, 1977)

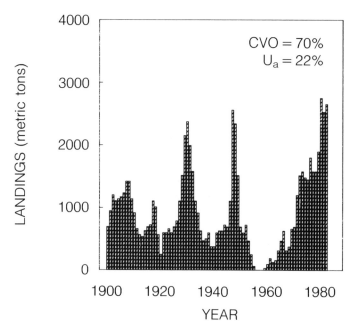

Fig. 28.8 Annual yield of lake whitefish, *Coregonus clupeaformis*, from Lake Michigan, 1900–82 (Taylor *et al.*, 1987)

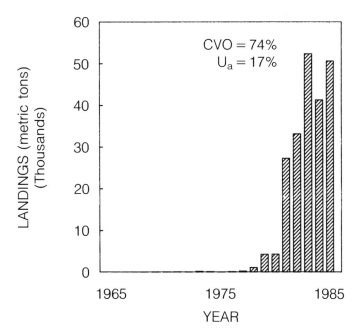

Fig. 28.9 Annual yield of Nile perch, *Lates niloticus*, from Lake Victoria in East Africa, 1968−85 (CIFA, 1988)

possible that the development in the yield will be similar to that for *Tilapia rendalli* in Lake Alaotra (Lévèque *et al.*, 1988) (Fig. 28.10). The yield of *T. rendalli* after its introduction into Lake Alaotra (Madagascar) showed a sharp initial increase and a low stabilization level after peak catches were recorded. This pattern is seen regularly in the annual yield of reservoir fisheries. By contrast, a continuous downward trend was observed in the fishery for sauger (*Stizostedion canadense*) in the Ohio waters of Lake Erie (Rawson & Scholl, 1978) (Fig. 28.11).

Both sauger and Nile perch showed significant decreases in the *CV* after fitting a polynomial but the case of *T. rendalli*, is not well described by either a 1st or a 2nd order polynomial. This is also true for other studies. For instance the rise, fall and rise again of walleye catches in North America (Hatch *et al.*, 1990; Schlick, 1978) cannot be described by this analysis. An indication whether the long-term trend in catches is negative or positive can be found by investigating column dY/Y in Table 28.1. A negative sign means a decline of the catch rates during the investigated period; a positive one an increase.

A number of cases come from water bodies in which more than one species is caught with the same gear. One example is the gill-net fishery (101 mm stretched mesh) for pikeperch (*Stizostedion lucioperca*) and perch (*Perca fluviatilis*) in Lake IJssel, the Netherlands (Densen *et al.*, 1990). The 101 mm mesh size is set to avoid catching pikeperch until they reach maturity (*c.* 40 cm). The pikeperch are caught from 40 cm onwards at 2−3 years of age. The fishing effort and thus the fishing mortality is very high. Landings consist of age-2 and age-3 group pikeperch.

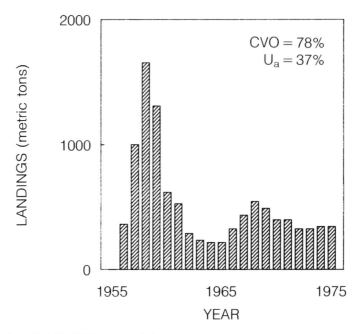

Fig. 28.10   Annual yield of *Tilapia rendalli* from Lake Alaotra, Madagascar, 1955–75 (Moreau *et al.*, 1988)

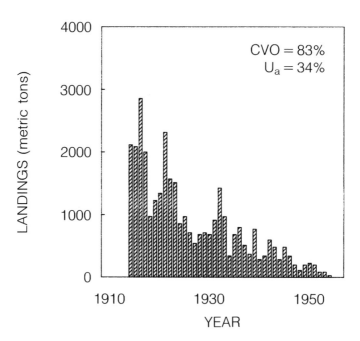

Fig. 28.11   Annual yield of sauger, *Stizostedion canadense*, from Lake Erie, 1914–58 (Rawson & Scholl, 1978)

Estimates of total removal for these age-groups are 77% and 76% per year. The slower growing perch are caught from 25 cm onwards at 3−4 years old. Landings comprise 3- to 6-age group fish. Estimates of total removal of perch are 27%, 58%, 63% and 46% for the consecutive age-groups (Buijse, unpublished data). The *CV0*s of the annual yield of pikeperch and perch are high, 95% and 76% respectively. The much lower U_a for perch (27%) than for pikeperch (65%) is attributed to the buffering effect of the simultaneous exploitation of more year-classes. The relatively slow growing perch gradually become vulnerable to the 101 mm gill nets, thus keeping fishing mortality low, whereas the faster growing pikeperch fully recruit to the catchable stock at age 3 or 4. This results in a sudden and high fishing mortality exerted on the pikeperch stock. Although the variation in annual juvenile recruitment is higher for perch than for pikeperch, the more rational exploitation of perch results in a more even distribution of landings.

28.4   Discussion

In some cases mention has been made of the factors causing the variation in yield. Occasionally these factors are completely man-induced, e.g. the ban on fishing for walleye in Lake Erie (Hatch *et al.*, 1990) and in Lake Winnipeg (Schlick, 1978) both because of mercury pollution. In many cases, however, the causes are not yet identified but the fishermen should be aware of typical catch variations in their fishery. In cases of very high variation, as was demonstrated for blue pike in Lake Erie, the fishermen will probably be willing to communicate with the fisheries managers and accept ideas on stabilization of the catch by lowering fishing effort more easily. By contrast, when the variation is high but trends are less easily detected and are only clear over very long periods, e.g. sauger in Lake Erie, the fishermen are less likely to be cooperative.

Caddy and Gulland (1983) categorized the natural variation in fish stocks, and thus in the annual catch by merely qualifying the variation into steady, cyclical, irregular or spasmodic and stressed the importance of taking these patterns into account when management strategies are developed. The categorization given here might be helpful in developing criteria for stability as a management objective and systems can be evaluated for their rationality of exploitation in this respect. High variation in recruitment and high fishing effort will result in a high uncertainty. Therefore, since recruitment is often difficult to manage because of uncertain stock-recruitment relationships, the stabilizing management strategy will be mainly to reduce the fishing effort.

Computer simulation can indicate the differential ways in which catches are optimized for maximum sustainable yield or for low variability. For example, a simulation of harvest strategies for lake whitefish in Lake Michigan showed that a constant effort policy produced relatively larger sustainable yields, whereas a constant quota policy was more effective in reducing the variability in annual yield (Jacobson & Taylor, 1985).

The variation in income experienced by the commercial fishermen is expected to be less than depicted in this study because most fisheries are multi-species and earnings are compensated for by higher prices when landings are low. It is important to point out that variation in catchable stock size is probably larger than variation in the yield because fishermen increase their effort when catches are dropping. Therefore variation in the economic status of a fishery tends to be less and the biological variation tends to be more than the variation in the yield elaborated in this chapter.

References

Aass, P. (1990) Management of Arctic charr (*Salvelinus alpinus* L.) and brown trout (*Salmo trutta* L.) fisheries in Lake Tunhovdfjord, a Norwegian hydro-electric reservoir. In *Management of Freshwater Fisheries* (Ed. by W.L.T. van Densen, B. Steinmetz and R.H. Hughes) Wageningen: PUDOC. pp. 382–389.

Ben-Tuvia, A. (1978) Fishes of Lake Kinneret In *Lake Kinneret* (Ed. by C. Serruvia). Dr W. Junk, Pub, The Hague, pp. 407–430.

Bhukaswan, T. (1985) The Nam Pong Basin (Thailand). In *Inland fisheries in multi-purpose river basin planning and development in tropical Asian countries: 3 case studies.* (Ed. by T. Petr). *FAO Fish Tech Paper* No. 265, pp 55–90.

Biro, P. (1977) Effects of exploitation, introductions, and eutrophication on percids in Lake Balaton. *J. Fish. Res. Board Can.* **34**: 1678–1683.

Biro, P. (1990) Population parameters and yield-per-recruit estimates for pikeperch (*Stizostedion lucioperca* L.) in Lake Balaton, Hungary. In *Management of Freshwater Fisheries* (Ed. by W.L.T. van Densen, B. Steinmetz and R.H. Hughes) Wageningen: PUDOC. pp. 248–261.

Brown, E.H., Argyle, R.L., Payne, N.R. and Holey, M.E. (1987) Yield and dynamics of destabilized chub (*Coregonus* spp.) populations in Lake Michigan and Huron, 1950–84. *Can. J. Fish. Aquat. Sci.* **44** (Suppl. 2): 371–383.

Busch, W.-D.N., Scholl, R.L. and Hartman, W.L. (1975) Environmental factors affecting the strength of walleye (*Stizostedion vitreum vitreum*) year-classes in Western Lake Erie, 1960–70. *J. Fish. Res. Board Can.* **32**: 1733–1743.

Caddy, J.F. and Gulland, J.A. (1983) Historical patterns of fish stocks. *Marine Policy* 7: 267–278.

CIFA (1988) Report of the fourth session of the sub-committee for the development and management of the fisheries of Lake Victoria. Kisumu, Kenya, 6–10 April 1987. *FAO Fish. Rep. No.* **388**: 112 p.

Crivelli, A.J. (1990) Fisheries decline in the freshwater lakes of northern Greece with special attention to Lake Mikri Prespa. In *Management of Freshwater Fisheries* (Ed. by W.L.T. van Densen, B. Steinmetz and R.H. Hughes) Wageningen: PUDOC. pp. 230–247.

Densen, W.L.T. van, Cazemier, W.G., Dekker, W. and Oudelaar, H.G.J. (1990) Management of the fish stocks in Lake IJssel, The Netherlands. In *Management of Freshwater Fisheries* (Ed. by W.L.T. van Densen, B. Steinmetz and R.H. Hughes) Wageningen: PUDOC. pp. 313–327.

Eck, G.W. and Wells, L. (1987) Recent changes in Lake Michigan's fish community and their probable causes, with emphasis on the role of the alewife (*Alosa pseudoharengus*). *Can. J. Fish. Aquat. Sci.* **44** (Suppl. 2): 53–60.

Gerdeaux, D. (1990) Fisheries management in an international lake: Lake Geneva. In *Management of Freshwater Fisheries* (Ed. by W.L.T. van Densen, B. Steinmetz and R.H. Hughes) Wageningen: PUDOC. pp. 161–181.

Grimas, U., Nilsson, N.-A., Toivonen, J. and Wendt, C. (1972) The future of salmonid communities in Fennoscandian Lakes. *J. Fish. Res. Board Can.* **29**: 937–940.

Groot, S.J. de (1989) Literature survey into the possibility of restocking the River Rhine and its tributaries with Atlantic salmon (*Salmo salar*). RIZA, RIVM, RIVO. Publications and reports of the project 'Ecological rehabilitation of the River Rhine'. 1989–11. 56 p.

Groot, S.J. de (1989) Literature survey into the possibility of restocking the River Rhine and its tributaries with sea trout (*Salmo trutta trutta*). RIZA, RIVM, RIVO. Publications and reports of the project 'Ecological rehabilitation of the River Rhine'. 1989–12. 11 p.

Hanson, F.H. (1987) Bioeconomic model of the Lake Michigan Alewife (*Alosa pseudoharengus*) fishery. *Can. J. Fish. Aquat. Sci.* **44** (Suppl. 2): 298–305.

Hatch, R.W., Nepszy, S.J. and Rawson, M.R. (1990) Management of percids in Lake Erie, North America. In *Management of Freshwater Fisheries* (Ed. by W.L.T. van Densen, B. Steinmetz and R.H. Hughes) Wageningen: PUDOC. 624–636.

Jacobson, L.D., MacCallum, W.R. and Spangler, G.R. (1987) Biomass dynamics of Lake Superior lake herring (*Coregonus artedii*): application of Schnute's difference model. *Can. J. Fish. Aquat. Sci.* **44** (Suppl. 2): 275–288.

Jacobson, P.C. and Taylor, W.W. (1985) Simulation of harvest strategies for a fluctuating population of lake whitefish. *N. Am. J. Fish. Manage.* **5**: 537–546.

Johnston, D.A. (1977) Population dynamics of walleye (*Stizostedion vitreum vitreum*) and yellow perch (*Perca flavescens*) in Lake St. Clair, especially during 1970–76. *J. Fish. Res. Board Can.* **34**: 1869– 1877.

Lehtonen, H. and Hudd, R. (1990) The importance of estuaries for the reproduction of freshwater fish in the Gulf of Bothnia. In *Management of Freshwater Fisheries* (Ed. by W.L.T. van Densen, B. Steinmetz and R.H. Hughes) Wageningen: PUDOC. pp. 82–89.

Ligtvoet, W. and Mkumba, M.C. (1991) A Pilot Sampling Survey For Monitoring The Artisanal Nile Perch (*Lates niloticus*) Fishery In Southern Lake Victoria (East Africa). In: *Catch effort sampling strategies: their application in freshwater fisheries management* (Ed. by I.G. Cowx). Oxford: Fishing News Books, Blackwell Scientific Publications Ltd.

Löffler, H. (1990) Fisheries management of Lake Constance: an example of international co-operation. In *Management of Freshwater Fisheries* (Ed. by W.L.T. van Densen, B. Steinmetz and R.H. Hughes) Wageningen: PUDOC. pp. 38–52.

MacCallum, W.R. and Selgeby, J.H. (1987) Lake Superior revisited 1984. *Can. J. Fish. Aquat. Sci.* **44** (Suppl. 2): 23–36.

Milliman, S.R., Bishop, R.C. and Johnson, B.L. (1987) Economic analysis of fishery rehabilitation under biological uncertainty: a conceptual framework and application. *Can. J. Fish. Aquat. Sci.* **44** (Suppl. 2): 289–297.

Moreau, J., Arrignon, J. and Jubb, R.A. (1988) Introduction of foreign fishes in African inland waters, suitability and problems. In *Biology and ecology of African freshwater fishes*. (Ed. by C. Lévêque, M.N. Bruton, and G.W. Ssentongo). ORSTROM Paris. Traveaux et documents. No. 216. pp. 395–425.

Moriarty, C. (1990) Eel management practice in three lake systems in Ireland. In *Management of Freshwater Fisheries* (Ed. by W.L.T. van Densen, B. Steinmetz and R.H. Hughes) Wageningen: PUDOC. pp. 262–269.

Müller, R. (1990) Management practices for lake fisheries in Switzerland. In *Management of Freshwater Fisheries* (Ed. by W.L.T. van Densen, B. Steinmetz and R.H. Hughes) Wageningen: PUDOC. pp. 477–492.

Nagiec, M. (1977) Pikeperch (*Stizostedion lucioperca*) in its natural habitats in Poland. *J. Fish. Res. Board Can.* **34**: 1581–1585.

Nepszy, S.J. (1977) Changes in percid populations and species interactions in Lake Erie. *J. Fish. Res. Board Can.* **34**: 1861–1868.

Nümann, W. (1972) The Bodensee: effects of exploitation and eutrophication on the salmonid community. *J. Fish. Res. Board Can.* **29**: 833–847.

Petr, T. (1985) Inland fisheries in multi-purpose river basin planning and development in tropical Asian countries: three case studies. *FAO Fish. Tech.* **265**.

Poddubnyi, A.G. (1979) The ichthyofauna of the Volga. In Mordukhai-Boltovskoi, Ph.D., 1979. *The River Volga and its life*. Dr W. Junk Publishers, The Hague-Boston-London. pp. 304–339.

Rawson, M.R. and Scholl, R.L. (1978) Re-establishment of sauger in Western Lake Erie. *Am. Fish. Soc. Spec. Publ.* **11**: 261–265.

Roest, F.C. (1988) Predator–prey relation in Northern Lake Tanganyika and fluctuations in the pelagic fish stocks. In *Predator–prey relationships, population dynamics and fisheries productivities of large African lakes.* (Ed. by D. Lewis). CIFA Occasional Paper. No. **15**. pp. 104–129.

Rühle, Ch. (1977) Biologie und Bewirtschaftung des Seesaiblings (*Salvelinus alpinus* L.) im Zugersee. *Schweizer Zeitschrift für Hydrologie* **39**: 12–45.

Rundberg, H. (1977) Trends in harvests of pikeperch (*Stizostedion lucioperca*), Eurasian perch (*Perca fluviatilis*), and northern pike (*Esox lucius*) and associated environmental changes in lakes Mälaren and Hjälmaren, 1914–74. *J. Fish. Res. Board Can.* **34**: 1720–1724.

Schlick, R.O. (1978) Management for walleye or sauger, South Basin, Lake Winnipeg. *Am. Fish. Soc. Spec. Publ.* **11**: 266–269.

Schneider, J.C. and Leach, J.H. (1977) Walleye (*Stizostedion vitreum vitreum*) fluctuations in the Great Lakes and possible causes, 1800–1975. *J. Fish. Res. Board Can.* **34**: 1878–1889.

Schupp, D.H. and Macins, V. (1977) Trends in percid yields from Lake of the Woods, 1888–1973. *J. Fish. Res. Board Can.* **34**: 1784–1791.

Silvert, W. (1982) Optimal utilization of a variable fish supply. *Can. J. Fish. Aquat. Sci.* **39**: 462–468.

Smith, C.L. and McKelvey, R. (1986) Specialists and generalists: roles for coping with variability. *N. Am. J. Fish. Manag.* **6**: 88–99.

Smith Jr., L.L. (1977) Walleye (*Stizostedion vitreum vitreum*) and yellow perch (*Perca flavescens*) populations and fisheries of the Red Lakes, Minnesota, 1930–75. *J. Fish. Res. Board Can.* **34**: 1774–1783.

Ssentongo, G.W. and Welcomme, R.L. (1985) Past history and current trends in the fisheries of Lake Victoria. *FAO Fish. Rep.* **335**. 123–138.

Svärdson, G. (1976) Interspecific population dominance in fish communities of Scandinavian lakes. *Rep. Inst. Freshw. Res. Drottningholm* **55**: 144–171.

Taylor, W.W., Smale, M.A. and Freeberg, M.H. (1987) Biotic and abiotic determinants of lake whitefish (*Coregonus clupeaformis*) recruitment in north-eastern Lake Michigan. *Can. J. Fish. Aquat. Sci.* **44** (Suppl.2): 313–323.

Wells, L. (1977) Changes in yellow perch (*Perca flavescens*) populations of Lake Michigan, 1954–75. *J. Fish. Res. Board Can.* **34**: 1821–1829.

Wolfert, D.R. (1981) The commercial fishery for walleyes in New York waters of Lake Erie 1959–1978. *N. Am. J. Fish. Manag.* **1**: 112–126.

Chapter 29
Statistical sampling methods for improving the catch assessment of lake fisheries

F.L. ORACH-MEZA *Fisheries Department, Ministry of Animal Industry and Fisheries, P.O. Box 4, Entebbe, Uganda*

Several catch and effort sampling techniques have been proposed for improving the catch assessment of large water bodies. However, most have not been successfully implemented on a sustained basis. Solutions to the possible constraints and problems encountered in designing catch assessment sampling techniques are suggested, and a method based on stratified simple random sampling is proposed.

It is shown that the proposed method, once the constraints are removed, can provide the desired precision of the estimates at the lowest possible cost besides possessing a high degree of accuracy. Accurate data collection of the randomly selected statistical units is emphasized. The design and technique are highly simplified for ease of application in the field and for ready adoption or modification for use on any inland water fisheries.

29.1   Introduction

Wise management of fisheries resources hinges on the knowledge of their magnitude, of their spatial and temporal distribution, variations in their annual recruitment levels and their general ecology. However, existing methods of statistical data collection from artisanal commercial lake fisheries are inadequate. Estimates of annual catches which often depend heavily on samples of catches from fish landings may have a large margin of error; and researchers have often doubted the accuracy of such estimates. Since not only more precise, but also more accurate, catch statistics are important in fisheries management, it becomes imperative to update some of the existing sampling systems. Any improvement in sampling design must, however, be achieved at the least possible cost. Thus the sampling scheme should be designed to minimize the sampling error for a given cost or alternatively to minimize costs for a given allowable sampling error.

The objective of this chapter is to present and illustrate a sampling method for improving the catch assessment of lake fisheries, which can provide the desired precision of the estimates at the lowest possible cost and yet possess a higher degree of accuracy. The design and technique, which are illustrated with the fisheries of Lake Kyoga, Uganda, are made as simple as possible for ease of application so that it can be readily adopted or modified for use on any lake with relatively well developed fisheries.

29.2   Lake Kyoga Fisheries

29.2.1   *The Lake*

Lake Kyoga (Fig. 29.1), located in the centre of Uganda, has an open water area of approximately 2354 km², and can be broken down into the Kyoga, Bisina, and Kwania complex. It is adjoined by many seasonal and permanent swamps of approximately similar dimensions. Its main source of water is the River Nile which originates from Lake Victoria, but there are also a series of smaller rivers and streams originating from its vast watershed covering an area of about 56 125 km². Its average depth is 7·8 metres. The long meandering shoreline of about 2000 km is fringed with papyrus and other swamp vegetation. Shore width varies from a few metres along the more exposed shores to several kilometres in the bays and river mouths which are often entirely chocked with sudd. Many of the sudds are often blown into the open water as floating papyrus islands.

29.2.2   *Existing monitoring scheme*

The commercial fisheries of Lake Kyoga are based on 15 species of fish of which only six are relatively abundant: *Lates* spp., *Clarias* spp., *Bagrus* spp., *Tilapia* and *Oreochromis* spp., and *Protopterus* spp. The fisheries, like those of many other inland waters, have developed mainly as a subsistence occupation. It is only recently that commercialization has started (see the annual landings since 1959, Fig. 29.2). However, fishing on the lake is not yet a fully organized industry; production units are small and often family owned. Fish landings are scattered along the meandering shoreline in a random fashion. Any village with passable shelter for canoes, or the mouth of a small stream, may become a fish landing. More open shores with good access roads have relatively settled fishing communities. Many temporary fishing camps are located on floating islands.

Several fishery surveys were carried out on the lake by the Uganda Fisheries Department (UFD) between 1949 and 1973 (Rhodes & Newton, 1958; Proude & Newton, 1960; Rogers, 1970; Orach-Meza, 1972; Dhatemwa, 1972). Many of the essential survey items such as number, size, and distribution of the fishing sites and landing centres have become known, and the changes in the number of fishing boats, fishing gears, and fishermen over time were noted (Fig. 29.2(b)). On the basis of what was gained from these frame surveys, the landing centres were grouped into statistical collection zones within each political boundary (districts) as shown in Table 29.1.

The sampling method to date involved the visiting of a high proportion of fish landing points on a regular and *ad hoc* basis. Preparation of the programme involves matching at random each of the landing centres in a survey zone to a

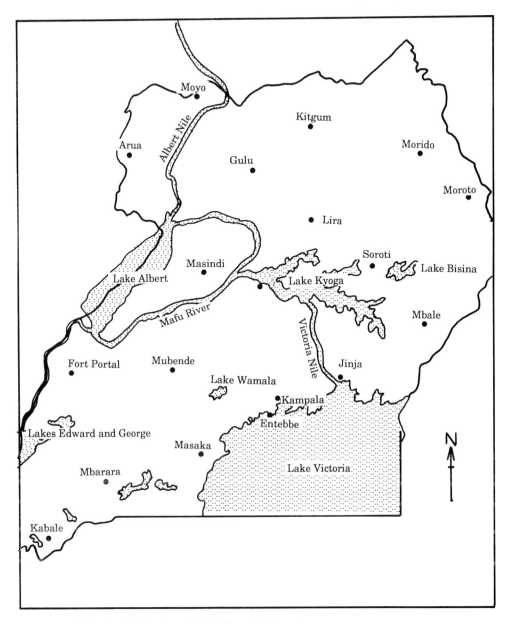

Fig. 29.1 Map showing location of Lake Kyoga in Uganda

working day in a month (at times using random numbers). Depending on the number of landing centres in the zone, some were visited only once and others twice a month without giving statistical weight to the size of the landing centre. At a landing centre, a surveyor selects (at random) a number of fishing vessels from which the catch is identified, counted and weighed. The total landings for the day

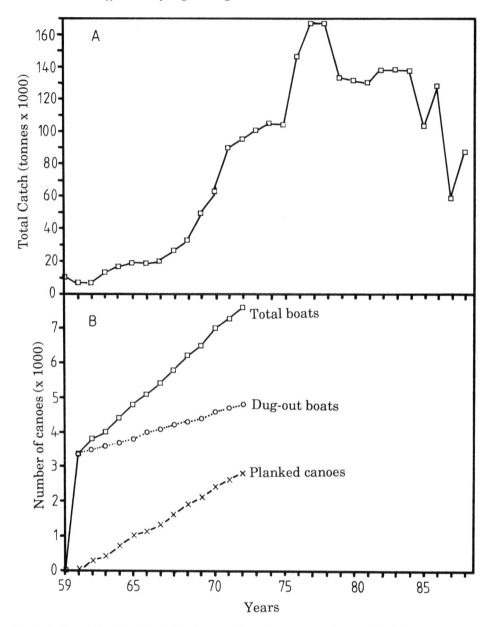

Fig. 29.2 Trends in fish catches (a) and the number of canoes operating (b) in Lake Kyoga between 1959 and 1988

are estimated by extrapolation of the catch from the selected canoes to the total number of canoes operating that day. The monthly and annual catch for each zone are estimated from the data collated. The trends in total catch and number of canoes operating since 1959 are shown in Fig. 29.2.

Table 29.1 Distribution of landing centres and fishing vessels by the existing sampling zones as of 1973. Figure in brackets gives the number of isolated landing centres with less than four canoes each and those spread along the shoreline. These are rarely surveyed due to lack of access roads. Many of them are often observed landing their catches at the nearby main landing centre.

Survey centres by districts	Landing centres	Number of fishing vessels		
		Planked/ motor	Planked	Dugout
Teso				
Kaberamaido	7 + (4)	17	146	198
Soroti	3 + (1)	1	103	167
Bugondo	5 + (6)	44	93	81
Labori	7 + (2)	12	104	188
Serere	6 + (6)	1	39	507
Ngora	3 + (10)	1	30	336
Kumi	10 + (3)	2	132	275
Lango				
Maruzi	6 + (5)	12	381	313
Kwania	10 + (10)	5	325	421
Namasale	6 + (3)	28	39	62
Amolatar	8 + (3)	19	43	22
East Buganda				
Bugerere	10 + (4)	22	149	221
Buruli	10 + (3)	96	386	94
Busoga				
Kagulu	11 + (2)	58	98	265
Kidera	8 + (4)	65	149	262
Bukedi				
Passisa	5 + (9)	0	28	283
Bunyoro				
Kibanda	3 + (1)	1	6	51
Total	118 + (67)	384	2284	3746

29.3   The proposed statistical sampling method

29.3.1   *Introduction*

Ecological diversity of tropical lakes presents several constraints that often render the design of a single suitable statistical system for lake fisheries rather difficult. The lakes are mostly characterized by having multiple stocks of fish and fisheries. Human settlements all along the meandering shorelines, full and part-time fishermen, multiple types and sizes of fishing gears and canoes including outboard engines, multiple roles for the canoes, variation in setting of gears and landing of catches in space and time and diversity in fishing skills. Total enumeration of catches is not possible for many tropical lakes because fish are landed by large numbers of small canoes at isolated places and at different times of the day and night. For example, in Lake Kyoga a total of 185 landing centres (excluding minor sites of < 5 canoes) scattered along the entire shoreline, were identified in 1972 (Dhatemwa, 1973).

Total enumeration, when applied intensively, gives numbers of fishermen, fishing boats, fishing gears, fish landing places, and other factors which do not vary greatly over a year. Each of these can be further sub-divided for statistical purposes. This shows how complex the fishery can be and the difficulty in assessing it accurately by a properly executed sampling or total enumeration system.

If a total enumeration of the catches were to be attempted, a much bigger staff than is physically possible would be needed; if only a large proportion of the landing centres were to be adequately covered the number of staff required would still be prohibitive. These limitations preclude the reliability and applicability of simple frame surveys for direct enumeration of the status of the fisheries and warrant the use of a stratified sampling procedure (Bazigos, 1974).

29.3.2   *Proposed stratified random sampling procedure*

In the stratification procedure, the fishing area of Lake Kyoga is divided into nine strata (Fig. 29.3) on the basis of similarity in the habitat characteristics (Worthington, 1929; Orach-Meza, 1973). This stratification also reflects similarity

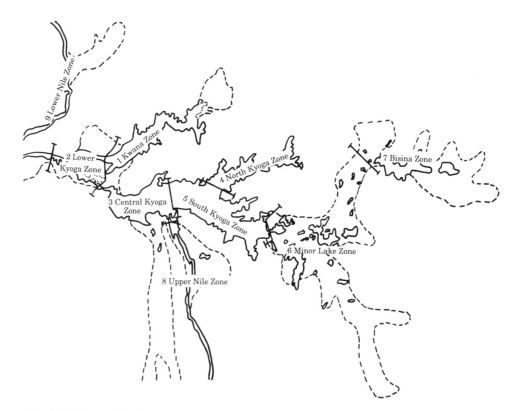

Fig. 29.3 Proposed stratification (zonation) of the fishing areas in Lake Kyoga

of the type of fishing canoes used in each stratum (Table 29.2). Each stratum is assumed to be homogeneous, and the variability of a particular parameter of the fishery to be minimal in that region. Thus a precise estimate of any stratum parameter mean can be obtained from a small sample in that stratum.

The optimum number of fish landing centres (n_k) which need to be sampled to estimate the number of canoes within each stratum, with a given relative precision, is determined by:

$$n_k = n \ (N_k \ S_k)/(\Sigma N_k \ S_k) \tag{1}$$

where n is calculated as:

$$n = (\Sigma N_k \ S_k)^2/(V_o + \Sigma \ N_k \ S_k^2) \tag{2}$$

N_k is the number of landing centres in the kth stratum and S_k^2 the variance determined as:

$$S_k^2 = [\Sigma B_{ki}^2 - (\Sigma B_{ki})^2/N_k]/(N_k - 1) \tag{3}$$

and B_{ki} is the number of boats counted at the ith landing centre and in the kth stratum. (A small value for S_k^2 implies that the probability of a large deviation of the means of the landing centres from the stratum mean is small.)

V_o, the expected precision of the overall estimate, is calculated as:

$$V_o = (CV \times \Sigma \ B_{ki})^2 \tag{4}$$

where CV, the coefficient of variation, is:

$$CV = (S_{st})/(\Sigma \ B_{ki}) = (\overline{B}_k) \tag{5}$$

$\overline{B}_k$ is the mean number of boats per landing centre and S_{st}^2 the stratified variance determined as:

$$S_{st}^2 = \Sigma \ (W_k^2 \ S_k^2/n_k) \tag{6}$$

where W_k is N_k/N and N is the total number of landings.

Table 29.2 Zonation and the division of landing centres and fishing vessels (Brackets as for Table 29.1)

		Fishing vessels	
Zones	Landing centres	Planked	Dugout
1 Lake Kwania	23 + (0)	453	484
2 Lake Kyoga (Lower)	4 + (1)	278	147
3 Lake Kyoga (Central)	19 + (8)	630	177
4 Lake Kyoga (North)	12 + (11)	292	407
5 Lake Kyoga (South)	18 + (12)	575	526
6 Minor Lakes	21 + (24)	144	1057
7 Lake Bisina	10 + (3)	136	321
8 Upper Victoria Nile	7 + (5)	133	326
9 Lower Victoria Nile	4 + (3)	27	301
Total (9 zones)	118 + (67)	2668	3746

The total number of landing sites (n) which need to be sampled for a relative precision of 50% is given by equation (2). For Lake Kyoga, using the data from Table 29.3, this would be:

$$n = (4140.7)^2/(102\,848.5 + 148\,775.8) = 68 \text{ landing sites.}$$

Therefore only 68 of the 118 landing centres need to be surveyed to give a statistically accurate result with the same level of precision. Furthermore, if these 68 centres are optimally allocated to the strata, it would give the smallest variance for the estimated total fishing vessels (Bazigos, 1974). Results of the allocation are listed in Table 29.3. If all the within-stratum variances are equal, proportional allocation could be employed (Cochran, 1963).

In order to reduce the possibility of variation within each stratum further, the allocated sample size can be distributed using random and unequal probability selection which is proportional to size (Cochran, 1963) before actual counts are made. For example, Stratum 1, with a total of 23 landing centres of which only nine are allocated to be sampled, can be subdivided as follows and as shown in Table 29.4:

Table 29.3 Estimate of sample size and computation units

Stratum	N_k	S_k	$N_k S^2_k$	$N_k S_k$	n_k
1	23	23.93	13 173.3	550.4	9
2	4	41.98	7 049.3	167.9	3
3	19	29.52	16 557.2	560.9	9
4	12	34.01	13 880.2	408.1	7
5	18	31.12	17 432.2	560.2	9
6	21	48.76	49 928.3	1024.3	17
7	10	27.64	7 639.7	276.4	5
8	7	43.10	13 003.3	301.7	6*
9	4	50.28	10 112.3	201.1	3
Total	118		148 775.8	4140.7	68

* Due to the large variability, the extra 1 landing centre to be sampled is added to stratum 8 so that $5 + 1 = 6$*

Table 29.4 Sub-division of stratum = 1: (Lake Kwania zone)

Sub-stratum	N_k	S_k	$N_k S^2_k$	$N_k S_k$	n_k
1	4	2.08	17.33	8.32	1*
2	4	2.06	17.00	8.24	1*
3	15	21.17	6723.86	317.55	9
Total	23		6758.19	334.11	9

* By computation $n_1 = 0.224$ and $n_2 = 0.222$ and $n_3 = 8.554$. For sampling purpose let $n_1 = 1.0$ and $n_2 = 1.0$ and $n_3 = 9$ so that $n = 11$ instead of 9 as optimally allocated.

Small centres — landings with 15 or less canoes
Medium centres — landings with 16 to 30 canoes
Large centres — landings with more than 30 canoes

This would assure the optimization of (a) accuracy required for estimates of the number of counts within the stratum, and (b) convenience and practicability with respect to management and the number of enumerates available.

Should money be the limiting factor, the sample size can be reduced accordingly, at the expense of increasing the estimate error, since the standard error of the estimates depends closely on the number of samples. Where the cost of enumerating canoes at respective landing centres is known, the expression given below (7) can be used for estimating the sample size (n) for a fixed cost (Cochran, 1963).

$$n = \frac{(C - C_o) \; \Sigma \; (N_k \; S_k / \sqrt{C_k})}{\Sigma \; (N_k \; S_k \; \sqrt{C_k})} \tag{7}$$

where C is the total cost, C_o the overhead costs and C_k the cost per landing centre which may vary from stratum to stratum and in relation to the sample size.

The estimated total number of boats for the fisheries is expressed by:

$$\hat{B}_{st} = \sum_1^k N_k \; \overline{B}_k \tag{8}$$

and the variance of the estimated total from optimum allocation is:

$$S_{st}^2 = \frac{1}{n} \; (\sum_1^k W_k \; S_k)^2 - (\sum_1^k W_k \; S_k^2)/N \tag{9}$$

with a coefficient of variation (the standard error expressed as a fraction of the total counts) given by:

$$(CV)_{st} = S_{st}/(\Sigma \; B_{ki}/n) \tag{10}$$

29.3.3   *Catch assessment*

The above sampling design can be usefully employed for estimating the total number and weight of fish caught by simply stratifying canoes at any sampled landing centre into types of fishing boats (i.e. planked canoe with outboard motor, planked canoe without outboard motor and dug-out canoe), each of which have optimal foraging distances. If detail distribution of the catch by species and sizes is also to be analysed, further stratification into fishing gear type is necessary. As with landing centres, not all the stratified units (boats, gears) can be inspected for the estimation of total catch of fish. Since the number and types of fishing boats in use at each landing centre are known, the best that can be done is to ensure that every boat within each sub-stratum (gear and boat type) has an equal chance of being included in the optimum sample size taken. Under the existing frame survey methodology it can only be assumed that the fishing gears in use are proportionally represented in the canoes sampled.

Considering that only the fish caught are brought to the landing centre, and very rarely the gears used in capturing the fish, it may be just as well to record the catches by canoes rather than placing statistical reliance on reports by fishermen. When a reasonable distribution of gears by canoes is eventually known, or can be determined, a two-stage or three-stage stratified sampling design, with the respective optimum allocation of the sample sizes, can be appropriately applied for the estimation of catches by sizes, species, etc.

One of the best methods of sampling from the sub-strata (the stratified boats by types) is to systematically sample the boats by their landing order, so that boats fishing near and far from the landing centre, or the crews which were idle and got up late and those that went out early all get an equal chance of being selected. Since the optimum number of boats to be selected from each sub-stratum would be known, the rank of arrival to be picked for inspection can also be selected in advance by random sampling in any of the following ways:

(1)   The lottery method – by drawing lots.
(2)   The use of random numbers from published tables of random digits.
(3)   Systematically following the order of arrivals of canoes at fish landing.

Ideally, each landing should be visited each day, but this is impractical for many lake fisheries; the landings can be visited only a few times each month. It is therefore necessary to randomize the order of visits because catches could easily be influenced by the moon phase, tidal strength, or some periodic biological or physical factor, so that the collection of landing figures in the same order on every occasion could give erroneous results which lead to wrong conclusions.

The catch from each boat sampled is sorted, identified by species, counted and weighed. The total catch for each sub-stratum (boat type) is calculated from the average catch of the respective sampled canoes. Summing these gives the total catch for the landing centre sampled that day (i.e. one day). Total catch in a month for the centre is calculated from the number of sampling days. Assuming a 26-day fishing month and 4 sampling days, the monthly catch for a landing centre is calculated using:

$$Y_{jki} = (26/4)\Sigma X_i \tag{11}$$

where X_i is the sum of the catches for the total number of sampling days spent at landing centre i (in this case 4).

The monthly landings for each of the ecological strata are obtained from:

$$Y_{jk} = (N_k/n_k) \sum_1^n Y_{jki} \tag{12}$$

where Y_{jk} is the estimated catch for the jth month and k^{th} stratum, and Y_{jki} is the estimated catch for the ith landing centre for the jth month and k^{th} stratum. Totalling the above estimates gives the monthly fish landings for the fisheries as a whole as:

$$\hat{Y}. = \sum_1^k \hat{Y}_{jk} \tag{13}$$

The annual estimates are similarly obtained by pooling the monthly catches as:

$$Y.. = \sum_{1}^{12} \hat{Y} \tag{14}$$

The variance of the monthly catch estimates is calculated by using similar methods as in equation (9). The precision of the total catch Y is:

$$\hat{S}_{jk}^2 = (N_k^2/n_k) \, S_k^2 \tag{15}$$

where S_k^2 is the estimate of the stratum variance which is given by:

$$S_k^2 = [\Sigma \, Y_{jki}^2 - (\Sigma \, Y_{jki})^2/n_k]/(n_k - 1) \tag{16}$$

The estimated variance of $Y.$ and $Y..$ are given respectively by:

$$\hat{S}.^2 = \sum^{k} S_{jk}^2 \tag{17}$$

and

$$\hat{S}.. = \sum^{k} \sum^{j} S_{jk}^2 \tag{18}$$

29.4   Summary and recommendations

Stratifying heterogeneous fishing areas and fishing vessels (as well as fishing gears when statistically feasible) before sampling the stratum catch increases the precision and accuracy of the total estimates greatly. The division makes each stratum homogeneous for independent sampling and gives estimates which can be combined to provide an assessment of the overall fishery. Unfortunately this requires prior knowledge concerning the ecology of the fish and the habits of the fishermen. The variance of the estimate is also obtained by combining the variance of the stratum estimates. Since each stratum is relatively homogeneous, variation of catch estimates within each stratum is expected to be small, so that the variance of the final combined estimate will also be small – much less than without stratification.

Where the landing centres are diffused or scattered along the shoreline, every attempt should be made to encourage the fishermen to land at a designated point nearby. If this is not possible, the sampler may have to cover a strip of shoreline designated as a sampling or landing centre. This may be applicable in strata 8 and 9 in Lake Kyoga. It is suggested that for stratum 9 (minor lakes) near Lake Kyoga, each small lake should constitute a landing centre except for Lake Nakuwa which already has five main landing centres. The use of a motorized canoe may be necessary in such cases.

An ultimate objective of any sampling design is to obtain accurate estimates on the characteristics of the population, when a complete coverage is either physically impossible or economically impracticable. Accurate data collection on the selected statistical units is therefore of prime importance, because management planning for the future is dependent upon a good knowledge of the amount of fish being caught in various zones by different fishing techniques. Unless this work is done carefully, honestly, and thoroughly, the returns will give a false impression of the state of the fisheries.

The sampling design described assumes that there is no difference between the catches of sampled canoes within a landing centre and between landing centres within a real stratum at the selected level of significance (normally $= 0.05$). The number of canoes sampled for inspection from sampled landing centres within each stratum is expected to ensure this high level of precision. An analysis of variance or regression analysis coupled with experimental fishing can be applied to verify the assumptions.

References

Bazigos, G.P. (1974) The design of fisheries statistical surveys – inland water. *F.A.O., Fish Tech. Pap.* **133**, 122 pp.

Cochran, W.G. (1963) *Sampling Techniques*. New York: John Wiley & Sons. Inc., 413 pp.

Dhatemwa, C. (1973) Aerial counts for fishing vessels on the lakes of Uganda. Manuscript – A report to the Fish. Dept., Uganda.

Orach-Meza, F.L. (1972) Survey of production factors of Lake Kyoga fisheries. Manuscript – A report to the Fisheries Department, Uganda.

Orach-Meza, F.L. (1973) Annual Report – 1972. Manuscript – A report to the Fish. Dept., Uganda.

Proude, P.D. and Newton, R.P. (1960) Aerial survey of fishing vessels of Lake Kyoga region. Manuscript – A report to the Fish. Dept., Uganda.

Rogers, P. (1970) Annual Report, 1969. Manuscript – A report to the Fish. Dept., Uganda.

Snedecor, G.W. and Cochran, W.G. (1973) *Statistical Methods*. Iowa State Univ. Press. Ames. 593 pp.

Worthington (1929) A report on the Fishing Survey of Lakes Albert and Kyoga. Crown Agents, London, S.A.L., 136 pp.

Chapter 30
A catch effort data recording system for the fishery on the small pelagic *Rastrineobola argentea* in the southern part of Lake Victoria

P.J. MOUS *Haplochromis Ecology Survey Team, P.O. Box 9516, 2300 RA Leiden, The Netherlands,*
Y.L. BUDEBA and M.M. TEMU *Tanzanian Fisheries Research Institute, P.O. Box 475, Mwanza, Tanzania,*
W.L.T. VAN DENSEN *Department of Fish Culture and Fisheries, Agricultural University Wageningen, P.O. Box 338, 6700AH Wageningen, The Netherlands.*

Dagaa, *Rastrineobola argentea* (Cyprinidae), began to appear prominently on the markets of Kenya and Tanzania in the 1960s. The dagaa is exploited by an artisanal light fishery using beach seines, scoop nets and lift nets. The dagaa is dried ashore by the fishermen, who sell the fish locally, both on the islands and the mainland. About 50% of the dagaa catch from Lake Victoria is transported to other regions and countries through the harbour of Tanzania's second largest town, Mwanza. A specific catch effort data recording system was set up for dagaa and major constraints were identified. The total dagaa catch from the Tanzanian side of Lake Victoria can be estimated by recording the total dagaa landings at the harbour of Mwanza. Information on the catch per unit of effort can be obtained from catch-effort recordings at the sites where the fishermen land the fresh dagaa. For the unit of effort the unit lamp burning hours, which is equivalent to lamp fuel consumption, was adopted.

30.1 Introduction

Lake Victoria, the largest lake of Africa with a surface area of about $70\,000\,km^2$, has recently drawn much attention because of the introduction of *Lates niloticus*, the Nile perch and the developing fishery on this large predator (CIFA, 1988). Before the increase of the Nile perch stock, a light-attraction fishery was developing on a small pelagic cyprinid, *Rastrineobola argentea*. *Rastrineobola argentea* is known as 'dagaa' in Tanzania, as 'mukene' in Uganda and as 'omena' in Kenya. The total length of dagaa rarely exceeds 10 cm and it has a short life span of 1–2 years (Wanink, 1989).

Dagaa began to appear prominently on the markets of Kenya and Tanzania in the 1960s (Okedi, 1974). To gain more information about the fishery a beach sampling programme was conducted in the Ukerewe district in Tanzania in 1978 (Okedi, 1980). The Ukerewe district is a group of islands in the south-eastern part of Lake Victoria where the dagaa fishery is concentrated (Fig. 30.1). In this area an annual yield of 3500 metric tonnes fresh weight was recorded at the end of the 1970s (Okedi, 1980). In 1989 a dagaa trader of Mwanza (Fig. 30.1), the major trading place for dagaa in the Tanzanian part of the lake, estimated that about 6000 metric tonnes of dried dagaa were transported through the harbour of Mwanza (Kirumba). An evaluation of the catch effort data recording systems for Uganda, Kenya and Tanzania revealed that dagaa catch effort statistics were not collected on a regular basis by the Fisheries Departments of the three countries (Bernacsek, 1986). The District Fisheries Office of Ukerewe also had not collected catch effort data of the dagaa fishery.

To follow trends in the stocks of dagaa crucial for the management of the fishery, catch and effort data need to be recorded. The aims of the present study were:

(1) To characterize the dagaa fishery in the Mwanza region.
(2) To design a possible catch effort data recording system for the dagaa fishery that takes into account some features of the fishery, e.g. different gear types, different landing times, different catch sizes and behaviour of the fishermen in relation to their mobility.

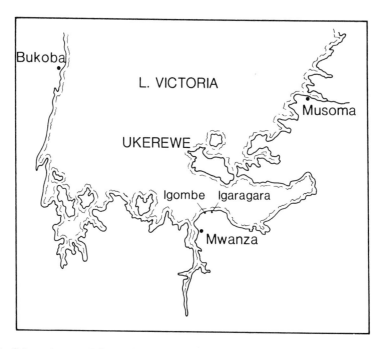

Fig. 30.1 Schematic map of the southern part of Lake Victoria

30.2   Description of the dagaa fishery

In the Tanzanian part of Lake Victoria, three main types of dagaa fishery exist:

- Beach seine fishery.
- Scoop net fishery.
- Lift net fishery.

Specific details relating to each of these fisheries are given in Tables 30.1, 30.2 and 30.3, but they have the following features in common:

(1)   They are based on the attraction of dagaa by an artificial light source during the night. The lamps used are pressure lamps, locally known as 'karabai'. The fuel consumption is about $0.21 \, hr^{-1} \, lamp^{-1}$.

(2)   Since fishing is based on light attraction, fishing is not conducted for a period of seven to ten days around the full moon each month.

Table 30.1   Composition of artisanal dagaa fishing groups per type of fishery of the southern part of Lake Victoria (Mwanza region). n = number of observations.

	Beach seine n = 21		Scoop net n = 7		Lift net n = 10	
	av.	sd	av.	sd	av.	sd
Number of fishermen per group	4.4	1	2.9	0.6	4.0	0.0
Number of lamps per group	3.8	1	2.9	0.5	2.9	0.2

Table 30.2   Characteristics of dagaa fishing gear and susceptibility to weather conditions for each type of fishery

	Beach seine	Scoop net	Lift net
Mesh size (stretched mesh in mm)	5–8	8	8
Maximum distance from shore (lamps) in km	0.5	1	20
Susceptibility to weather conditions	high	moderate	low

Table 30.3   Average and maximum catch for each type of fishery. These figures apply to the period October–December 1989 only. Catches of 0 kg were not included in the calculations. n = number of observations.

	Average catch (kg group^{-1} night^{-1})	Maximum catch (kg group^{-1} night^{-1})	n
Beach seine	80	360	71
Scoop net	77	640	24
Lift net	480	1950	55

(3) A minor part, about 20%, of the dagaa catch is sold fresh. The remainder is spread on the beach to dry for one to three days, depending on the weather conditions. In the rainy season the catch does not dry properly. This results in low quality dagaa which is sold as chicken feed at about half the price of high quality dagaa. A 30 kg gunny bag of high quality dagaa cost *c.* Tsh 2800 in February 1990 (190 Tsh = 1 US Dollar).

(4) The unit of measurement for the quantity of fresh dagaa is the 'debe', a tin which can contain about 20 kg of dagaa. Dried dagaa is sold per gunny bag of *c.* 30 kg (high quality) or of *c.* 30−40 kg (low quality). The latter is heavier because of the higher water and sand content of the product.

(5) The fishermen operate in groups, the size of which depends on the gear type (Table 30.1). The entrepreneur receives about $\frac{1}{2}$ to $\frac{3}{4}$ of the revenues. Several groups may live together in one camp and sometimes they even share a net and/or a boat in case of a beach seine or a scoop net fishery, but the catches are always kept separate.

(6) Most fishermen are transitory. They tend to shift to other sites when catches are not satisfactory or if conditions are not favourable, for instance rough water.

In the beach seine fishery, the lamps are attached to locally made rafts which are anchored in a straight line approximately perpendicular to the shoreline. The distance between the lamps varies between 10 and 20 m. Approximately three hours after setting, the lamps are hauled slowly, until they are grouped close to the shore. Next, the beach seine is set around the lamps and the net is hauled. The length of the beach seine varies between 40 and 100 m, the net is pulled by three to six men. If the fishermen are operating on their home beach, the catch is spread on the beach for drying immediately after a haul. This type of fishery is only conducted on sandy beaches. In some cases the fishermen make several hauls in different places. In this case the catch is transported by a canoe equipped with an outboard engine to the home beach where it is dried. Beach seining is the most common type of fishery in the Mwanza and Ukerewe regions.

In the scoop net fishery, the lamps are set on rafts similar to those used in the beach seine fishery. In this case, however, the lamps are concentrated near the canoe and the fish are scooped with a hand net into the canoe. The diameter of the scoop net is *c.* 1.3 m, and the total length of the net is *c.* 4.5 m. One scoop net is handled by one man.

The lift net vessel is composed of two canoes, connected to each other as a catamaran. The net is lowered into the water between the two canoes and is kept open by outriggers on the catamaran. The lamps are attached to the catamaran. When in operation the lamps are set in the middle of the two boats. The depth of the net is about 10 m and the surface area of the water column above the net is about 25 m² (Nedelec, 1975). About half of the lift net groups encountered used outboard engines (7 hp).

Catches from the scoop net fishing groups and of the beach seine fishermen are of about the same order of magnitude (Fig. 30.2). The mode for both types of fishery is 50–100 kg night^{-1} group^{-1}. The daily average catch of a lift net group is considerably higher with approximately 40% of the recorded catches being higher than 500 kg. The average catch per group (Table 30.3) confirmed these differences.

30.3   Catch effort sampling methods

The Fisheries Department of Tanzania conducts a yearly canoe census and this gives an indication of the fishing activities and their intensity in different parts of

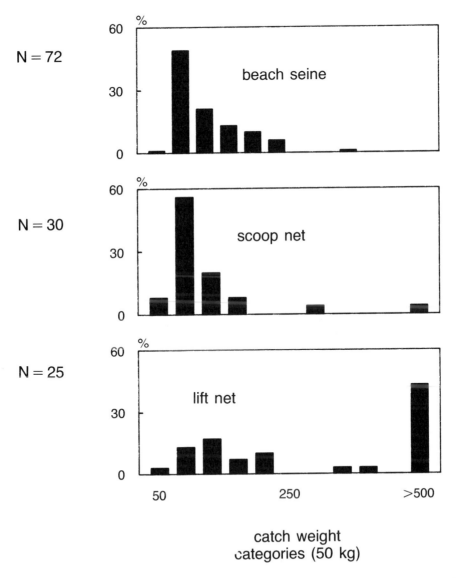

Fig. 30.2 Percentual frequency distribution of catch per group and per night fishing in 50 kg categories for each gear type

the lake. This information was cross-checked by asking local dagaa fishermen about other centres associated with the dagaa fishery. From these preliminary findings, visits were paid to seven landing sites to assess their suitability for a continuous monitoring programme. The following criteria were considered the most important for the programme:

- Accessibility.
- Number of fishermen.
- Types of fishery present.

It was decided to concentrate on two landing sites, Igaragara and Igombe (Fig. 30.1). Both landing sites have beaches with some scattered vegetation. Igaragara is located in a small bay and is more sheltered, while Igombe is more exposed to the wind. Beach seine fishing was conducted at both Igaragara and Igombe, scoop netting was practiced only at Igaragara and lift net fishing was practised only at Igombe.

During the months October–February 1989 a team of three persons paid regular visits to Igaragara and Igombe to collect information on catch and effort. This was carried out by interview (fishermen's estimate approach) and by actual measurements ('real measurement approach', (Bazigos, 1974)).

To initiate the exercise, during the first visit time was taken to explain the objectives of the study to win the cooperation of the fishermen. A number of precautions were taken to avoid potential conflict with the fishermen.

(1)   'Personal' questions, such as boat registration number, name of the fisherman and costs and earnings were avoided. In Tanzania a considerable number of the fishing vessels are not licensed; thus fishermen would not feel that they would be reported to the authorities.
(2)   The team also tried to give a neutral but not disinterested impression.

The results from interviewing were compared with the results from the 'real measurement approach'. This was done by comparing the spreading densities of estimated catches and measured catches on the drying fields. The spreading density was defined as the amount of fresh dagaa per unit weight (kg) spread on an area of $1 \, m^2$. The densities were t-tested against each other. A difference in variance was tested by the F-test for equal variances. The F-statistic was calculated as:

$$F = \textit{the larger } s^2 / \textit{the smaller } s^2$$

where s^2 is the variance of the spreading density in $(kg \, m^{-2})^2$. If the variance according to the 'fisherman's estimate approach' is significantly larger, it is likely that this is caused by less accurate estimates.

The spreading densities can also be compared by t-tests. A significant difference in spreading density would indicate that the fishermen are consistently over- or underestimating their catches.

No significant difference was found for the densities (Table 30.4; t-test; $P > 0.05$). The variances also did not differ significantly ($P > 0.05$). It was concluded therefore that the estimates of the fishermen could be used for total catch estimates.

The unit of effort was evaluated by calculating the correlation coefficient and probability for the catch effort relationship in one lunar month for one type of fishery. The variability in the catch per unit of effort (CPUE) was assessed by calculating the coefficient of variation (CV):

$$CV = 100 \; SD_y / \overline{Y} \hspace{3cm} \text{(Steel and Torrie, 1980)}$$

where SD_y is the standard deviation and $\overline{Y}$ is the average of variable Y. The information on variability can be used to estimate the sample size needed for a certain maximum relative error of the mean using the formula:

$$n = CV^2 / (a/2)^2 \hspace{3cm} \text{(Caddy and Bazigos, 1985)}$$

where n is sample size, a is the percentual maximum relative error (referred to as percentage accuracy of the mean in Caddy & Bazigos (1985) and Bazigos (1974)) at the 5% significance level.

The Fisheries Department also collect data on the dagaa landings at Kirumba (Mwanza), the main trading centre of dagaa in Tanzania, and at the Tanzania Railways Corporation (Mwanza). An estimation was made of the proportion of the catch of dagaa from the Tanzanian part of Lake Victoria which is transported through Kirumba.

30.4   Results

30.4.1   *Definition of the unit of effort*

The catch (C) per unit effort (f) is influenced by the average abundance of dagaa (N) during the same unit time period and the catchability (q) according to:

$$C/f = q \times N \hspace{3cm} \text{(Gulland, 1983)}$$

If the unit of effort is appropriate, q is constant during a short time period. Thus for the same time period a given effort should catch the same proportion from the stock. The unit of effort can be tested by plotting the catch on effort applied for short time periods.

The unit of effort was taken as lamp burning hours, which was assumed to be proportional to lamp fuel consumption. The use of lamp hours as the unit of

Table 30.4 Summary statistics of spreading density of catch data collected following the real measurement approach and fishermen's estimate approach. (Density = average spreading density in kg m^{-2}, SD = Standard deviation, n = number of observations.)

	Density	SD	n
Real measurement approach	2.21	0.87	18
Fishermen's estimate approach	2.34	0.89	64

effort is based on the following assumptions. Since the fish concentrates under the lamp, the total amount of concentrated fish around a set of lamps will be determined by the total number of lamps. If the lamps are set for a longer time, more fish will be attracted to the lamp. The success of the fishing depends largely on the weather conditions and the moon phase. If the weather gets too rough during the night or the lunar cycle makes the lamps less effective in concentrating fish, fishing is ended.

To test the unit of effort, regression analysis was performed on the catch effort data of the beach seine fishery (Table 30.5) and the scoop net and lift net fishery (Table 30.6) at Igaragara and Igombe. A significant and positive correlation between catch and lamp hours would provide evidence for the appropriateness of lamp hours as a unit of effort.

For the beach seine fishery, in all cases except one (Igaragara, November/December), a large part of the variance in the catch was explained by the variance in effort. In the scoop net fishery, variance in the catch was explained partly by the variance in effort only in the December/January sample.

The non-significant catch-effort regression for the lift net data of November/December is probably due to the low variability in effort. The coefficient of variation in effort applied was only 27%. The coefficient of variation of effort was more than 50% in all other cases. Conversely, the non-significant catch-effort regression of the lift net fishery in December/January could have been caused by low dagaa abundance. This is supported by a significantly ($p < 0.05$) higher CPUE in November/December (44 kg lamp hour^{-1}; n = 14) than in December/January (7 kg lamp hour^{-1}). The non-significant catch-effort regression for the beach seine fishery at Igaragara in November/December is probably due to a strong fluctuation in abundance of dagaa which also caused the fishermen to migrate to and from Igaragara within one month.

Table 30.5 Correlation coefficients (r^2) for the regression of catch on effort for beach seine fishery. Observations are pooled per lunar month. n = number of observations, P = P-value (significance of regression).

| | November/December | | | December/January | | |
	n	r^2	P	n	r^2	P
Igaragara	12	0.19	0.14	16	0.56	0.00
Igombe	7	0.59	0.03	6	0.82	0.01

Table 30.6 Correlation coefficients (r^2) for regression of catch on effort for the scoop net and lift net fisheries. Observations are pooled per lunar month. n = number of observations, P = P-value.

| | November/December | | | December/January | | |
	n	r^2	P	n	r^2	P
Scoop net/Igaragara	13	0.03	0.8	10	0.52	0.01
Lift net/Igombe	14	0.07	0.4	10	0.11	0.41

30.4.2   *Effect of place and of type of fishery on CPUE*

The average CPUE and the *CV* of the CPUE of each type of fishery per lunar month per site is presented in Table 30.7. A t-test was used to explain any significant differences in CPUE between the sites and the types of fishery. Only for the beach seining could a location effect be investigated but the relationship was non-significant (P > 0.05). Differences between beach seining and lift netting and between scoop netting and lift netting were significant for November/December (P < 0.01) but not for December/January catches.

During routine data collection, the lift net fishery should be assessed separately. Lift net fishery is more efficient, and it takes place in more off-shore waters. There is no reason to assess scoop net and beach seine fishery separately, since their CPUE did not differ significantly. However, the variation in CPUE was high, which means that the power of the test was low. It can be concluded that stratification of the data collection (Bazigos, 1974) within a limited area of about 10–20 km shoreline could be carried out according to type of fishery.

30.4.3   *Collection of CPUE data at the landing site*

The relationship between sample size and percentual maximum relative error is presented in Fig. 30.3. For *CV* the unweighted mean of all three types of fishery at both landing sites (*CV* = 80.4) was adopted. It can be concluded that to acquire a maximum relative error of 15% a sample size of 115 is needed. For a maximum relative error of 30% and 50% sample sizes of respectively 29 and 10 are needed. These figures refer to an estimate of the CPUE of a single type of fishery in one month at one site.

The problems with routine collection of data stem from the following realizations:

(1)   The fishermen can land their catch at any time of the night and the catch is processed soon after landing. Therefore, planning of a catch assessment survey (CAS) is difficult. By the time the recorder arrives, the catch may already have been landed, and after the catch is spread on the beach, it is no longer possible to weigh it.

Table 30.7   CPUE statistics (X indicates that the fishery is not present at that site).

| | Beach seine | | | Scoop net | | | Lift net | | |
	Mean	CV	n	Mean	CV	n	Mean	CV	n
Nov./Dec.									
Igaragara	5.5	85.6	12	9.5	137.7	13	X	X	X
Igombe	7.8	102.3	7	X	X	X	44.1	69.5	14
Dec./Jan.									
Igaragara	6.4	64.2	16	4.3	47.9	10	X	X	X
Igombe	3.6	53.5	6	X	X	X	6.7	77.5	10

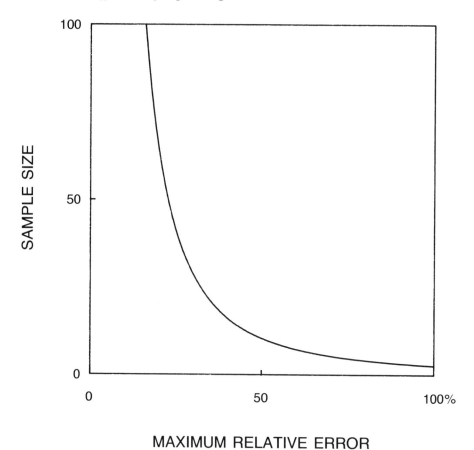

MAXIMUM RELATIVE ERROR

Fig. 30.3 Plot of sample size against percentual maximum relative error according to the formula $Y = 80.4^2/(X/2)^2$, where Y is the number of catch effort observations needed to obtain a maximum relative error of X%

(2) Since the catch does not have to be sold soon after landing there is no need for a centralized marketing site, which makes it difficult to contact a reasonable number of groups.
(3) The fishermen are mobile, changing their fishing sites regularly. This means that many enumerators are needed to cover many fishing and landing sites.

The beach seine catches are the most difficult to record, since in most cases, the catch of one haul is spread directly after landing. Thus, by the time the recorder arrives only the catch from the last haul can be measured.

The catches from the lift net fishermen are less difficult to record and since fisheries are less susceptible to bad weather conditions the time of landing is more predictable. Lift net fishermen operate all night, only landing their catch the following morning. As soon as they land their catch the weight can be assessed using boxes of known capacity. This process can be time consuming if the catch is high.

During this study a team of three researchers managed to measure, at best, the catch of *c.* three fishing groups at Igombe during one morning. These were the catches of two lift net groups and one beach seine group. However, on average less than one measurement per visit could be taken. Thus in a regular catch effort data recording system it is not possible to collect sufficient information by independent measurements and the recorder has to rely on fishermen's estimates which have been proven reliable (Table 30.4). The scoop net fishermen and the beach seine fishermen, who have on average smaller catches than the lift net fishermen, usually relate their catch to number of tins. The lift net fishermen, however, have problems with estimating their catch, especially if it is big, but their catch can be measured more easily by the recorder himself.

30.4.4   *Total catches from Kirumba and Mwanza railway station*

From Kirumba, the harbour of Mwanza, traders go by boat ('dhow') to the main dagaa fishing areas around Ukerewe Island where they collect the dried dagaa from the fishermen. Each dhow can take up to 250 gunny bags of about 30 kg each. At Kirumba the dagaa can be stored for a few hours to several days. The dagaa is then transported by lorry or by train to other regions in Tanzania and to other countries. Before distribution, more dagaa is sometimes added to each gunny bag, so that the weight increases up to 40 kg.

At Kirumba, enumerators from the Fisheries Department have to register the number of bags which are brought to Kirumba. However, the enumerators do not get exact figures from the traders, since the traders have to pay transport fees per gunny bag (80 Tsh/gunny bag). Thus the data from Kirumba is often an underestimate of the real landings.

It is also difficult to estimate which part of the total dagaa catch is transported through Kirumba. There are no other big dagaa trading centres at the Tanzanian side of Lake Victoria, but the fishermen sell part of the fresh dagaa to local people, often in exchange for other food. Also, part of the dried dagaa is bought from the fishermen by small-scale traders, who do not bring their dried dagaa to Kirumba or any other trade centre. From the measured catches of the lift net fishermen an average of 20% of the catch was sold fresh off the vessel. It was not possible to assess by the real measurement approach how much of the catch from the beach seine and scoop net fishermen was sold fresh, since the local people take the fish from the drying field during the day. According to the fishermen they sold *c.* 0.5 to 1 tin of fresh dagaa per day. Assuming an average catch/night/group of 80 kg, this would be about 20% of their catch. Thus, it can be concluded that for all three types of fishery, about 20% is sold fresh. A rough estimate of the part of the catch bought by small-scale traders would be about 20%, and a rough estimate of the total catch transported through minor trading centres would be about 10% of the total catch. Therefore only about 50% of the catch from the Tanzanian waters is going through Kirumba.

Since February 1990 a recorder from the Fisheries Department has been appointed to register the number of gunny bags of dagaa which are transported by train. The Tanzania Railways Corporation (TRC) keeps a record of the number of gunny bags, value, weight and destination of the dried dagaa which is transported. The figures concerning number of gunny bags, weight and destination are quite accurate, since a gunny bag of dagaa is not transported unless it is paid for and registered. It is difficult to assess which part of the dagaa from Kirumba is transported by train. Factors influencing the proportion of the total catch transported by train are the costs and the destination of the dagaa. Transport by train is preferred to lorry because it is cheaper, but not all destinations can be reached by train. However, the information from TRC can provide an indication of the total catch of dagaa.

30.5   A dagaa catch effort recording system (CAS)

The dagaa artisanal fishery is difficult to monitor because of the many small fishing groups spread over many areas, the variable landing time, the movement of the fishermen, the remote landing sites and the diverse fishery. It is therefore recommended that the collection of data on CPUE is based on landing site catches and those data on the total catch collected at Kirumba, the dagaa trading centre, and at the Tanzania Railways Corporation.

The CPUE data should be collected in a CAS at a limited number of indicator areas covering approximately 15 km of shore length and encompassing at least two concentrations of dagaa fishing groups. For example:

- three indicator areas at Ukerewe (dagaa fishery centre);
- one indicator area at Mwanza, 1 at Bukoba and 1 at Musoma (Fig. 30.1).

In this way information is obtained on the fluctuations in CPUE throughout time in the Tanzanian part of Lake Victoria. The CAS should be conducted by well-trained enumerators who operate in groups of two persons, consisting of one enumerator and one assistant. They should be equipped with transport, at least bicycles, and with boxes which can contain about 50 kg of dagaa. At first the CAS should be conducted on a bi-monthly basis. After assessment of the seasonal fluctuations, the frequency of the CAS can be adjusted. To estimate the mean catch with a relative error of not more than 30%, the enumerators must collect at least 30 catch effort estimates for each type of fishery per indicator area per month. This means that in a working month of 22 days of which on about four days no work can be done due to the full moon, they have to collect about five catch-effort measurements each day, which is not unrealistic.

The enumerator should collect the following parameters:

(1)   Date, place.
(2)   Type of fishery (beach seine, scoop net, lift net, other).
(3)   Catch weight per group per night (first in number of tins and then converted to kg fresh weight).

(4) Effort (lamp hours) by which the catch was realized.
(5) Mesh size (stretched mesh, in mm).
(6) Price of fresh dagaa (per tin).

This information can only be obtained by asking the fishermen questions. Only if lift net units are present the enumerator can measure the catch himself by supplying the fishermen with boxes of known volume.

Most of the catch-effort data can be acquired by the enumerator via interviewing the fishermen. Enforcement of restrictive measures, taxing etc. should not be the task of the enumerators, but of other officials. To avoid fishermen's suspicion, it is better not to ask for the name of the fisherman and the registration number of the boat.

The total dagaa catch is estimated best by enumeration at Kirumba (Mwanza). The enumerators should count the number of gunny bags of dagaa which are unloaded from the dagaa transport ships. The total catch from the Tanzanian part of Lake Victoria can be estimated by multiplying the total landings at Kirumba by a factor of two. Conversion from dried dagaa weight to fresh dagaa weight can be done by multiplying by a factor of four (Wanink, 1989). The information from Kirumba can be related to the information from the Tanzania Railways Corporation (Mwanza) on dagaa shipments. Data from the Tanzania Railways Corporation can be collected against little cost and are reliable.

Acknowledgements

We thank Dr F. Witte for his helpful comments on the manuscript. The efforts of Mr M. Brittijn and Mr H. Heijn in preparing Fig. 30.1 is acknowledged. The research for this study was financed by the Netherlands Minister for Development Coöperation.

References

Bazigos, G.P. (1974) The design of fisheries statistical surveys — inland waters. *FAO Fisheries Technical Paper No.* **133**. 122 p.

Bernacsek, G.M. (1986) Kenya, Tanzania and Uganda. Evaluation of statistical services for Lake Victoria fisheries. Mission report. FAO/CIFA/Sub-Committee for the development and management of the fisheries of Lake Victoria. September 1986. 25 pp. + addenda.

Caddy, J.F. and Bazigos, G.P. (1985) Practical guidelines for statistical monitoring of fisheries in manpower limited situations. *FAO Fisheries Technical Paper No.* **257**. 86 pp.

CIFA (1988) Report of the fourth session of the sub-committee for the development and management of the fisheries of Lake Victoria. *FAO Fish. Rep. No.* **338**: 112 pp.

Gulland, J.A. (1983) *Fish stock assessment. A manual of basic methods.* Chichester: John Wiley & Sons. 223 pp.

Nedelec, C.(Ed.) (1975) Catalogue of small-scale fishing gear. West Byfleet: Fishing News Books Ltd. 191 pp.

Okedi, J. (1974) Preliminary observations on *Engraulicypris argenteus* (Pellegrin) 1904 from Lake Victoria. In *Annual report of the East African Freshwater Fisheries Research Organization 1973.* Jinja, Uganda. pp. 39–42.

Okedi, J. (1981) The *Engraulicypris* 'dagaa' fishery of Lake Victoria: with special reference to the southern waters of the lake. In *Proceedings of the workshop of the Kenya Marine and Fisheries Research Institute on Aquatic Resources of Kenya*, 13–19 July 1981. pp. 445–484.

Steel and Torrie (1980) *Principles and procedures of statistics, a biometrical approach*. McGraw-Hill Book Company. 633 p.

Wanink, J.H. (1989) The ecology and fishery of dagaa, *Rastrineobola argentea* (Pellegrin) 1904. In *Fish stocks and fisheries in Lake Victoria*. A handbook to the HEST/TAFIRI & FAO/DANIDA regional seminar, Mwanza January/February 1989. Report of the Haplochromis Ecology Survey Team (HEST) and the Tanzanian Fisheries Research Institute (TAFIRI) no. 53. Leiden, The Netherlands, RUL.

Chapter 31
A pilot sampling survey for monitoring the artisanal Nile perch (*Lates niloticus*) fishery in southern Lake Victoria (East Africa)

W. LIGTVOET *Haplochromis Ecology Survey Team, Leiden, The Netherlands. Present address: Witteveen & Bos, Consulting Engineers, Postbus 233, 7400 AE Deventer, The Netherlands.*

O.C. MKUMBO *Tanzanian Fisheries Research Institute, P.O. Box 475, Mwanza, Tanzania.*

In the early 1960s Nile perch (*Lates niloticus*), a large predator, was introduced into Lake Victoria. Following its establishment and expansion by the end of the 1970s an artisanal Nile perch fishery developed. Currently, Nile perch is the dominant demersal fish in the lake, supporting an important commercial fishery. In order to manage any fishery there is a basic requirement to collect length frequency and catch effort data on a regular basis. For the Lake Victoria Nile perch fishery, these data have been collected at monthly intervals from six landing beaches on the southern part of the lake between 1987 and 1988. The data have been used to describe the types of Nile perch fishery developing and are to be used in establishing a viable sampling programme for future monitoring. In the present study, the different usage and selectivity of three main gear types (gill nets, long lines and beach seines) are described and the unit of fishing effort is discussed. Important trends in the CPUE and modal length of the Nile perch catches are also given.

31.1   Introduction

In the early 1960s, Nile perch (*Lates niloticus*; Centropomidae), a predatory fish, was introduced into Lake Victoria. Following its establishment and rapid expansion, an artisanal fishery had developed by the end of the 1970s (Arunga, 1981; Okemwa, 1979, 1984; Barel *et al.*, 1985). Nowadays this fishery constitutes the most important commercial fishery in the lake (CIFA, 1988; Ligtvoet *et al.*, 1988). Catch data obtained by the statistical data collection services of the riparian states, Kenya, Uganda and Tanzania, estimate the annual yields at between 200 000 and 300 000 tonnes. The validity of these data, however, need verification (Bernacsek, 1986; CIFA, 1988).

For adequate management of this fishery, an increased research effort directed towards describing and monitoring the developments in both the Nile perch stock and its fishery is needed (CIFA, 1985; 1988). Earlier research effort was geared towards describing the trophic dynamics (Ogari, 1985; Hughes, 1986; Ogari & Dadzie, 1988) and population parameters (Acere, 1985; Asila & Ogari, 1987) of the Nile perch. Preliminary descriptions of the changes taking place in the fisheries are found in Goudswaard and Ligtvoet (1988) and Ligtvoet *et al.* (1988). Reynolds and Greboval (1988) presented an extensive overview of the socio-economic changes due to the impact of the Nile perch upon the fisheries of Lake Victoria.

In Tanzania, a fishery statistics collection system exists under the authority of the Fishery Division. Enumerators record the daily catch per species, per canoe and per type of gear for as many days per month as possible. However, on an annual basis only a 'mean catch per canoe' is calculated, irrespective of gear type. Due to various problems, such as lack of incentives for enumerators, lack of transport and equipment, the reliability of official statistics are disputed (Bernacsek, 1986; CIFA, 1988). Length measurements are not incorporated in the data collection system, so changes in the exploitation pattern of the fisheries cannot be directly detected from changes in the mean size of the Nile perch in the catch.

In the design of an efficient sampling system, the basic characteristics of the respective fisheries, particularly variations in the catches, need to be known (cf. Bazigos, 1974; Caddy & Bazigos, 1985). A sampling programme, covering a limited area of the Tanzanian part of Lake Victoria, was set up which had the following objectives:

(1) To provide a basic description of the types of Nile perch fisheries.
(2) To collect preliminary data on catch and effort (CPUE) and exploitation pattern of the different fisheries.
(3) To provide basic information required for designing an efficient monitoring system for the Nile perch fishery.

The present paper describes the data collection scheme used during 1987 and 1988 for the three major Nile perch fisheries (gill netting, long lining, beach seining) and illustrates the value of incorporating length measurements of the fish into the sampling routine.

31.2   Methods

The Tanzanian shoreline of Lake Victoria covers *c.* 2900 km (Fig. 31.1). Landings are made at numerous (possibly hundreds) of relatively small sites servicing both inshore and off-shore fishing grounds. The Mwanza part of the lake is the most densely fished area in Tanzanian waters. For this study six landing sites were chosen in this region for regular monitoring. These were distributed over *c.* 100 km shoreline. The fishing grounds associated with these landing sites ranged from shallow inshore waters (depth < 4 m) inside the Mwanza Gulf, to off-shore waters up to 40 m deep. The basic characteristics of the landing sites sampled are given in Table 31.1. The inshore fishing grounds of the gill net and long line fishery comprise shallow and sheltered waters inside the Mwanza Gulf, an elongated

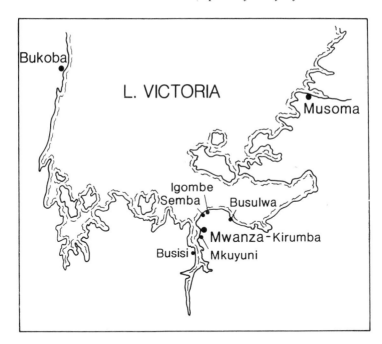

Fig. 31.1 The southern part of Lake Victoria with Mwanza and the landing sites indicated

Table 31.1 Summary of basic characteristics of six sites in the Mwanza region monitored over the years 1987–88

Landing site	Fishing ground	Average no. of canoes	Target species	Dominant gear
Busisi	inshore depth −4 m	5–10	Nile perch Tilapia lungfish	gill nets beach seine long-line
Mkuyuni	inshore depth −15 m	10–15	Nile perch	gill nets
Kirumba	in- and off-shore depth 10–40 m	10–20	Nile perch	gill nets
Semba	inshore depth −ca. 10 m	1	Nile perch	beach seine
Igombe	off-shore depth −40 m	10–35	Nile perch	gill nets
Busulwa	off-shore depth −30 m	10–30	Nile perch	gill nets

bay. The inshore fishing ground of the beach seine at Semba is also shallow but exposed and along the open waters of the lake.

At each landing site, defined as a 'Primary Sampling Unit' (Bazigos, 1974), 5–10 boats were randomly sampled on each occasion. The 'sampling unit' was one boat (canoe). This concurs with the 'Fishing Economic Unit' consisting of a fishing craft, fishing gear and fishermen of Bazigos (1974). The sampling was carried out monthly by a team of two persons.

Information with respect to each sampling unit was obtained from direct observation and from questionnaires. The total number of fish from each sampling unit was counted and Nile perch lengths (TL) to the nearest cm were also obtained for the whole catch or a sub-sample thereof. Weight of the catch was estimated using the length-weight relationship $W = 0.000006 \times L^{3.17}$ (Ligtvoet & Mkumbo, 1990). The questionnaire sought information on: number of fishermen; type of fishing ground; frequency of fishing (daily or seasonal); type, characteristics and number of gears used, as well as the relative mobility of the canoe (paddles, sails, motored).

The variation in the catches is expressed as the 'coefficient of variation' (CV). The 'maximum relative error' (MRE) is used as a measure of accuracy; this expression MRE (Sparre *et al.*, 1989) is equivalent to the 'accuracy of the mean' (Caddy & Bazigos, 1985).

31.3   Description of the artisanal Nile perch fishery

31.3.1   *General*

Three types of Nile perch fishery, gill netting, long lining and beach seining, are evident. The fishermen generally use planked canoes of up to *c.* 10 m long, mostly powered by paddles or sails. The gill net fishery is the most important at all fishing grounds (inshore and off-shore). Long lining is largely restricted to the shallow inshore waters (e.g. Busisi; Table 31.1) where other commercial fish species (e.g. lungfish, *Protopterus aethiopicus*) are also caught. In these areas about 40% of the recorded landings were from long line fishermen. At the other landing sites long lining played an insignificant role. Beach seining for Nile perch is restricted to inshore waters on gentle slopes. The relative importance of this fishery was difficult to assess, because landings were normally sold on the spot. They were not brought to the same sites where the gill net and long line fishermen landed their catches. In the vicinity of Igombe, three beach seines are known to operate.

31.3.2   *Gill netting*

A gill net fishing unit consists of one canoe operated by 3–5 fishermen. The number of gill nets per canoe varies between 10 and 100 plus. At Igombe, the majority of the canoes were found to operate with 30–50 nets (mean: 40), while in the shallower areas fleets of 5–40 nets (mean: 25) were common. The gill nets were not seen to have standard dimensions; reported lengths varied between 15 and 90 m, with lengths between 30 and 60 m being the most common. The nets were normally set in the late afternoon and hauled at dawn the next morning. Fishermen operating in off-shore fishing areas often collect only the fish and leave the nets set; they only bring the nets ashore when they need repairing.

Due to shortage of ordinary netting materials, other types of material, e.g. split nylon ropes, polyethylene fibres (obtained from fertilizer bags) and twine used in the manufacture of car tyres were common substitutes for making gill nets (Ligtvoet *et al.*, 1988).

The gill net fishery mainly uses the 7−8 inch mesh size which exploits Nile perch in the total length groups of 60−80 cm (Fig. 31.2(a)). However, in the early exploitation of the fishery, meshes of 9−10 inches and occasionally 12−16 inches were also used (Fig. 31.3). Nile perch strongly dominates the catch, contributing 98% of the number of fish caught. The length frequency distributions of the gill net landed Nile perch from Igombe in 1987 and 1988 are depicted in Fig. 31.2(a). There was slight decrease in modal length of the captured Nile perch from 1987 to 1988. This coincides with a shift in the mesh sizes used (Fig. 31.3): the proportion of fishermen using gill nets with 9-inch mesh size declined in 1988, whilst the proportion using 7-inch mesh nets increased. The small Nile perches (30−55 cm) recorded in 1987 were caught by fishermen who had not turned to the Nile perch fishery and still operated with 4- and 5-inch mesh sized gill nets (Fig. 31.3). Formerly, the latter mesh sizes were used for fishing the catfishes *Clarias gariepinus* and *Bagrus docmac*; however, in 1987 Nile perch was the main species caught.

The average catches per canoe between January 1987 and October 1988 at Igombe (off-shore fishing grounds) and Busisi (inshore fishing ground) are shown in Fig. 31.4. The average landings per canoe were also lower for the inshore fishery (Fig. 31.5; cf. Table 31.2). At both landing sites, the catches fluctuate over the years but reveal no particular trend (analysis of variance for Igombe: df = 17, F = 1.55, P = 0.09; for Busisi: df = 18, F = 0.78, P = 0.7).

31.3.3   *Long lining*

A long line fishing unit comprises one boat with 2−4 fishermen. Usually, 500−1200 hooks are used. Live haplochromine cichlids, obtained by angling on rocky shores, fresh dead dagaa (*Rastrineobola argentea*) or fresh pieces of Nile perch are normally used as baits. Long lines are set either near the bottom or in mid-water in the late afternoon and hauled the following morning. Long line fishermen also catch lungfish *Protopterus* (inshore waters), and the catfishes *Bagrus* and *Clarias* (off-shore fishing grounds). At Busisi, these species comprised not more than 25% of the total catch by number.

Long lines exploit a broad variation in length but size group 40−80 cm TL contributed most to the catch (Fig. 31.2(b)). In the long line catches an increase in modal length from 1987 to 1988 was found. This increase can probably be attributed to the growth of the large 1987 cohort. This follows from the assumption that long lines are less selective and the increase in size composition is reflected in the catches.

The average catch landed per canoe over 1987 and 1988 amounted to 60−70 kg (Table 31.2). When compared with the gill net fisheries (Fig. 31.4), the seasonal trends in long line catches were at an intermediate level. Catch fluctuations were

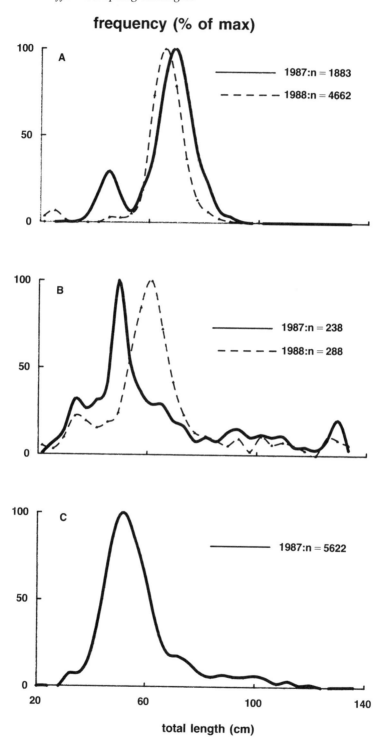

Fig. 31.2 Smoothed length frequency distributions of the catch of the three major Nile perch fisheries. A: gill net fishery (Igombe); B: long line fishery (Busisi); C: beach seine fishery (Semba). The gill net and long line fishery distributions for 1987 and 1988 (broken line) are given separately. The distribution of the beach seine represents the catch in 1987

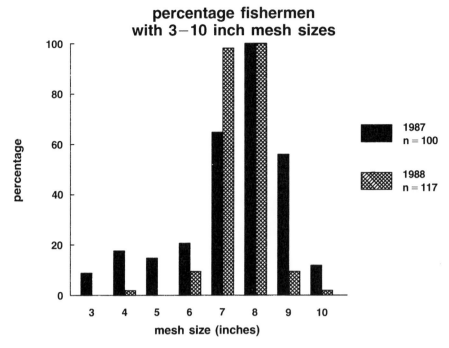

Fig. 31.3 Percentage of fishermen using gill nets with mesh sizes from 3–10 inches in the Nile perch fishery in 1987 (cross-hatched) and 1988 (solid bars)

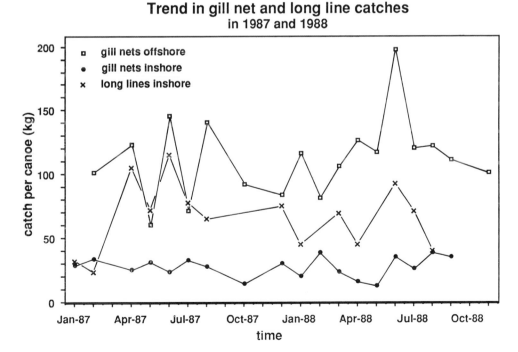

Fig. 31.4 The mean catch (kg) per canoe in the Nile perch fishery over 1987 and 1988. Bold line: off-shore gill net fishery; normal line: inshore gill net fishery; broken line: inshore long line fishery. For convenience, the large standard deviations (cf. table 31.2) around the means are not indicated

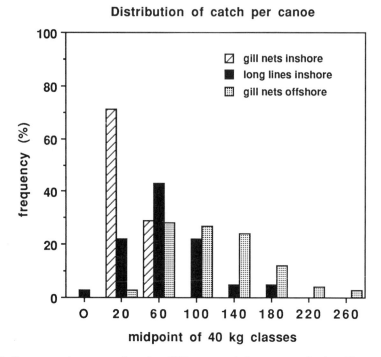

Fig. 31.5 Percentage frequency of catches (40 kg categories) per canoe in the gill net fishery of inshore waters (hatched bars), gill net fishery of off-shore waters (dotted bars) and long line fishery of inshore waters (black bars). Data for 1987 and 1988 are combined

Table 31.2 Coefficient of variation (CV in %) and maximum relative error (MRE; 95% probability) of estimates of mean total length (TL) and catch weight (CW) in Nile perch catches recorded from different types of fisheries in 1987 and 1988.

Fishery	Fishing ground	year	n	TL mean	σ	n	CW mean	σ	CV TL	CV CW	MRE TL	MRE CW
Gill net	offshore	1987	1883	70.5	6.7	41	107	55	10	51	<0.01	0.16
		1988	4662	67.6	5.8	71	118	50	9	42	<0.01	0.10
Gill net	inshore	1987	202	68.3	7.1	31	31	21	10	69	0.01	0.25
		1988	407	63.7	8.5	28	32	22	13	68	0.01	0.26
Long line	inshore	1987	238	65.2	23.5	20	69	44	36	64	0.05	0.29
		1988	288	71.0	25.6	17	60	40	36	67	0.04	0.34
Beach seine	inshore open water	1987/ 1988	6940	56.2	13.9	12	1140	582	25	51	<0.01	0.32

again evident but, like the gill net fisheries, no obvious trend was evident (df = 14, F = 1.23, P = 0.3).

31.3.4   *Beach seining*

A beach seine fishing unit consists of one or two wooden canoes, a large beach seine, the owner, a number of permanently employed fishermen and a varying number of persons assisting in pulling the net. The total number of fishermen operating a beach seine may be 30−40. The beach seines are of the type with a bag (or codend) mounted between the two identical wings. Poles are used at either end of the net to keep the wing netting vertically open. Mesh sizes used in the wings range from 3−6 inches; the codend mesh size varies between 1−1.5 inches. The beach seines are between 800 and 1000 m in length and have pulling ropes of c. 800 m. One haul per night is usual, the time for a single haul varying between 8 and 10 hours.

The total length of Nile perch captured varied from 25 to over 140 cm TL, with the majority falling between 40 and 70 cm (Fig. 31.2(c)). The catch per haul during the period March 1987 to June 1988 varied between 400 and 2000 kg with an average of 1140 kg ($\sigma = 580$ kg).

31.3.5   *Variation in the Nile perch catches*

Since no evidence was found for differences between the monthly catches over the two years sampled, the recorded landings were combined for each year and fishing method. In Table 31.2, the annual variation in the catches is compared for the gill net fishery from Igombe (off-shore fishing ground), the gill net and long line fisheries of Busisi (inshore fishing grounds) and the beach seine operating near Igombe. As might be expected, for the mean length the CV (averaging *c.* 10%) and MRE (*c.* 0.01) in the comparatively selective gill net fishery are orders of magnitude lower than in the long line and beach seine fisheries.

Overall, the coefficient of variation (CV) and the maximum relative error (MRE) for the estimate of average total length are low compared with those for the estimates of the catch weight. The number of canoes that need to be sampled, therefore, has to be based on the variation in the catch weight. The realized sampling intensity in the study gives an MRE of 0.25−0.35 with respect to the estimates of average catch in the littoral gill net, long line and beach seine fisheries and a lower range in the off-shore gill net fishery (MRE 0.1−0.16). The required sample sizes for the three types of fisheries for various levels of MRE are given in Table 31.3.

31.4   **Discussion**

In gill net fisheries, the unit of effort is usually the total surface area of the net fleet operated per fishing unit (Gulland, 1983). The number of nets is only a proper index of effort when fishermen use gill nets of standard dimensions. In this study, the fishing economic sampling unit, one canoe, had to be taken as unit of effort. Preliminary analysis showed no relationship between the catch per canoe and the number of nets per canoe, indicating that the variation in the gill net catches was not primarily determined by the number of nets. This seems to be supported by the fact that the variation in the catches landed by individual canoes was high (CV = 42%). This could be attributed to an irregular movement of the Nile perch stock between shallow and deeper water as noted in the Mwanza area

Table 31.3 Maximum relative error and the required number of canoes to be sampled in estimations of mean catch weight with respect to the different types of the Nile perch fishery. The figures are calculated using the mean values over 1987 and 1988 by type of fishery.

Fishery	Fishing ground	0.05	0.1	0.15	0.2	0.25	0.5
				Maximum relative error			
Gill nets	offshore	346	86	38	22	14	4
Gill nets	inshore	776	194	86	49	31	8
Long line	inshore	755	189	84	47	30	8
Beach seine	inshore	495	123	55	31	20	5

(P.C. Goudswaard *pers. comm.*; Ligtvoet & Mkumbo, 1990). This results in substantial fluctuations in the exploitable biomass of Nile perch at the different fishing grounds and complicates the establishment of an adequate unit of effort. Other factors that hamper the determination of a possible relationship between catch and number of nets relate to inaccurate information from the fishermen with respect to the number of nets and the fact that most of the nets are home-made and not of standard dimensions.

CPUE depends on the abundance of the target species. If well chosen, and the catchability coefficient is presumed to be constant, the fluctuations in the CPUE could reflect the stock abundance. To this end trawl surveys in the Mwanza area revealed a substantial lower exploitable biomass of Nile perch in the shallow waters (P.C. Goudswaard *pers. comm.*). The lesser abundance of Nile perch and the lower effort per canoe in terms of number of nets in inshore waters (25 versus 40 nets per canoe off-shore) may indicate why the catches in this gill net fishery are lower than the off-shore equivalent (Table 31.2, Figs. 31.4 and 31.5).

When overexploitation occurs, the first signs are usually decreasing mean individual size in the catch to which the fishermen often react by reducing their mesh size. In this study a slight shift in the modal length of the gill net catches was noted (Fig. 31.2a) and this was attributed to the use of smaller mesh sizes (Fig. 31.3). Beside the use of smaller mesh nets, overfishing could have crept in. Trawl surveys in the area indicated a substantial decrease in the exploitable biomass of Nile perch in the early part of 1990 (P.C. Goudswaard, *pers. comm.*). If the trend as found persists over the years, this supports the view that overfishing could be taking place.

The sampling programme revealed significant differences between the CPUE of the three major Nile perch fisheries. This implies the monitoring of the CPUE between different fisheries should be carried out separately. The annual average catch per canoe, which is the statistic used by the Tanzanian Fishery Division and which does not take into account gear type and fishing area is not, therefore, likely to be representative of the trends in the fishery.

Furthermore, data on length frequency distribution of the catch are not provided. Incorporation of length measurements in the data collection system of the Fisheries

Division is strongly recommended as these data could easily be collected at the same time as catch data.

In situations where resources are limited, CPUE and length frequency data from a small sector of the fishing fleet are extremely valuable (Shepperd, 1988). The CPUE data preferably should be obtained from each different type of fleet separately. The data in Table 31.3 may provide the basic information needed in designing an effective beach sampling programme covering the different Nile perch fisheries. It should be realized that the sample sizes given indicate the sampling intensity required at one landing site over one year. In addition, catches from the minimum number of canoes required (worked out by the MRE method) should be collected regularly over the year to give a representative annual pattern.

Acknowledgements

Thanks are due to Drs W.L.T. van Densen and F. Witte for their valuable comments on the manuscript, and Dr M.A.M. Machiels, Mr M. Brittijn and Mr H. Heijn for preparing the figures. The research was financed by the Netherlands Minister for Development Cooperation.

References

Acere, T.O. (1985) Observations on the biology, age and growth and sexuality of Nile perch *Lates niloticus* (linne) and the growth of its fisheries in northern waters of Lake Victoria. In Report of the Third Session of the Sub-Committee for the Development and Management of the Fisheries of Lake Victoria. Jinja, Uganda, 4–5 October 1984. *FAO Fish. Rep.* **335**: 42–61.

Arunga, J. (1981) A case study of the Lake Victoria Nile perch *Lates niloticus* fishery. In Proc. Workshop Kenya Marine and Fishery Research Institute. Mombassa, Kenya, July 1981. Nairobi, Kenya National Academy for the Advancement of Arts and Science: 165–183.

Asila, A.A. and Ogari, J. (1987) Growth parameters and mortality rates of Nile perch (*Lates niloticus*) estimated from length-frequency data in the Nyanza Gulf (Lake Victoria). In *Contributions to Tropical Fisheries Biology* (Ed. by S. Venema, J. Moller Christensen and D. Pauly): Papers by the Participants of FAO/DANIDA Follow-up Training Courses, *FAO Fish. Rep.* **389**: 272–287.

Barel, C.D.N., Dorit, R., Greenwood, P.H., Fryer, G., Hughes, N., Jackson, P.B.N., Kawanabe, H., Lowe–McConnel, R.H., Nagoshi, N., Ribbink, A.J., Trewawas, E., Witte, F. and Yamaoka, K. (1985) Destruction of fisheries in Africa's lakes. *Nature* **315**: 19–20.

Bazigos, G.P. (1974) The design of fisheries statistical surveys – inland waters. *FAO Fish. Tech. Pap.* **133**: 122 p.

Bernacsek, G.M. (1986) Kenya, Tanzania and Uganda evaluation of statistical services for Lake Victoria fisheries. Mission Report. Document of the Fourth Session of the Sub-Committee for the Development and Management of the Fisheries of Lake Victoria. CIFA:DM/LV/87/6: 62 p.

Caddy, J.F. and Bazigos, G.P. (1985) Practical guidelines for statistical monitoring of fisheries in manpower limited situations. *FAO Fish. Tech. Pap.* **257**: 86 p.

CIFA, (1985) Report of the Third Session of the Sub-Committee for the Development and Management of the Fisheries of Lake Victoria. Jinja, Uganda, 4–5 October 1984. *FAO Fish. Rep.* **335**: 145 p.

CIFA, (1988) Report of the Fourth Session of the Sub-Committee for the Development and Management of the Fisheries of Lake Victoria. Kisumu, Kenya, 6–10 April 1987. *FAO Fish. Rep.* **338**: 112 p.

Goudswaard, P.C. and Ligtvoet, W. (1988) Recent developments in the fishery for haplochromines (Pisces: Cichlidae) and Nile perch, *Lates niloticus* (L.), (Pisces: Centropomidae) in Lake Victoria. In Report of the Fourth Session of the Sub-Committee for the Development and Management of the Fisheries of Lake Victoria. Kisumu, Kenya, 6–10 April 1987. *FAO Fish. Rep.* **338**: 101–112.

Gulland, J.A. (1983) *Fish stock assessment. A manual of basic methods*. Chichester: John Wiley & Sons 223 p.

Hughes, N.F. (1986) Changes in the feeding biology of the Nile perch, *Lates niloticus* (L.) (Pisces: Centropomidae), in Lake Victoria, East Africa, since its introduction in 1960 and its impact on the native fish community of the Nyanza Gulf. *J. Fish. Biol.* **19**: 541–548.

Ligtvoet, W., Chande, A.I. and Mosille, O.I.I.W. (1988) Preliminary description of the artisanal Nile perch (*Lates niloticus*) fishery in Southern Lake Victoria. Report of the Fourth Session of the Sub-Committee for the Development and Management of the Fisheries of Lake Victoria. Kisumu, Kenya, 6–10 April 1987. *FAO Fish. Rep.* **338**: 72–85.

Ligtvoet, W. and Mkumbo, O.C. (1990) Synopsis of ecological and fishery research on Nile perch (*Lates niloticus*) in Lake Victoria, conducted by HEST/TAFIRI. Report of the Haplochromis Ecology Survey Team (HEST) and the Tanzanian Fisheries Research Institute (TAFIRI) no. 50: Leiden, The Netherlands, 40 p.

Okemwa, E.N. (1979) Changes in fish species composition of Nyanza Gulf, Lake Victoria. In Proc. Workshop Kenya Marine and Fishery Research Institute. Mombassa, Kenya, July 1981. Nairobi, Kenya National Academy for the Advancement of Arts and Science: 138–156.

Okemwa, E.N. (1984) Potential fishery of Nile perch, *Lates niloticus* (Pisces: Centropomidae) in the Nyanza Gulf of Lake Victoria, East Africa. *Hydrobiologia* **108**: 121–126.

Reynolds, J.E. and Greboval, D.F. (1988) Socio-economic effects of the evolution of the Nile perch fisheries in Lake Victoria: a preliminary assessment. FAO report RAF/87/099/TECH/02: 167 p.

Shepperd, J.G. (1988) Fish stock assessment and their data requirements. In *Fish population dynamics*. (Ed. by J.A. Gulland). Chichester: John Wiley & Sons. 35–62.

Sparre, P., Ursin E. and Venema, S.C. (1989) Introduction to tropical fish stock assessment. Part 1. Manual. *FAO Fish. Tech. Pap.* **306.1**: 337 p.

Chapter 32
Catch effort sampling data involving indigenous gears in Bukit Merah Reservoir, Malaysia: a re-evaluation

YAP SIAW-YANG *University of Malaya, 59100 Kuala Lumpur, Malaysia*

Analysis of the catch data from the Bukit Merah Reservoir, Malaysia, reveals the multi-species nature of the fisheries. Fishing takes place every-day, except on festive days, on the reservoir. Variability in numbers of fishermen and canoes operating varies little thus negating the use of catch per unit effort (CPUE) in terms of catch per fisherman and catch per canoe for the Bukit Merah Reservoir. Effort was therefore expressed in terms of net/gear-pieces employed.

Concomitant with the multi-species nature, a rich variety of traditional fishing gears are used to exploit the fish community. This multi-gear situation posed a problem for estimating and standardizing the fishing effort. Thus the three different gear-groups (gill net/cast net, bamboo traps, and stake/troll/stow net) were evaluated separately. Earlier work had shown that the catchability at Bukit Merah Reservoir varied with season. Therefore, the sampling data (catches and unit scores of gear-groups) were stratified into wet and dry seasons. This enabled a comparison of the variability in catch-effort estimates using both the unstratified and stratified sampling data.

Although there are few similarities between these expressions of effort for the non-mechanized gears, standardized catch per gill net/cast net against the number of nets per month showed a declining trend, while the catch per stake/troll/stow net with the numbers of stake/troll/stow net also showed a slight decline. The catch per trap appeared to change insignificantly with an increase in the number of traps per month. Therefore, catch per bamboo trap was selected as the best measure of catch fishing effort.

Using this assessment of fishing effort, total annual yield from Bukit Merah Reservoir was estimated at about $38\,kg\,ha^{-1}\,yr^{-1}$, ranging from 15.69 to $60.28\,kg\,ha^{-1}\,yr^{-1}$.

32.1   Introduction

A subsistence fishery has existed at Bukit Merah Reservoir, Malaysia, and its associated swamps for approximately 80 years. An early census on the size of the fishery, the methods of capture, composition and extent of the catch, weight and price of each fish species sold to the dealers, and the fishing effort (Yap, 1982), indicted an underexploited situation. In addition Yap (1983a and b) has described the various aspects of these multi-species and multi-gear fisheries. In this paper, the catch effort sampling strategies correlated to the diversity of gears are discussed and re-evaluated, with the aim of identifying an appropriate measure of fishing effort which best reflects fishing mortality.

32.2   Materials and methods

32.2.1   *Study area and fisheries*

Bukit Merah Reservoir (Fig. 32.1), Malaysia (5° 01 N, 100° 40 E) is a lowland impoundment. The reservoir is 9.65 km wide, 4.83 km long and with a depth ranging from 1.52–9.00 m. Since its construction in 1904, the perimeter bund of the reservoir has been further raised in 1964 to increase storage capacity: the average storage capacity is now $1852.66 \times 10^6 \, m^3$. The reservoir has an approximate surface area of 3495 ha, and is bisected by a railway track running east-west with the northern region covering an area of about 1036 ha, and the southern region 2459 ha. The shore slope is 1.48:1. A spillway was constructed to the west of the reservoir leading to the Kurau River. The water intake point for irrigation was diverted into a canal named Selinsing Canal (Fig. 32.1).

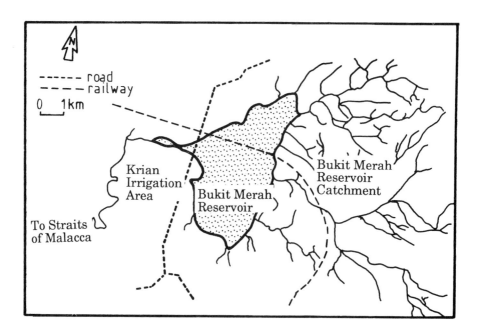

Fig. 32.1 Map showing the region around Bukit Merah reservoir

A small township, Bukit Merah Town, has become established on the north-western shore of the reservoir. It has about 150 families with an average of 10 individuals per family. Fishing by these local residents contributes about M\$4.50 per day (US\$1.00 = M\$2.30) or about 37% of the annual income of fishermen, with rice cultivation and rubber tapping contributing the rest. The reservoir fisheries have been examined by Yap (1983a, 1983b, 1984a, 1984b, 1987, 1988a and 1988b) who recorded low fish yields of $37\,kg\,ha^{-1}\,yr^{-1}$.

Bukit Merah Reservoir and its associated swamps hold a rich fish fauna, comprising 48 species (representing 14 families; Yap, 1983b). The ichthyofauna is mainly represented by indigenous species of cyprinid and anabantid complexes. All these species are of commercial importance in Bukit Merah Reservoir, regardless of their size (Yap, 1984a, 1987 and 1988a). With the exception of an introduced carp *Puntius gonionotus*, exotic species have neither established successfully, nor contributed significantly to the exploitable stocks.

32.2.2 *Unit scores of gear-type and measure of catch-effort*

The present study is based upon data collected between September 1978 and February 1981. A total of 904 observation days were recorded during the 30-month period. Catch and effort statistics were collected at three major trading sites during transactions of daily catch immediately after fishing (at 6.00−9.00 a.m.) and 5.00−6.30 p.m. hours daily, except on festive days). The trading sites are at reservoir landing sheds, on the roadside when the spillway gates are opened, and at the outflow leading to the diversion canal.

The main gears used by the local fishermen in the Bukit Merah Reservoir are stakes, trolling, stow nets, bamboo traps, gill nets and cast nets (Yap 1983a). The stakes are used in smooth-flowing high water, sometimes baited with palm-oil seeds or rubber seeds in the trap. The troll specializes in catching predatory fish with hooks baited with animal feed. The stow net is operated only by traditional or average income fishermen at high water and is losing its popularity. The non-exit bamboo traps are operated mainly among submerged macrophytes in the reservoir and near the shore, at swampy pools in running waters and in any shallow water. The trap fishermen have developed special tactics for scattering these individualistic traps in rows along the bank among submerged macrophytes. The gill nets are approximately 25 m in length and 1−2.5 m in depth, with a stretched mesh size of 50, 100 or 200 mm. This method of fishing has gained in popularity since its introduction in the late 1960s. The cast nets augment catches of other major gears or are used by anglers (besides hook-and-line) to catch schooling species. All gears are set at dusk and/or dawn and the fish are collected the next day or after 12 hours. Usually more than one type of gear is used, with one gear complementing the other.

An initial one-month evaluation of canoes/boats and fishermen was carried out by one to three enumerators. These were local residents who have been involved in fishing and who were personally known to fishermen in the Bukit Merah

Township. A form was distributed asking for the following information: date, name of fishermen, types and pieces/units of gears used, number of canoes/boats owned, landing stations, numbers of hours/days at fishing ground each trip, manpower working from each canoe. The number of fishermen/canoes was determined in this preliminary survey. A total of 93 regular reservoir fishermen and 47 running-water (inflow, spillway and diversion canal) fishermen, operating a total of 138 canoes, were found. This suggests that each canoe is usually operated by one man. Details of the number of gear-types, pieces of nets and traps per canoe, the species composition, and the weight of the catch by the 138 canoes/fishermen were recorded. In addition, the manner and combination of gear-type put into operation by randomly selected fishermen were determined by inspection and interview.

Since the number of canoes and fishermen were constant, the catch per canoe per day and the catch per fisherman per day were not attempted. It was assumed that catch variability changes with the gear-type and gear-piece per day. Hence catch per unit effort (CPUE) was formulated as catch per gear piece per month and standardized for different catch levels by expressing the relative CPUE as a percentage of the maximum in each stratum of the data.

In view of the diversity of gear, the number of pieces of each gear-type was scored separately. Three groups of gear-type were obvious (Yap, 1983a) and were differentially preferred for use in the wet and dry seasons. The data for the gears were thus combined into three gear groups: (1) gill nets and cast nets, (2) bamboo traps, and (3) stakes, trolling and stow nets. To consider the effects of seasonal variations in reservoir water level (m) and rainfall (mm) on catch, the CPUE data were stratified into wet and dry seasons and standardized into relative CPUE.

The catch efficiency of each gear-group in each season was analysed by plotting the number of gear-pieces employed by the fishermen during the study period against the respective relative CPUE (%). Fluctuations in the catch were compared to select the most appropriate measure of fishing effort which showed the least change and the most consistent relationship with gear-pieces per month. Such a measure should minimize the data required yet still be applicable to the management of the subsistence fisheries of the Bukit Merah Reservoir. The other criterion defined in this selection was that the number of gears set should be directly related to the catching power experienced by the exploitable fish stock.

32.3   Results

32.3.1   *Catches and score of fishing gears*

Peak catches for the Bukit Merah Reservoir fisheries were during the wet seasons (north-east and south-west monsoon seasons). Catches of 5.28 kg per unit effort (based on the average of overall gears) were recorded in October, 1979, coincident with the major north-east monsoon (July–November; Fig. 32.2). These five wet months were treated as replicates, and compared with the dry period months

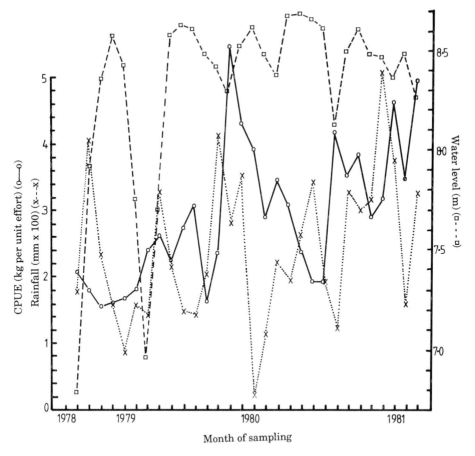

Fig. 32.2 Fluctuations in CPUE in relation to monthly water level (m) and rainfall (mm)

(January–June excluding a minor wet month of April). December, when the rainfall was receding, was omitted.

The proportional contribution of trophic-bonded fish species to the catch during the study period (September 1978–February 1981; Yap, 1988a) was: 42% detritivore (*Labiobarbus festiva, Cyclocheilichthys apogon, Puntius schwanenfeldii* and *Puntius strigatus*), 30% herbivore (*Osteochilus hasselti, Puntius gonionotus* and *Trichogaster pectoralis*), 19% piscivore (*Oxyeleotris marmorata, Clarias batrachus* and *Clarias macrophalus, Ophicephalus striatus, Ophicephalus micropeltes* and *Ophicephalus lucius, Hampala macrolepidota* and *Wallago leerii*) and 9% invertebrate-feeders (*Mystus planicep, Mystus nemurus, Mystus nigriceps, Pristolepis fasciatus, Notopterus notopterus Notopterus chitala*). Species dominance was not obvious and therefore management based on single-species methods was not satisfactory for such multi-species and multi-gear fisheries.

Among the 138 canoes (3.66–5.49 m long) inspected, 95% of the canoes used gill nets and 98% bamboo traps. Other gears commonly used included stakes

(65%), cast nets (45%) and trolling (30%). Stow nets were employed by only 10% of the canoes. Table 32.1 presents the monthly catches of the five main gear types used during the study period, based on 28–31 fishing days per month. Traps, stakes, gill nets and casting nets were predominantly used. Disregarding the unequal efficiencies of gears, the overall catch per month ranged from 4.35 tonnes (December 1978) to 33.19 tonnes (July 1980).

32.3.2   *Catch per unit effort (CPUE)*

Unstratified/unadjusted CPUE values

Based on unstratified sampling data the catch efficiencies of the three gear groups are shown in Fig. 32.3. The catch per gill net/cast net declines as the number of

Table 32.1 Fishing days, monthly catch, unit score of the main gears at Bukit Merah Reservoir, Malaysia (source − Yap, 1982)

Month	Fishing days	Catch (per month)	Gill net/ cast net	Bamboo trap	Stake	Troll	Stow net
				Unit scores (×1000 units)			
Sept 1978	30	8.150	6.750	12.570	3.000	0.900	0.090
Oct	30	7.544	7.920	12.060	3.000	0.900	0.090
Nov	30	5.746	6.510	10.980	3.000	0.840	0.090
Dec	31	4.348	3.255	9.021	2.201	1.085	0.062
Jan 1979	29	5.872	2.001	14.964	2.320	0.725	0.029
Feb	28	7.021	3.108	11.928	1.540	0.840	−
Mar	30	11.162	2.040	15.420	1.950	0.900	−
Apr	30	15.320	2.400	19.620	2.700	0.750	−
May	31	8.948	1.488	12.865	2.480	0.868	−
June	30	7.549	3.000	9.150	1.800	0.930	0.030
July	31	9.249	2.914	10.075	2.790	0.310	0.030
Aug	31	7.063	17.081	4.588	3.100	0.310	0.093
Sept	30	10.827	13.530	8.070	3.300	0.150	0.090
Oct	30	13.012	2.280	6.270	3.450	0.240	0.090
Nov	30	12.564	5.160	6.240	3.450	0.300	0.090
Dec	31	20.789	3.503	19.995	3.720	0.713	0.093
Jan 1980	30	10.683	6.000	10.200	2.700	0.900	0.060
Feb	29	15.431	5.452	15.486	1.740	0.928	0.029
Mar	31	19.486	3.844	20.398	1.860	0.992	−
Apr	30	11.446	2.130	16.020	1.950	0.900	−
May	31	12.891	7.502	19.840	1.643	0.900	−
June	30	13.858	5.940	23.250	1.500	1.200	−
July	31	33.186	17.068	21.064	3.000	0.600	0.030
Aug	29	17.345	9.249	13.050	3.000	0.750	0.087
Sept	30	17.099	6.550	13.500	2.520	0.747	0.090
Oct	30	16.094	8.150	18.000	2.710	0.750	0.090
Nov	29	14.563	7.860	13.047	2.610	0.740	0.087
Dec	31	15.900	0.550	13.950	2.705	0.650	0.093
Jan 1981	29	14.270	5.255	13.047	2.620	0.750	0.090
Feb	28	19.542	5.550	12.600	1.512	0.748	0.084

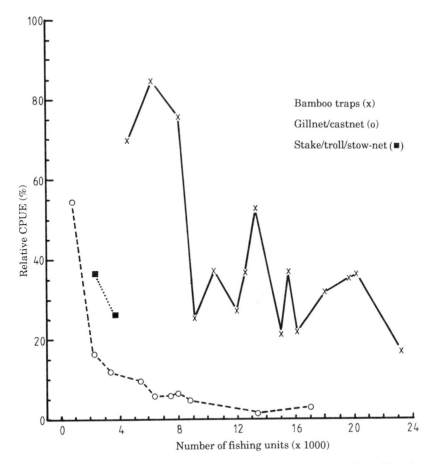

Fig. 32.3 Relative catch per unit effort (per cent) in relation to the number of (a) gill net/cast nets (×1000), (b) bamboo traps (×1000) and (c) stake/troll/stow nets

nets increases. Such a decline suggests gear-selectivity or gear-avoidance effects, and therefore negates this measure as a reliable catch-effort statistic for the overall fisheries. The catch per stake/troll/stow net also shows a declining trend with an increase in the number of units used in the range 2000–4000. However, the narrow range of stake/troll/stow net data available does not allow conclusive indication on the catch efficiency of these high-water gear-groups.

The pattern of catch per trap fluctuates without any obvious trend but stabilizes around 35% as the number of traps increases. The initial decline associated with the lower number of traps (between 4000 and 8000 unit score) is probably due to statistical error associated with single or small sample data (Fig. 32.3). The relatively consistent pattern of catch per trap suggests that the catch-effort remains more or less stable, despite fishermen manipulating the number and distribution patterns of the traps in the water. This in turn suggests the non-selectivity of the bamboo traps; any fluctuation in the catchability observed is considered to be attributable to variability in fish stock density.

Stratified for effects of seasonal fluctuations

Catch per net per month stratified for wet and dry seasons (Figs. 32.4 and 32.5) again shows similar and more obvious declining trends. The sampling data for gear group comprising stake/troll/stow net is inadequate for any conclusion to be drawn. The catch per trap for the wet season fluctuates constantly around 51% as the number of traps increases, whilst for the dry season it fluctuates around 61%. This observation justifies the stratification of CPUE and the grouping of gear type to reflect the seasonal fluctuations in the catch efficiencies.

Comparison of the various categories of CPUE (unstratified, wet season and dry season) using the three gear groups, highlights the limitations of catch effort sampling techniques, despite the availability of actual catch data and unit scores of gear types. However, after eliminating the cumulative interaction of variables/effects of gear interference (gear saturation, gear-selectivity and gear-avoidance)

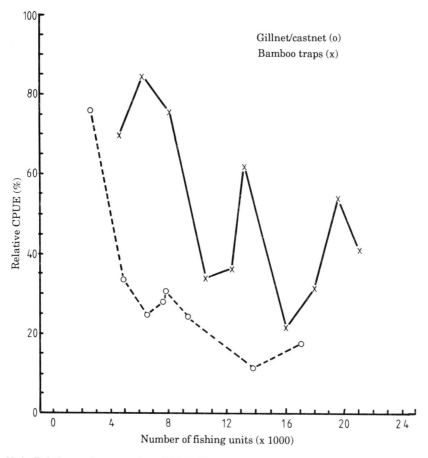

Fig. 32.4   Relative catch per net piece (CPUE %) against (a) number of gill net/cast nets (b) traps for the wet season

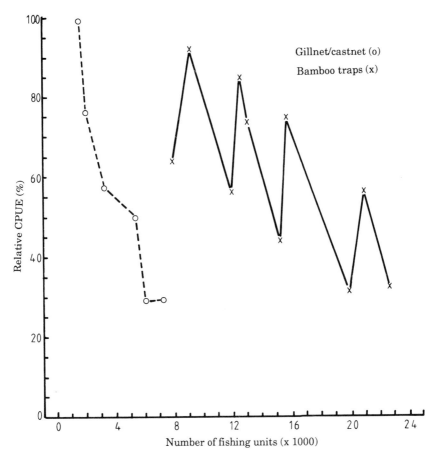

Fig. 32.5   Relative catch per net piece (CPUE %) against (a) number of gill net/cast nets (b) traps for the dry season

and fishing time, catch per trap appears to be a useful catch effort measure for the Bukit Merah Reservoir fisheries. This resultant assessment of total effective fishing effort (Ricker, 1975) should be representative of the entire fishing grounds and independent of changes in fishing power, fishing time, vulnerability, and relative distribution of fish and fishing. It should also be proportional to the instantaneous rate of fishing mortality for all values of effort (Morgan, 1979; Simpson, 1978; Gulland, 1956).

32.3.3   *Yield estimate application to the fishery*

The measure of total effective fishing effort has been calculated for the study period (Table 32.2). Adjusted for the effects of wet and dry season, the fish yield for Bukit Merah Reservoir was 38 kg ha^{-1} yr^{-1}, (ranging from 15.69 to 60.28 kg ha^{-1} yr^{-1}; 95% confidence intervals). This estimate approximates to the earlier

Table 32.2   Yield estimate using the catch per trap effort (CPUE) for Bukit Merah Reservoir fisheries

	Wet season				Dry season		
Month	CPUE (kg/unit)	Unit trap (×1000)	Yield (t)	Month	CPUE (kg/unit)	Unit trap (×1000)	Yield (t)
Sept 1978	0.4020	12.57	5.0531	Jan 1979	0.2228	14.96	3.3331
Oct	0.3708	12.06	4.4718	Feb	0.2872	11.93	3.4263
Nov	0.2814	10.98	3.0898	March	0.4723	15.42	7.2829
Apr 1979	0.5764	19.62	11.3090	May	0.3748	12.86	4.8199
July	0.4434	10.08	4.4695	June	0.3312	9.15	3.0305
Aug	0.7448	4.59	3.4186	Jan 1980	0.4674	10.20	4.7675
Sept	0.8093	8.07	6.5310	Feb	0.3153	15.49	4.8840
Oct	0.7398	6.27	4.6385	March	0.3060	20.40	6.2424
Nov	1.0684	6.24	6.6668	May	0.1728	19.84	3.4284
Apr 1980	0.2284	16.02	3.6590	June	0.1783	23.25	4.1455
July	0.4389	21.06	9.2432	Jan 1981	0.3780	13.05	4.9329
Aug	0.6043	13.05	7.8861	Feb	0.5209	12.60	6.5633
Sept	0.5577	13.50	7.5290				
Oct	0.3360	18.00	6.0480				
Nov	0.8327	13.05	10.8667				
Mean	0.5623 ±0.4657	12.34 ±9.61	6.3253 ±5.1182	Mean	0.3356 ±0.2222	14.93 ±8.40	4.7380 ±1.3756

value (i.e. $37 \, \text{kg} \, \text{ha}^{-1} \, \text{yr}^{-1}$) derived from a catch census on daily fishing operations and transactions at the reservoir site.

32.4   Discussion

In the Bukit Merah Reservoir fishing effort based on the number of a particular type of fishing gear is a more informative measure than numbers of canoes or fishermen. It is directly related to the catching power and thus exploitation of the fish stock. In the present study, values based on a field observation of the number of units of a particular gear used and the catch of the canoes/fishermen are more likely to be accurate than fishery statistics compiled from estimated values for the mean catch per boat multiplied by the number of registered boats. Moreover, data on the numbers of the predominant gear (e.g. bamboo traps) are in some ways easier to gather accurately at landing sites at the margins of the reservoir than numbers of fishermen in each canoe, since some of the fishermen may have left the shore when interviews and inspection takes place.

The diverse variety of fishing gears used in the Bukit Merah Reservoir fisheries, in terms of their structure and operation, makes it impossible to visualize a standardized unit for fishing effort. This is because of the uncertainty of the particular gear mainly responsible for the catch; stake and gill nets are extensively used, but bamboo traps predominate. Thus, even if a particular fishing gear is chosen as a standard for fishing effort, a substantial amount of information may be overlooked unless a large part of the catch is due to a particular gear-type (e.g.

in the Parakrama Samudra tilapia fisheries, Amarasinghe & Pitcher, 1986). Such problems are indeterminable in the Bukit Merah Reservoir multi-species fisheries, hence it is impossible to obtain a really satisfactory comparative measure of total fishing effort. Such a difficulty in standardization in relation to a diversity of fishing effort has been previously noted by Ricker (1975). The present study approaches this problem by classification into three gear groups embodying more or less similar operation procedures and by stratification of catch levels in relation to these gear groups.

To some extent, the catch efficiencies of the diversified gears are confounded by the effects of season at Bukit Merah Reservoir. Thus seasonal variation in catch necessitates stratified analysis. This seasonal variation in vulnerability is important in the measurement of stock abundance and fishing mortality rate, particularly when the seasonal distribution of effort is changing. Stratified estimates using catch per gear piece per day in the wet season avoid an underestimation when fishermen tend to compensate for poor catch by using a larger number of barriers (stakes and stow net) at inflows and high-water levels. The reverse probably holds true for an overestimation in the dry season.

Fishing effort, measured as standardized catch efficiency per net piece, declines with gill nets and casting nets, and probably with stake/troll/stow net suggesting a possibility of gear-avoidance, gear-saturation and other gear-interference phenomenon. Thus the gathering of the additional data on these gear groups is necessary. Conversely, fishing effort expressed as catch per trap do not alter significantly with the number of traps; this is probably because a trap's catch depends indirectly upon the fisherman's special operational tactics. Such comparisons of various versions of CPUE values and catch efficiencies of gears employed at the reservoir site suggest that catch per bamboo trap is the most reliable yield assessment of seasonal fisheries. Gulland (1979) noted an insignificant change from year to year in the amount of fishing conducted per year, type and size of gears used in many small-scale fisheries in the developing countries.

Acknowledgements

I thank the University of Malaya, Malaysia, for the permission to attend the Catch-effort sampling Techniques Symposium and Workshop, the Commonwealth Foundation and the Organizing Committee of the Symposium (particularly Dr I. Cowx, Chairman of the Organizing Committee) for the financial support.

References

Amarasinghe, U.S. and Pitcher, T.J. (1986) Assessment of fishing effort in Parakrama Samudra, an ancient man-made lake in Sri Lanka. *Fisheries Research* **4**: 271–282.

Gulland, J.A. (1956) On the fishing effort in the English demersal fisheries. *Fishery Invest.*, London. Ser. **2**. 20 (5), 41 pp.

Gulland, J.A. (1979) Stock assessment in tropical fisheries: Past and present practices in developing countries. In *Stock assessment for tropical small-scale fisheries*. (Ed. by P.M. Roedel and S.B.

Saila). Proceedings of an International Workshop, September 19–21, 1979. The University of Rhode Island, Kingston.

Morgan, G.R. (1979) Trap response and the measurement of effort in the fishery for the Western rock lobster. *Rapp. p-v Reun Cons. int. Explor. Mer.* **175**: 197–203.

Ricker, W.E. (1975) Computation and interpretation of biological statistics of fish populations. *Bulletin of Fisheries Research Board Canada* **191**: 382 p.

Simpson, A.C. (1975) Effort measurement in the trap fisheries for crustacea. *Rapp. p-v Reun Cons. int. Explor. Mer.* **168**: 50–53.

Yap, S.Y. (1982) Fish resources of Bukit Merah Reservoir, Malaysia. PhD Thesis, University of Malaya, Malaysia. 400 pp.

Yap, S.Y. (1983a) Some fishing gears in use in Malaysia inland waters. *Malaysian Agriculture Journal* **54(1)**: 10–17.

Yap, S.Y. (1983b) A holistic ecosystem approach to investigating tropical multi-species reservoir fisheries. *ICLARM Newsletter* **6(2)**: 10–11.

Yap, S.Y. (1984a) Cohort analysis on a freshwater fish *Osteochilus hasselti* C & V (Cyprinidae) at Bukit Merah Reservoir, Malaysia. *Fisheries Research* **2**: 299–314.

Yap, S.Y. (1984b) A summary of 'An overview on reservoir fisheries in Tropical Asia' by Yap, S.Y. and Furtado, J.I. The winning entry to the 1983 ICLARM Review of the Year Competition. *ICLARM Newsletter* **7(2)**: 29.

Yap, S.Y. (1987) Recent developments in reservoir fisheries research in tropical Asia. *Arch Hydrobiol Beih Ergebn Limnol* **28**: 295–303.

Yap, S.Y. (1988a) Food resource utilization partitioning of fifteen species at Bukit Merah Reservoir, Malaysia. *Hydrobiologia* **157**: 143–160.

Yap, S.Y. (1988b) Ecological basis of Asian reservoir fishery management and development: present status and future trends. *Wallaceana* **54**: 1–7.

Chapter 33
Catch and effort sampling techniques and their application in freshwater fisheries management: with specific reference to Lake Victoria, Kenyan waters

C.O. RABUOR *Kenya Marine and Fisheries Research Institute, Kisumu Laboratory, P.O. Box 1881, Kisumu, Kenya*

A catch and effort assessment survey in the Kenyan waters of Lake Victoria is conducted as part of a permanent monitoring programme for the artisanal fishery. The primary objective of the survey is to obtain reliable yield estimates of the fish harvested by the fishermen within this part of the lake. Secondary objectives include determination of the species composition of the catch, and the fishing effort involved in obtaining the catch. Sampling is by stratification of the primary landing sites and sub-sampling according to the type of fishing operation. The catch effort data have been used to determine the management policy for the fishery, and for planning, marketing and other aspects of fishery development.

33.1   Introduction

In almost all African countries, inland fisheries have an old established tradition. The main characteristics associated with lake fisheries in Africa are that they are extremely diffuse, volatile, disordered and in the first stages of spontaneous evolution. Further, in the big lakes, several thousand fishermen are involved in the fisheries on a private basis. A good proportion of the fishermen live in fishing villages of different sizes ranging from a single family unit with one canoe to more than a hundred persons with many canoes (Bazigos, 1970).

In order to manage these fisheries catch and effort assessment survey programmes have been established. The primary objective of these surveys is to obtain reliable yield estimates of the fish harvested by the fishermen. Secondary objectives include determination of the species composition of the catch, and the fishing effort involved in obtaining the catch. This chapter describes the catch assessment survey programme for the Kenyan waters of Lake Victoria, and how it has been used to determine the management policy for the fishery.

33.2   Materials and methods

33.2.1   *Description of the study area*

The Kenyan portion of Lake Victoria, which is generally referred to as the Nyanza (or Winam) Gulf (Fig. 33.1), comprises only 6% (3755 km^2) of the entire lake (68 000 km^2). It is a shallow bay that lies at an altitude of 1134 m above sea level on the north-east shore of the lake. It has an area of about 192 000 ha and an average depth of 6–8 m. The bottom consists of mud, sand, gravel and rock. The equatorial location of the Gulf provides a relatively constant climatic regime throughout the year. The south-east trade winds blow during most of the year, causing a daily rise and fall of tides within the Gulf. The water of the Gulf has a mean pH of 8.75, a mean water conductivity of 145 μS cm^{-1}, and a mean annual rainfall of 1159 mm (Okach & Dadzie, 1987).

Fig. 33.1 Map showing the Kenyan waters of Lake Victoria and the locations of the primary sampling units (landing sites)

33.2.2 *Description of the fishery*

The Lake Victoria fishery in general, and that of the Nyanza Gulf in particular, is basically artisanal, having all the characteristics described above. An estimated 6000–7000 boats, using several different types of gears, operate mainly from 65 principal landing beaches (Ogutu, 1988) out of an estimated 200 (Bernacsek, 1987) along the 760 km shoreline of the Nyanza Gulf. Thus approximately eight boats operate per kilometre of shoreline. Ssentongo and Welcomme (1985) estimated that between 25 000 and 30 000 fishermen are operating in the Nyanza Gulf which represents about 50% of all the fishermen in the entire lake.

33.2.3 *Sampling techniques*

Landing of fish takes place all the year round at the principal landing beaches, making it impossible to sample all these on a continuous basis. The catch and effort survey methodology, therefore, follows the procedure designed by Bazigos (1974), where the geographical area was stratified with respect to different types of fisheries in terms of species and gears, although consideration was also given to existing district boundaries. In each stratified area a number of primary sampling units were selected for continual monitoring. These were again selected to ensure adequate coverage of all the different fisheries in the zone. Other factors given consideration in the selection of the Primary Fishing Units were number of boats landing per day, infrastructure, and availability of dwelling houses for the beach recorders.

The next stage of sampling involves the identification of the Fishing Economic Units in the Primary Sampling Unit which need to be sub-sampled. The Fishing Economic Units are considered the reporting units, and they fall into two categories:

(1) Usual Fishing Unit (UFU), which is composed of the boat, fishing gear and the fishermen necessary to carry out the fishing operation.
(2) Minor Fishing Unit (MFU), which is composed of fishing gear and fishermen (without a fishing boat) necessary to carry out fishing operations, (Bazigos, 1970).

An average of three beach recorders operate at each primary sampling unit (landing beach). They are supplied with weighing balances and measuring boards to record length-weight frequency data for all the major species landed, except *Rastrineobola argentea* (sardines) and juvenile *Lates niloticus* (Nile perch), which are mass weighed in buckets. The data are collected on a monthly basis during the normal work days. From the Fishing Economic Units sampled the following information is also registered:

(1) Types and characteristics of the gears used (gill nets, long lines, mosquito seines, beach seines, etc.).

(2) Types and characteristics of the fishing vessel used (motorized, Sesse boats, Karua, etc.).
(3) Number of crew per boat.
(4) Number of hours spent fishing.
(5) Number of hauls.

33.2.4   *Estimation of annual catch*

The boat is considered the main component in the Fishing Economic Unit and hence the main factor in estimating the total catch. Using the data from the Primary Sampling Units, the mean catch per boat day (*MCBD*) and the catch per boat per year (*CPBY*), based on 250 boat fishing days per year, are calculated. The annual production (*Y*) is estimated as:

$$Y = MCBD \times Fd \ (Rf) \times Bn$$

where *Fd* is the number of fishing days, *Rf* the raising factor and *Bn* the number of boats, which is between 5000 and 7000 (Bernacsek, 1986).

33.3   **Results**

33.3.1   *Percentage compositions and estimated catch*

Table 33.1 summarizes the estimated total catch and the percentage composition of the dominant species landed in the Kenyan waters of Lake Victoria between

Table 33.1 Percentage species composition and the estimated total catch (metric tonnes) for the dominant species

	1986		1987		1988	
	MT	%	MT	%	MT	%
Lates niloticus	64 929	63.5	86 681	69.0	82 019	59.3
Rastrineobola argentea	31 084	30.4	30 778	24.5	50 512	36.5
Oreochromis spp.	2 781	2.7	3 266	2.6	3 361	2.4
Others	3 456	3.4	4 724	3.8	2 420	1.8
Total catch (MT)	103 163		125 625		138 312	

Table 33.2 Contribution to total catch made by various fishing gears

	Mainga (1985) 1981 & 1982	This study		
		1986	1987	1988
Gill nets	72	63.1	55.8	45.8
Beach seines	7	8.2	8.8	9.6
Long lines	4	9.2	9.7	9.6
Mosquito seines	17	19.55	25.65	35.0

Table 33.3 Statistical presentation of catch per unit effort (CPUE) characteristics of the fishery

Year	Sample weight kg	Boat day			Net day			Crew day			Boat haul			Boat hours		
		x̄	σ	CV	x̄	σ	CV	x̄	σ	CV	x̄	σ	CV	x̄	σ	CV
1986	4 372 014	88	23	28	5	2	35	24	5	19	31	16	52	6	2	28
1987	5 514 682	102	18	18	6	2	31	31	3	10	34	24	71	8	2	27
1988	5 527 235	113	11	10	6	1	16	35	3	9	25	8	41	8	2	15

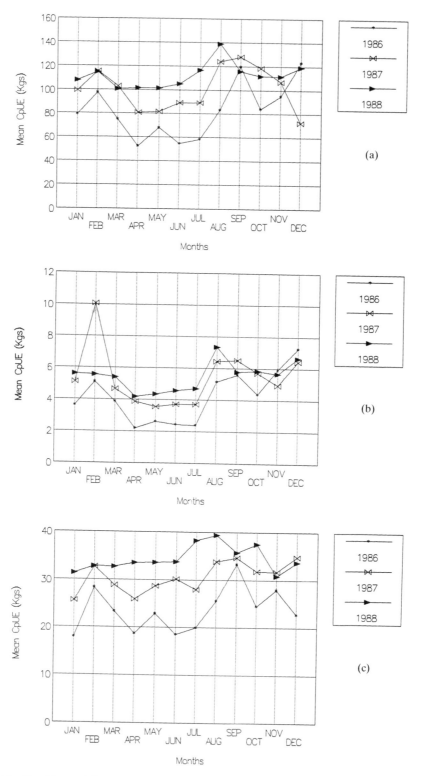

Fig. 33.2 Monthly trends in mean catch per unit effort for various parameters: (a) CPUE boat day^{-1}; (b) CPUE net day^{-1}; (c) CPUE crew day^{-1}; (d) CPUE boat haul^{-1}; (e) CPUE boat hour^{-1}

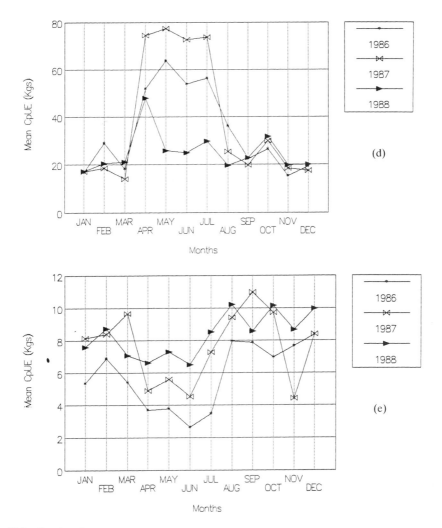

Fig. 33.2 Continued

1986 and 1988. The contribution of the various gears to these catches is given in Table 33.2 and is compared with an earlier study by Mainga (1985).

33.3.2   *Statistical analysis of CPUE*

The mean catch per unit effort for the various parameters recorded is presented in Table 33.3. The mean catch per boat day shows an upward trend with a decreasing variance and coefficient of variation for all the years. The mean catch per net day, mean catch per crew day and mean catch per boat hour all show a stable state with minimal variance, indicating that there has been very little change in the number of nets and crews over the study period. However, the mean catch per haul day exhibits a marked change due to the change in attitude towards the *Rastrineobola argentea* fishery.

33.3.3   *Trends in the fishery*

Figs. 33.2(a)-(e) show the seasonal trends in the fishery determined from CPUE statistics. The CPUEs are, with the exception of catch per boat haul, lowest between April and July. This period is the 'closed-season', a time when the Fisheries Department impose a ban on mosquito and beach seines to ensure breeding of *Rastrineobola argentea* and growth of juvenile *Lates niloticus*. This is also the reproductive period for the *Oreochromis* spp. and the banning of the beach seines is aimed at relieving the pressure exerted by the gears on their breeding grounds. The peaks in August are indicative of a large number of fishermen resuming fishing after harvesting of the agricultural crops in July. The peaks may also be attributed to an increase in stock abundance following closure of the fishery between April and July.

It is important to note there has been a general increase in the exploitation of the Nyanza Gulf fishery over the three-year sampling period. However, this trend was not exhibited in the catch per boat haul because the usual ban on the *Rastrineobola argentea* fishery between April and July was waived in 1988.

33.4   **Discussion**

From a sampling point of view, information on a population may be gathered either on the basis of complete enumeration or stratified sampling surveys. It is more advantageous to gather such information by stratified sampling because it is less costly, the collection and processing of data takes a shorter time and the survey, if well planned and executed, will have more consistent results. However, any statistical investigation is likely to have some degree of error and bias. Bias in fisheries statistical surveys can be eliminated by choosing the proper estimator, by increasing the size of the sample or by using the method of stratification after selection and weighting with the real proportions of strata. Secondly, bias by means of incomplete sampling frames can be eliminated by introducing correction factors in the estimation procedure (Bazigos, 1974).

The catch and effort assessment survey described was initiated in 1980 as a permanent monitoring programme on the fish stock of Lake Victoria, Kenyan Waters. Prior to this, catch and effort assessment surveys on this artisanal fishery were conducted under the auspices of the defunct East African Freshwater Fisheries Research Organization (Wetherall, 1973; Wanjala & Martens, 1974). These surveys, coupled with fishery independent stock assessment studies (Asila & Ogari, 1988) have proved a reliable source of information of the fish stocks of the Nyanza Gulf.

From these data, it is evident that there are 13 families and 29 genera of fishes in Lake Victoria. The benthic biomass is dominated by *Lates niloticus* (family of Centropomidae): Ssentongo and Welcomme (1985) estimated the standing stock of this exotic species to be of 120 000 tonnes. The largest number of species is found in the family Cichlidae, containing seven genera, the most important of which are the *Oreochromis* and *Haplochromis* species. It is estimated that there are more than 250 species of this fish in Lake Victoria, illustrating the wide ecological diversity of the lake. Kudhogania and Cordone (1974) estimated from bottom trawls that, prior to 1974, 83% (564 000 metric tonnes) of the demersal ichthyomass of the lake consisted of these cichlids. However, the fishery has changed in favour of Nile perch which has completely colonized the lake.

One point that should be highlighted is the need to properly train and renumerate field personnel to avoid inconsistency in data collection. Supervision should also be carried out at regular intervals to maintain collection of quality data.

Acknowledgements

I would like to express my sincere thanks to Mr S.O. Allela, the Director of KMFRI, for having accorded me the opportunity to travel and stay in Mombasa during the preparation of this paper and to Mr S.O. N'gete for assisting in the computer work. My special thanks are extended to all the beach recorders for their tireless work during the data collection exercise and to Messrs D. Owage, J. Nyanjwa and S. Osewe for assisting in editing the data. Mrs M.A. Opiyo is thanked for typing the manuscript.

References

Asila, A. and Ogari, J. (1988) Growth parameters and mortality rates of Nile perch (*Lates niloticus*) estimated from length frequency data in the Nyanza Gulf (Lake Victoria). *FAO Fish. Rep.* **389**: 519 p.

Bazigos, G.P. (1970) Sampling techniques in inland fisheries with special reference to Volta Lake. *FAO: F10: SF/GHA/10*: 40 p.

Bazigos, G.P. (1974) The design of fisheries statistical surveys – inland waters. *FAO. Fish. Tech. Pap.* **133**: 122 p.

Bernacsek, G.M. (1986) Kenya, Tanzania and Uganda: Evaluation of Statistical Services for Lake Victoria Fisheries. (Mission Report. Accra, Ghana, FAO Regional Office for Africa): CIFA Sub-Committee for the Development and Management of the Fisheries of Lake Victoria: Fourth Session Kisumu, Kenya, 6–10th April 1987. *CIFA: DM/LV/87. Inf.* **4**, 25 pp.

Kudhogania, A.W. and Cordone, A.J. (1974) Batho-spatial distribution patterns and biomass estimates of the major demersal fishes in Lake Victoria. *Afr. J. Trop. Hydrobiol. Fish.* **3** (1): 15–31.

Mainga, O.M. (1985) Preliminary results of an evaluation of fishing trends in the Kenyan Waters of Lake Victoria from 1981–82. *FAO. Fish. Rep.* **335**: 110–117.

Norman, T.J. and Bailey, M.A. (1984) *Statistical Methods in Biology*. London: Hodder and Stoughton. 216 p.

Okach, J.O. and Dadzie, S. (1988) The food, feeding habits and distribution of a Siluroid Catfish, *Bagrus docmac* (Forsskal), in the Kenyan Waters of Lake Victoria. *J. Fish Biol.* **32**: 85–94.

Ogutu, G.M. (1988) Artisanal fisheries of Lake Victoria, Kenya. Social and Economic Aspects of Production and Marketing. Nairobi: Int. Dev. Res. Centre, 45 p.

Ssentongo, G.W. and Welcomme, R.L. (1985) Past history and current trends in the Fisheries of Lake Victoria. *FAO Fish. Rep.* **335**: 123–138.

Wanjala, B. and Martens G. (1976) Survey of the Lake Victoria fishery in Kenya. EAFFRO Annual Report 1974: 81–85.

Wetherall, A.J. (1973) On the Catch Assessment Survey of Lake Victoria. EAFFRO Occ. Pap. 14.

Chapter 34
The fishery of Lake Naivasha, Kenya

S.M. MUCHIRI *Fisheries Department, Moi University, Eldoret, Kenya*
P. HICKLEY *NRA, Severn-Trent Region, 550 Streetsbrook Road, Solihull, B91 1QT, UK*

Lake Naivasha is a freshwater lake situated in the eastern rift valley of Kenya. Only five species of fish are present, all of which have been introduced. They are *Micropterus salmoides*, *Oreochromis leucostictus*, *Tilapia zillii*, *Barbus amphigramma* and *Lebistes reticulata*. The first three of these form the basis of a commercial gill net fishery. Bass are also taken by sport fishermen and *Barbus* are occasionally caught by dip net. Catch data for the Lake Naivasha fishery are presented. The fish populations are shown to be under pressure from overfishing, fluctuating water levels, changes in aquatic macrophyte densities and inadequate species diversity. Potential yields are discussed and some management options proposed.

34.1   Introduction

Lake Naivasha is a freshwater lake, about $150\,km^2$ in area, situated in the eastern rift valley of Kenya about 100 km north of Nairobi. It lies in a closed basin at an altitude of 1890 m above sea level and receives 90% of its water from the perennial River Malewa. The remaining input comes from two ephemeral streams, rainfall and ground seepage. The lake is shallow, having, for the most part, a maximum depth of about 8 m, and is subject to wide fluctuations in level. Marginal vegetation is dominated by papyrus (*Cyperus papyrus* L.). *Salvinia molesta* Mitch. covers large areas of water surface and submerged macrophytes occur to varying degrees, the principal species being *Najas pectinata* (Parl.). A detailed description of Lake Naivasha is given by Litterick *et al.* (1979) and updated by Harper *et al.* (1990).

Tropical lakes generally have a diverse fish fauna but, due to a probable history of drying out, Lake Naivasha had only one, the endemic *Aplocheilichthys antinorii* (Vinc.) which was last recorded in 1962 (Elder *et al.*, 1971). Various fish introductions have been made since 1925, the details of which are given in Table 34.1. The fish population of the present day comprises five species: *Micropterus salmoides* Lacepede (largemouth black bass), *Oreochromis leucostictus* (Trewewas), *Tilapia zillii* (Gervais), *Barbus amphigramma* Blgr. and *Lebistes reticulata* Peters (guppy). The two tilapias and the bass form the basis of an important commercial gill net fishery which opened in 1959. Bass are also caught by sport fishermen and, in recent years, some *Barbus* have been taken with dip nets.

Table 34.1 Summary of changes to the fish population of Lake Naivasha

Species	Date and success of introduction
Aplocheilichthys antinorii (Vinc.)	Endemic. Probably extinct; last reported in 1962.
Oreochromis spirulus niger (Gunther)	Introduced in 1925. Disappeared by 1971.
Micropterus salmoides (Lacepede)	Introduced in 1929, several times during 1940s and in 1951. Present today.
Tilapia zillii (Gervais)	Introduced in 1956. Present today.
Oreochromis leucostictus (Trewevas)	Introduced unintentionally in 1956 with *T. zillii*. Present today.
O. leucostictus x *O.s.niger* hybrid	Abundant in the early 1960s but due to back-crossing with *O. leucostictus* disappeared by 1972.
Oreochromis niloticus L.	Introduced in 1967. Disappeared by 1971.
Gambusia sp. and *Poecilia* sp.	Introduced but dates unknown. Absent since 1977.
Lebistes reticulata Peters	Introduced; date unknown. Recorded since 1982. Present today.
Oncorhyncus mykiss (Walbaum)	Introduced into the River Malewa; dates unknown. Caught in the lake on rare occasions.
Barbus amphigramma Blgr.	Natural invader from inflowing rivers. Recorded since 1982. Present today.

34.2   Methods

Fishing on Lake Naivasha is controlled by the Fisheries Department of the Kenya Government. For commercial fishing, gill nets are set from wooden or glass-fibre canoes which are driven either by sail and oar or by outboard motor. Every commercial fisherman is allowed to use up to 10 multi-filament gill nets of 100 m in length and with a minimum stretched mesh size of 100 mm. Netsmen must land their catch at a single landing station near Naivasha town where fisheries personnel record the weight of the catch from each vessel. Data collection commenced in 1963 but sorting of the catches into bass and tilapia has only taken place since 1974. Separate recording of *Oreochromis leucostictus* and *Tilapia zillii* began in 1987. In addition, bass are taken by sport fishermen using rod and line; the bait usually being artificial lure. *Barbus amphigramma* is captured by fishermen using locally made scooping nets when ripe fish attempt to ascend the River Malewa to spawn and are unable to cross a weir constructed about 5 km from the river mouth.

Gill net catch statistics were fitted to a version of the Schaefer model (Ricker, 1975; Pauly, 1983):

$$C = a + bE$$

where C is the catch per unit effort, E the effort (measured as the number of canoes licensed to fish in a given year), and a and b are constants.

Maximum sustainable yield (MSY), optimum effort and equilibrium yields were calculated as:

$$MSY = a^2/4b$$

$$optimum \; effort = a/2b$$

$$equilibrium \; yield = aE - bE^2$$

Theoretical potential yield of fish was estimated using, firstly, a morpho-edaphic index (MEI; the ratio of total dissolved salts to mean lake depth) and, secondly, a primary productivity model.

Yield based on MEI was calculated according to the regression equation of Henderson and Welcomme (1974):

$$Y = 8.7489 \; M^{0.3813} \qquad (r^2 = 0.5073)$$

where Y is the fish yield (kg ha^{-1} year^{-1}) and M the morpho-edaphic index. (N.B. In this study conductivity was used rather than total dissolved salts.)

The regression equation of Melack (1976) was used to predict yield from primary production data:

$$\log Y = 0.113 \; P + 0.91 \qquad (r^2 = 0.57)$$

where Y is the fish yield (kg ha^{-1} year^{-1}) and P the average daily photosynthetic rate.

34.3   The gill net fishery

34.3.1   *Fish catches*

The total annual fish catches landed from the gill net fishery of Lake Naivasha during the period 1963–1988, and the number of canoes that were used, when known, are given in Table 34.2. Over the years there have been great fluctuations in both the amount of fish landed and the number of vessels. The former ranged from 37 to 1150 t yr^{-1} and the latter from 6 to 104 canoes with catch per canoe varying between 1.1 and 24 t yr^{-1}. The overall species composition of the catch (since 1987) was *Oreochromis leucostictus* 84.7%, *Tilapia zillii* 0.5% and *Micropterus salmoides* 14.8%. The mean fork length of *Oreochromis* was 225 mm (S.D. = 16.18) and that of bass 298 mm (S.D. = 17.00). Figure 34.1 shows the regression plot of computed catch per unit effort on effort from 1974 to 1988, together with the resulting equilibrium yield curve.

A summary of events for the gill net fishery is presented in Fig. 34.2. The fishery can be divided into two main phases with the first being the period of development of the new fishery from 1959 to the time when it collapsed in the early 1970s. During this phase the maximum catch recorded was 1150 t yr^{-1}. The second, recovery, phase from 1975 to the present day is typified by marked fluctuations in fish catches. Some of these fluctuations can be attributed to three main causes; fishing pressure, fluctuating lake levels and, more recently, effects related to the loss and subsequent return of submerged vegetation.

Table 34.2 Fish catches from the commercial gill net fishery of Lake Naivasha for the years 1963–88 inclusive

Year	Number of canoes	Total catch tonnes	Catch per canoe tonnes
1963		183	
1964		550	
1965		650	
1966		950	
1967		955	
1968		885	
1969		929	
1970		1150	
1971		484	
1972		117	
1973		62	
1974	8	62	7.8
1975	6	144	24.0
1976	12	252	21.0
1977	10	67	6.7
1978	17	255	15.0
1979	33	529	16.0
1980	52	471	9.1
1981	32	269	8.4
1982	80	89	1.1
1983	76	576	7.6
1984	104	277	2.7
1985	49	206	4.2
1986	75	478	6.4
1987	85	224	2.6
1988	30	37	1.2
1989*		96	

* First seven months of 1989 only

34.3.2   *Fishing pressure*

Although there were restrictions imposed on the maximum number of gill nets allowed under each fishing licence, the number of licensed canoes was at times very large, the maximum being 104 (in theory a total length of 104 km of netting) in 1984. In practice, however, not all fishermen were able to use their full legal entitlement of net but the increase in the number of canoes engaged in fishing still led to a general decline in catch per canoe (Table 34.2). A similar trend in the relationship of fish catches with fishing effort has been described by Fryer and Iles (1972) for the fishery of Lake Victoria (Kenya waters).

Factors that led to the increase in the number of fishermen were two-fold. Firstly, there was a considerable demand for fish in the neighbouring urban centres whose populations have recently recognized the importance of including fish protein in their diet following the launch, in 1960, of a nationwide 'Eat more fish' campaign. Originally, the Maasai and Kikuyu inhabitants of the Naivasha

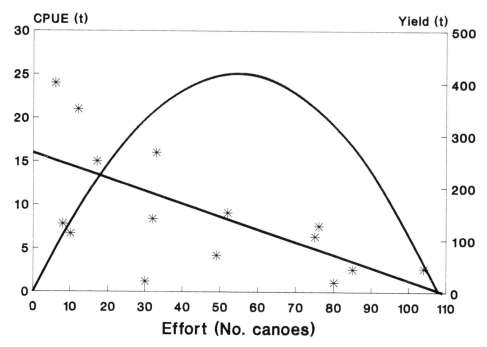

Fig. 34.1 Regression relationship (* and solid straight line) of catch per unit effort (CPUE) on effort (E) and the resulting equilibrium curve (solid curved line) for total gill net catches from Lake Naivasha for the period 1974–1988 inclusive (see also Table 34.3)

area were not fish eaters. Secondly, the fishing pressure on the Lake Naivasha fishery has increased for economic reasons. More and more people have sought employment, some of whom have been absorbed into the fishing industry.

Another detrimental influence on the lake fishery has been non-compliance with restrictions on gill net mesh size. Occasional inspection surveys of fishing gear indicate that there is often contravention of the minimum (100 mm) mesh size regulations by some licensed fishermen. Surveys carried out between 1984 and 1989 showed that nets with stretched mesh down to 75 mm were in common usage.

34.3.3   *Effects of lake level fluctuations*

Fish catches appear to be related to water level changes (Fig. 34.2). Rise in lake level is followed by increased catches whilst a fall in lake level is followed by a corresponding decline in fish catch. Lake level fluctuations influence fish numbers through effects on food, breeding grounds and predator-prey relationships. As food has been shown not to be a limiting factor for the fish in Lake Naivasha (Muchiri, 1990), a more probable effect is that on the breeding behaviour; tilapias being particularly sensitive to fluctuating water levels since they breed in shallow

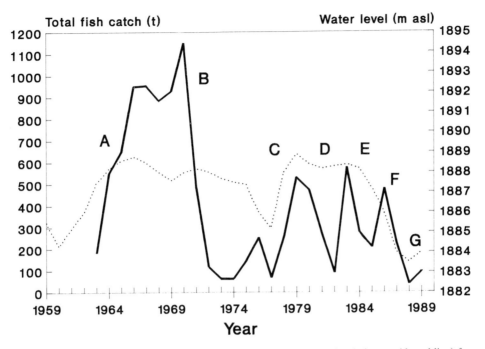

Fig. 34.2 Total catches (solid line) from the gill net fishery and water level changes (dotted line) for Lake Naivasha during the period 1963–1988 inclusive. Commercial fishing started in 1959 and records were kept from 1963. (A = period of development of the new fishery; B = period of receding lake level. Also, pressure to supply fish for processing factory led to the use of small meshed nets; C = small fishing effort (6–17 canoes) coupled with a rise in water level allowed the fishery to recover; D = period of high water but with loss of submerged plants; E = sudden decline in lake level combined with highest effort (104 canoes); F = recovery of aquatic plants but sustained high effort (49–75 canoes) and further decline in water level; G = small effort (30 canoes). Lake level starting to rise, flooding terrestrial vegetation and providing good nursery areas

water. Habitat drying and flooding were listed by Lowe-McConnell (1982) as factors controlling tilapia numbers within fish communities. Also cited was predation by piscivorous fish and birds. Fryer and Iles (1972) described how predation and fishing pressures on tilapias of Lake Victoria were minimized by flooding of marginal terrestrial vegetation and it appears likely that the fluctuating water levels of Lake Naivasha have similar effects.

Based on the past history of the lake it is unlikely that levels will stabilize in the near future and human interference is bound to exacerbate the problem. Lakeside farms irrigate with water from the lake and, although abstraction may not significantly affect the water levels, this lake level/farming relationship is potentially detrimental. When the lake recedes the farmers extend their boundaries thus acquiring more land for cultivation, consequently needing more water for irrigation. In addition, there is a proposal to dam the River Malewa which may create yet further instability in the future.

34.3.4   *Effects of submerged macrophytes*

Submerged macrophytes disappeared during the early 1980s and this appears to have contributed to the decline in fish catches. Following the recovery of submerged and swamp macrophytes in 1987 and 1988, in spite of the low lake level, catches in 1989 showed an upward trend which seems to be continuing in 1990. Since the three main commercial fish species use vegetation-rich spawning grounds, the loss of macrophytes in the early 1980s severely affected recruitment of new individuals to sustain the fishery. The return of aquatic plants provides extensive breeding and nursery grounds as well as offering abundant food and cover from predation.

34.3.5   *Fish yields*

The computed maximum sustainable yield (MSY) is 418.8 t yr^{-1} with an optimum effort of 54 canoes (Table 34.3). During the early years of the commercial fishery the catch was maintained at a much higher level until 1970 when the maximum

Table 34.3 Estimates of actual and theoretical fish yields calculated for Lake Naivasha:
(a) Regression of catch per unit effort (CPUE) on effort (E).
(b) Morpho-edaphic index (MEI) model.
(c) Gross primary production model.
Figures in brackets indicate sources of data, (1) Melack 1979; (2) Harper 1987; (3) Muchiri (1990); (4) Harper 1990, in prep.

(a)

Year	CPUE =	r^2	Optimum E (Canoes)	MSY t yr^{-1}
1974−88	15.541−0.14418E	0.456	54	418.8

(b)

Year	Area km^2	Mean Depth m (Z)	Conductivity (K) uS^{-1} cm^{-1}	MEI (K/Z)	Total Fish Yield t yr^{-1}
1974	160	4.7	345 (1)	73.4	720.0
1982−3	150	8	259 (2)	32.4	495.0
1984	150	6	350 (2)	58.3	618.4
1987−8	130	5	430 (3)	86	621.4

(c)

Year	Area km^2	Gross Production (g O$_2$ m^{-2} d^{-1})	Total Fish Yield t yr^{-1}
1973−4	160	4.08−6.84 (1)	573.5
1982	150	4.20−7.56 (2)	617.7
1988	120	15.6 (4)	5649.0

fish landing ($1150 \, t \, yr^{-1}$) was recorded. Since that time the fish catches have been close to or below the MSY (Table 34.2).

The fish yield estimates for Lake Naivasha calculated from published limnological data using the morpho-edaphic index and gross primary production models are given in Table 34.3. Most of the estimates show little variability, ranging between 495 and $720 \, t \, yr^{-1}$, except for 1988 when the yield calculated by the primary production model was particularly large at $5649 \, t \, yr^{-1}$.

34.4   The Barbus fishery

Exploitation of *Barbus amphigramma* on the River Malewa commenced in 1983. In 1983 and 1984 there were no records of catches but in 1985 60 t of *Barbus* were taken. In 1986 the catch amounted to 62.9 t but this decreased considerably in 1987 to 26.1 t. No fishing took place during 1988 because the annual spawning run did not occur. Only a small number of fish were caught in 1989.

34.5   Discussion

Since commercial fishing started, fish have been taken with gill nets which, in theory, should make regulating the fishery easier than in multi-gear fisheries. As gill nets are selective, desired mesh sizes can be imposed and effort controlled by limiting numbers and lengths of nets in addition to regulating the numbers of fishing vessels. Unfortunately, the task of enforcing such regulations is often the difficult part for the fisheries manager. This is particularly true for Lake Naivasha owing to its proximity to the towns of Nairobi and Nakuru with their great demands for fish. The most serious problem to be addressed is that of curbing illegal fishing by unlicensed fishermen who have no regard for the gill net regulations. This is, however, as much a social concern as a fisheries one and requires a combined effort by the Fisheries Department, the administrative officials and social workers in the Naivasha region in order to educate and rehabilitate those people involved.

With regard to the sport fishery, fishermen are supposed to make monthly returns of their catches, but this has met with little success. Spot checks by fisheries personnel reveal that substantial amounts of fish are landed by anglers but not reported.

Although fish caught by unlicensed fishermen and through sport fishing were mainly unrecorded, information on catches brought to the central landing station was more or less complete. Therefore, in spite of difficulties in obtaining accurate data, the commercial catch statistics can provide a good representation of the pattern of exploitation. In the future, the ratio of reported to unreported catch could be calibrated and yield figures could be adjusted accordingly. It may be that the catch per unit effort for illegal fishing is similar to that for legal fishing, in which case working on just the legal data by way of a sub-sample would still give a reliable insight into the status of the fishery.

The *Barbus* fishery is relatively easier to monitor since all the fishing activity is concentrated on one short section of the River Malewa just below the weir. There is difficulty, however, in predicting the consequences of this new fishery as little is known of the biology and ecology of *Barbus amphigramma* in Naivasha or indeed elsewhere. The intensive fishing that took place between 1985 and 1987 may have drastically reduced the population of breeding individuals to a level that will require several spawning seasons before recovery. In an effort to have as many fish as possible reach their breeding grounds upstream, fishermen are now required to release one scoop net-full of fish in five into the river on the upper side of the weir. A more lasting and dependable solution would be to construct fish passes to ensure adequate migration and escape from fishermen's nets.

When potential fish yields were calculated from limnological data (Table 34.3), in each case values were higher than the $418.8 \, t \, yr^{-1}$ MSY value obtained from catch data. In particular, the gross primary production model provided a very high estimate for 1988 ($5649 \, t \, yr^{-1}$). These high theoretical fish yield estimates suggest that Lake Naivasha may have a potential for a larger fisheries output than is realized at the present time. Of 31 lakes examined by Henderson and Welcomme (1974), 18 had more than one fisherman per km^2. By contrast, Naivasha has had less than 0.7 fisherman per km^2 throughout the history of its commercial fishery and yet has experienced overfishing from time to time. This is possibly due to the low number of target species. It is likely, therefore, that the potential fisheries resources of Lake Naivasha are at present underutilized. The hypothesis that fish yields are below potential is also supported by results obtained by the authors on the feeding of the two tilapias and the largemouth bass. It appears that zooplankton and off-shore benthic macro-invertebrates remain uncropped, yet studies on zooplankton (Mavuti & Litterick, 1981; Harper, 1984, 1987) and macro-invertebrates (Clark, *et al.*, 1989; Muchiri, 1990) indicate that the off-shore secondary production is high.

With the fishery based on species that depend to a large extent on the stability of the water levels, and which do not utilize all available resources in terms of food and space, it becomes tempting to support earlier suggestions (e.g. the Fisheries Department; Siddiqui, 1977) to introduce additional fish species to exploit the remaining niches. The option of further introductions is attractive not only in diversifying the ecosystem, but the subsequent increase in the commercial catches would provide greater availability of much needed protein and a useful increase in employment opportunity.

Early consideration should be given to assessing the suitability of those fish that could be contenders for introduction. Possible species are: *Limnothrissa miodon*, *Stolothrissa tanganyikae* or *Alestes* spp. to occupy the open water and feed on zooplankton; *Heterotis niloticus* to consume phytoplankton; *Mormyrus* spp. or *Haplochromis angustifrons* to feed on the benthos. The feasibility appraisals must carefully consider all aspects of biology and ecology such as recommended in the EIFAC (1988) code of practice.

Due to the very unstable conditions that continually affect the ecology of Lake

Naivasha there is need for continuous appraisal of the fishery in order to facilitate implementation of appropriate management measures. Meaningful assessment of the fishery calls for more precise data collection in relation to the commercial catch and the MSY and optimum effort should be determined regularly in order to maintain the fishing pressure within the capacity of the fishery. It is suggested that, in future, numbers of gill nets be the measure of effort when determining the optimum exploitation rather than the number of canoes licensed. As a result, the number of craft could remain high (thus accommodating more fishermen) whilst the number of actual gill nets would be maintained at the optimum level. It is more difficult to keep away a fisherman who depends on the fishery for his livelihood than it is to adjust the number of nets he is allowed to use.

Enforcement of the fisheries regulations is essential if accurate catch statistics are to be obtained. Also, until the problem of illegal fishing is resolved, there can never be effective management. Accordingly, attempts need to be made to increase the numbers of trained personnel engaged in the management of the fisheries resource, to upgrade transport, to improve communication and to ensure that fish catches are landed at the central landing station only.

Acknowledgements

This study is part of a wider research project at Naivasha undertaken by members of the Departments of Zoology of the Universities of Leicester, UK, and Nairobi, Kenya. We are grateful to the Office of the President of the Government of Kenya for research permission to Dr D.M. Harper for the project within which this work was carried out and to the Chairman of the Department of Zoology of the University of Nairobi for formal support and affiliation of the project. We are particularly grateful to Dr K.M. Mavuti of the Zoology Department at the University of Nairobi, the co-director of the overall project, for help and many hours of useful discussions. We are much indebted to the Fisheries Department in Naivasha for access to catch data for the fishery. Work in the field would not have been possible without the logistical assistance of Peter Magius, the late Roger Mennell, John and Jane Carver, Peter Robertson, Angus and Jill Simpson and all the staff at the Elsamere Conservation Centre. Fieldwork on Lake Naivasha was assisted by students from the University of Leicester Adult Education Department and by volunteers from Earthwatch, Boston, USA. The views expressed are those of the authors and not necessarily those of their parent organizations.

References

Clark, F., Beeby, A. and Kirby, P. (1989) A study of the macro-invertebrates of Lakes Naivasha, Oloidien and Sonachi, Kenya. *Rev. Hydrobiol. Trop.* **22**: 21–23.
Elder, H.Y., Garrod, D.T. and Whitehead, P.J.P. (1971) Natural hybrids of the African cichlid fishes *Tilapia spirulus nigra* and *T. leucosticta*, a case of hybrid introgression. *Biol. J. Linn. Soc.* **3**: 103–146.

EIFAC (European Inland Fisheries Advisory Commission) (1988) Code of practice and manual of procedures for consideration of introductions and transfers of marine and freshwater organisms. *FAO EIFAC Occ. Pap.* No. **23**, 23 pp.

Fryer, G. and Iles, T.D. (1972) *The cichlid fishes of the great lakes of Africa: their biology and evolution.* Neptune City, New Jersey, T.F.H. Publ.; also Edinburgh, Oliver and Boyd.

Harper, D.M. (1984) Recent changes in the ecology of Lake Naivasha, Kenya. *Verh. Internat. Verein. fur Theor. Ang. Limnol.* **22**, 1192−1197.

Harper, D.M. (1987) The ecology and distribution of the zooplankton in Lakes Naivasha and Oloidien. In D.M. Harper (Ed.) University of Leicester studies on the Lake Naivasha ecosystem, 1982−1984; Final Report to the Kenya Government, 101−121.

Harper, D.M. (1990) Primary production in Lake Naivasha, Kenya. In prep.

Harper, D.M., Mavuti, K.M. and Muchiri, S.M. (1990) Ecology and management of Lake Naivasha, Kenya, in relation to climatic change, alien species introductions and agricultural development. *Env. Conserv.*, in press.

Henderson, H.F. and Welcomme, R.L. (1974) The relationship of yield to morphoedaphic index and number of fishermen in African inland fisheries. *UN/FAO CIFA Occ. Pap.* No. **1**, 19 pp.

Litterick, M.R., Gaudet, J.J., Kalff, J. and Melack, J.M. (1979) The limnology of an African lake, Lake Naivasha, Kenya. Manuscript prepared for the SIL-UNEP Workshop on Tropical Limnology, Nairobi. 73 pp. Mimeo.

Lowe-McConnell, R.H. (1982) Tilapias in fish communities. In *The Biology and Culture of Tilapias* (Ed. by R.S.V. Pullin and R.H. Lowe-McConnell), ICLARM Conference Proceedings **7**, 83−113.

Mavuti, K. and Litterick, M.R. (1981) Species composition and distribution of zooplankton in a tropical lake, Lake Naivasha, Kenya. *Arch. Hydobiol.* **93**: 52−58.

Melack, J.M. (1976) Primary production and fish yields in tropical lakes. *Trans. Am. Fish. Soc.* **105**: 575−580.

Melack, J.M. (1979) Photosynthetic rates in four tropical African fresh waters. *Freshwat. Biol.* **9**: 444−571.

Muchiri, M. (1990) The feeding ecology of tilapias and the fishery of Lake Naivasha, Kenya. PhD Thesis (unpublished), University of Leicester, UK.

Pauly, D. (1983) Some simple methods of assessing fish stocks. *F.A.O. (Rome) Fish. Tech. Pap.* No. **234**, 52 pp.

Ricker, W.E. (1975) Computation and interpretation of biological statistics of fish populations. *Bull. Fish. Res. Bd Can.* **191**: 382 pp.

Siddiqui. A.Q. (1977) Lake Naivasha (Kenya, East Africa) fishery and its management together with a note on the food habits of fishes. *Biol. Conserv.* **12**: 217−218.

Chapter 35
Catch effort data and their use in the management of fisheries in Malawi

S.B. ALIMOSO *Fisheries Research Unit, Box 27, Monkey Bay, Malawi*

Catch and fishing effort data are available for most fisheries in Malawi. Acquisition of these data from artisanal fisheries is carried out through monthly catch assessment surveys and annual frame surveys. All commercial (industrial) fishing units are obliged by law to submit monthly returns of their daily fishing activities to the Malawi Fisheries Department. Information derived from the analysis of these data, using surplus yield models, provides useful guidelines for managing fisheries in the country. Dynamic pool models cannot be used because data required to fit these models are either inadequate or unavailable. Information on the Chambo (*Oreochromis* spp.) fishery of Lake Malawi is presented to demonstrate the use of catch-effort data for the management of the fishery.

35.1   Introduction

The human population of Malawi, approximately 8 million people, depends mainly on fish for its source of animal protein: fish makes up about 70% of the total animal protein consumed. The total annual fish production is about 70 000 tonnes, mainly from capture fisheries operating in lakes Malawi, Malombe, Chilwa and Chiuta, and from the rivers, marshes and lagoons associated with these lakes (Fig. 35.1). Artisanal fisheries land approximately 90% of the fish; the remainder are caught by the commercial fisheries who are only found in the southern part of Lake Malawi. Fish farming is relatively undeveloped in the country and does not contribute significantly to the estimated total fish production.

The two sectors of capture fisheries in Malawi, artisanal and commercial fisheries, are divided mainly by their scale of operation, but also for administrative convenience (Alimoso, 1988). Approximately 34 000 artisanal fishermen operate in the various national waters of Malawi using 12 000 dugout canoes and small plank boats with or without outboard engines. A variety of traditional fishing gears, such as gill nets, long lines, seine nets, traps, hand lines, cast nets, etc., are employed. Fish are landed at approximately 1000 fishing sites scattered around the water bodies. Commercial fisheries, on the other hand, include 15 pair-trawlers (44 hp−160 hp), five stern-trawlers and two pair-ring-netters (120 hp−235 hp). Less than 1000 people are employed in commercial fishing and operations are confined to the southern part of Lake Malawi.

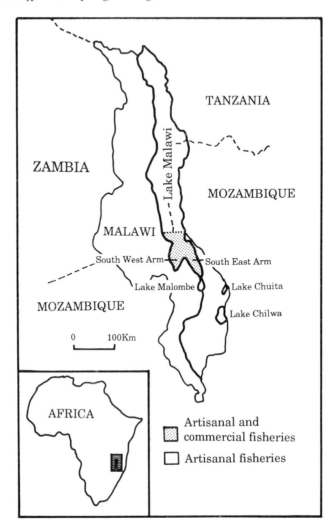

Fig. 35.1 Map showing Malawi and the lakes in which most fishing takes place. The extent of the commercial and artisanal fishing in Lake Malawi is also shown

In view of the importance of fish as a source of protein for human consumption, it is essential that the capture fisheries in Malawi be managed to maintain and, where possible, increase the fish supply. Fishery management based on dynamic pool models is unsuitable at present because the detailed information on growth, recruitment and natural mortality of the many species that comprise the fish stocks is either scanty or completely lacking. Surplus yield models, however, which require only catch and fishing effort data can be applied as a basis for formulating reasonable guidelines for managing fisheries in Malawi since sufficiently long time-series are available. In this chapter, the methods currently employed in collecting catch and fishing effort data in Malawi are presented. In addition, the

Chambo (*Oreochromis* spp.) fishery of the southern part of Lake Malawi is briefly described to explain the use of catch-effort data in managing the fishery.

35.2   Methods of data collection

As distinct differences exist between the commercial and artisanal fisheries two methods of collecting data, one for each fishery, are employed.

35.2.1   *Commercial fisheries*

As a result of studies conducted by FAO (1976), the southern part of Lake Malawi was partitioned into eight commercial fishing areas for stock assessment purposes (Fig. 35.2(a)). Operators of commercial fishing units are required, as a condition of their fishing licence, to submit daily fishing records to the Fisheries Department at the end of each calendar month. The records, which are submitted on a standard form, include the date when fishing took place, area(s) fished, number of hauls (for ring nets), duration of fishing operations (hours), wet weight of catch (in kg) by major category or species and depth of the water fished. From these returns, catch and effort time series are tabulated for each commercial fishing area and for each fishing unit.

35.2.2   *Artisanal fisheries*

The statistical method used in collecting data from the artisanal fisheries was developed by Bazigos (1972) and established in Malawi by Walker (1974). For the purpose of collecting statistics from the artisanal fisheries, the water areas are divided into major and minor strata (Fig. 35.2(b)). Stratification is based on the differences between the areas in such factors as intensity of fishing, fishing methods and physical characteristics of the area (Walker, *op. cit*). Minor strata boundaries in the south of Lake Malawi closely follow those of commercial fishing areas.

Two types of surveys, the frame survey (FS) and the catch assessment survey (CAS), are carried out in each minor stratum. The frame survey is conducted in August every year by field staff, on foot or bicycle. All the fishing sites along the coastline are visited, recording data on a standard form. The main details recorded are the number and type of fishing craft and gear owned by a fisherman, and the number of assistants he employs to operate the fishing gear. These data, therefore, provide some basic information on the size of the fishery in absolute terms.

Since there is a large number of fishing sites in each minor stratum with different types of fishing gears used at each site, it is impractical to collect data on total catches and fishing effort. Consequently, samples of catch and fishing effort are recorded in a more specific catch assessment survey. For this only four fishing sites are selected for sampling according to the following procedure. The sites are first classified into two groups, 'small' and 'large'. The size of a site is determined by the number of usable fishing craft counted on the site during the annual frame

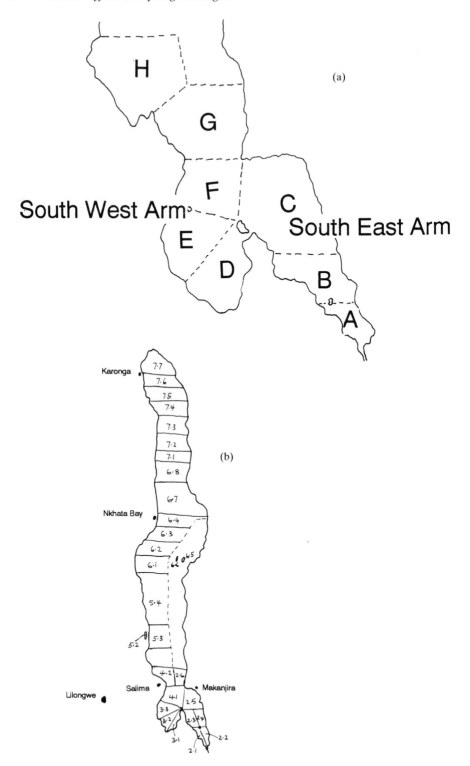

Fig. 35.2 (a) Southern Lake Malawi showing commercial fishing areas. (b) Lake Malawi showing minor strata from which artisanal fisheries statistics are gathered

survey. Size classification is based on the size-frequency distribution curve of the fishing sites. Two fishing sites from each size group are selected with probability proportional to their size, i.e. unequal probability sampling where a greater weighting is given to a site which has a larger number of fishing craft (Walker, 1974). Only four sites are sampled because of financial and resource retrictions and experiences elsewhere in Africa showed this number of sites provided reliable yield estimates. Sampling of catch and fishing effort is carried out on the selected sites, every month, until the results of a new frame survey are available. When new frame survey data have been obtained, selection of a new set of sampling sites (selection with replacement of the 'old' sampling sites) is carried out.

During each month, the four selected sampling sites are visited. The order in which the sites are visited is randomly predetermined before sampling begins in that month. Each sampling site is visited on four consecutive days during which time the following activities are carried out. On the first day, a complete count of all the fishing units existing on a fishing site is carried out, recording, for each fishing unit, the number of fishermen, their gear and the number of boats they own.

On the second day, actual recording of the catches begins. The operational fishing units are observed and listed in their order of leaving the fishing site. On their return, fishing units whose catch are to be sampled are selected systematically on the order of their landing at the fishing site. For example, if seven fishing units went out fishing, only four of the returning units − the first, the third, the fifth and the seventh − would be sampled. If more than eight fishing units operate only five would be sampled. For each unit sampled, the weight of the catch by species is recorded together with the fishing effort (number of gill nets or number of pulls (hauls) in the case of seine nets) expended to catch the fish. If the quantity of the catch is too large to be weighed, a sub-sample is taken and, based on its weight and the number of units (tins, crates of fish, etc.) actually counted from the catch, an estimate of the total catch by species is determined and recorded. This procedure is repeated on the third and fourth days − a total of three days of actual sampling − after which the recorder moves on to another selected sampling site.

Since it takes four days to collect records at each of the four selected fishing sites, the monthly catch assessment survey lasts for the first 16 days of each calendar month.

Estimation of total catch and fishing effort (E) from each minor stratum is based on the sample data collected in the catch assessment and frame surveys according to:

$$E = \frac{L}{S} \times \frac{M}{n} \times \frac{Z}{z}$$

where L is the number of fishing units which landed on the selected fishing site during *n* days of sampling during the catch assessment survey (three days in this case), *S* is the number of fishing units actually sampled during the three days of sampling, *M* is the number of days in a month, *Z* is the number of fishing craft counted from the minor stratum during an annual frame survey and z is the number of fishing craft counted from the selected fishing site during the annual frame survey.

The calculation outlined above is repeated for data collected at each selected fishing site so that a set of four values of E are obtained for the minor stratum. Total catch or fishing effort for a minor stratum is estimated by calculating the mean of the four values of E obtained.

35.3   Catch-effort data and fishery management in Malawi

Catch and fishing effort data have been used, quite successfully, as a useful tool for fishery management in Malawi and it would appear that these data will continue to provide useful management guidelines for a long time to come. The commercial demersal fishery of southern Lake Malawi, which produces about 25% (approximately 10 000 tonnes) of the total Lake Malawi catch, has been monitored and controlled by the Malawi Fisheries Department for the past two decades, using catch-effort data and the Schaefer (1954) surplus yield model (Turner, 1977; Tweddle Magasa, 1989). Analyses of catch-effort data collected from the various artisanal fisheries in Malawi are currently being carried out, using surplus yield models, with a view to recommending appropriate fishing effort quotas. The Chambo fishery of the south-east arm of Lake Malawi, probably the most valued fishery in Malawi, has also been managed on the basis of catch-effort information. This fishery is described below to serve as a typical example of how catch-effort data are used in managing fisheries in Malawi.

35.3.1   *Chambo fishery*

Chambo is a local name given to a species complex of the genus *Oreochromis*. The fish is the basis of important artisanal and commercial fisheries in the south-east arm of Lake Malawi. The commercial Chambo fishery, which started in 1943, exploits Chambo using two Chambo ring nets (mesh size 102 mm), an Utaka ring net (mesh size 38 mm), a Chambo midwater trawl (mesh size 102 mm) and an Utaka midwater trawl (mesh size 38 mm). Some Chambo also form part of the by-catch of demersal trawling (<1.0% of the demersal catch). A concise history of the commercial fishery is given by Lewis (1986).

Attempts to rationalize commercial exploitation of Chambo started early, in 1951, with the increase in the minimum mesh size from 51 mm to 102 mm and a one month closed season (December) which was later extended to two months (November to December) in 1960. These management measures were based on an earlier study, by Lowe (1952), on the biology of Chambo and its fisheries. The first analysis of catch-effort data was by Williamson (1966), who, by examining the trends in the Chambo catches, catch rates and amount of fishing, recommended an optimum fishing effort level of between 6000 and 7000 ring net pulls per year for the commercial Chambo fishery. FAO (1976) reviewed Williamson's assessments using a longer series of data and fitting the data to Schaefer's surplus yield model. The results of this assessment, which are the basis of present regulations of the commercial fishery, estimated maximum sustainable yield (*MSY*) to be 2000 tonnes per year at an optimum fishing effort (F_{opt}) of about 8000 ring net pulls per year. A recent analysis of the catch-effort data from the commercial fishery by Tweddle and Magasa (1989), which includes all the commercial data collected from 1951 to 1985 gave *MSY* and F_{opt} estimates similar to those obtained by FAO.

Artisanal fisheries also exploit Chambo in the same area as the commercial fisheries. The main gears that catch Chambo are gill nets (mesh size approximately 96 mm) and two types of beach seine nets: the Kambuzi seine nets (mesh size 19 mm to 25 mm) and Chambo seine nets (76 mm to 102 mm). About 6000 artisanal fishermen, using more than 1300 small plank boats and dugout canoes, are involved in the fishery. Assessments of the Chambo fishery by Williamson (1966) excluded the artisanal Chambo fisheries because there were no reliable data at that time. FAO (1976) also did not take into account artisanal fisheries because the data then available were not adequate. Although sufficient artisanal Chambo fishery catch-effort data were available, Tweddle and Magasa (1989) excluded the data on the assumption that artisanal fisheries did not exploit the same Chambo stock as the commercial fishery. An analysis carried out by Alimoso (1986) treated the Chambo fishery of the south-east arm of Lake Malawi as a multi-gear fishery in which both the artisanal and commercial fisheries were assumed to exploit the same Chambo stock. Results suggested that of the four types of surplus yield models tried, the Fox (1970) model fitted the data best. *MSY* was estimated at about 4000 tonnes per year at an optimum effort of approximately 900 000 gill net units. It was shown that each fishery contributed equally to the total annual catch and that there was a need to take into account the fishing effort exerted by the artisanal fishermen.

Table 35.1 shows the updated catch and effort data series collected from the commercial and artisanal fisheries exploiting Chambo in the south-east arm of Lake Malawi. Data from the demersal trawlers are not included because these do not exploit Chambo to any noticeable extent. The table also shows the total catch and the standardized annual effort for the multi-gear Chambo fishery. The effort was standardized by dividing the total Chambo catch by the catch per unit effort (CPUE) of the gill nets, assuming that this gear is a reliable sampling gear and that, therefore, its CPUE was the best representative of an index of abundance of Chambo in that area of the lake. Gill nets are the dominant gear in terms of the amount of Chambo they catch annually. A gill net is also a passive rather than an active gear and, therefore, is not influenced by human skill. Hence gill net catch per unit effort was expected to be more related to the Chambo abundance than the other gears. Alimoso (1986) used this method successfully to standardize fishing effort in the Chambo fishery of the south-east arm of Lake Malawi.

Fig. 35.3(a), whose curve was fitted from an exponential regression of CPUE on fishing effort (Fig. 35.3(b)) (Fox's model, 1970) suggests that maintaining an

Table 35.1 Chambo catch (tonnes) data from the commercial and artisanal fisheries of the South East Arm of Lake Malawi. Total standardized fishing effort, gill net CPUE and total Chambo catch are also tabulated. RN = ring net, MWT = midwater trawl, CSN = Chambo seine net, KSN = kambuzi seine net, GLN = gill net. The figures in brackets, e.g. MWT (38 mm) are the mesh sizes of the commercial gears. Gill net effort is in gill net units where a gill net unit is a net of length 100 yd (91.4 m) set overnight and checked the following morning.

Year	Standard RN (102 mm)	Commercial fisheries			Artisanal fisheries			Total (tonnes)	Gill net Catch (kg/net)	Total CPUE Effort (gill nets)
		RN (38 mm)	MWT (102 mm)	MWT (38 mm)	CSN	KSN	GLN			
1976	1151	0	0	683	106	57	1249	3246	1.7	1 909 412
1977	471	22	0	1146	162	277	3138	5216	3.06	1 704 575
1978	975	8	0	912	301	520	1614	4330	2.59	1 671 815
1979	504	9	0	858	100	85	1806	3362	3.75	896 533
1980	1481	1	0	873	51	34	1678	4118	3.04	1 354 605
1981	1056	159	80	768	208	78	1561	3910	3.12	1 253 205
1982	1070	335	90	714	442	37	2985	5673	5.31	1 068 362
1983	1838	327	9	533	359	47	3302	6415	7.51	854 194
1984	1293	628	0	368	1458	56	2444	6247	3.65	1 711 507
1985	1444	142	197	1250	799	17	2403	6252	3.66	1 708 197
1986	1265	222	388	287	583	203	2211	5159	3.53	1 461 473
1987	1201	17	602	120	560	266	826	3592	2.33	1 541 631

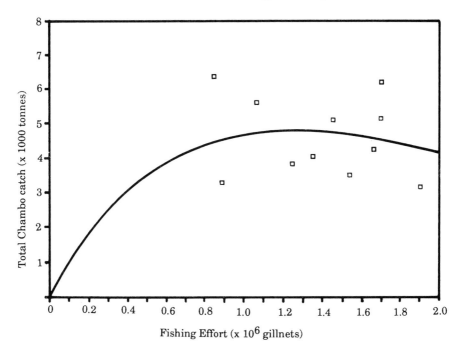

Fig. 35.3(a) The relationship between catch and total fishing effort in the multi-gear Chambo fishery of the south-east arm of Lake Malawi

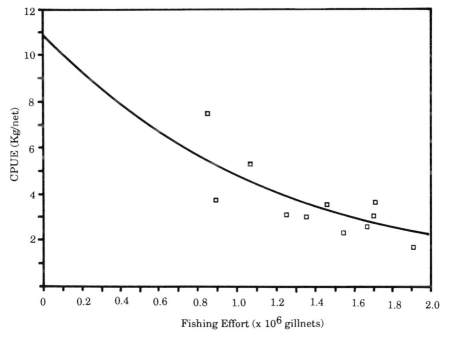

Fig. 35.3(b) The relationship between CPUE and total fishing effort in the multi-gear Chambo fishery of the south-east arm of Lake Malawi

annual total fishing effort of about 1×10^6 standard gill nets would result in an *MSY* of about 5000 tonnes per year. The data also suggest that the Chambo fishery is at present operating at or near the *MSY* or, indeed, may have been slightly over-fished. Action to limit fishing effort and to continuously monitor the fishery is recommended.

Implementation of recommendations from this work has been delayed awaiting the results of a more comprehensive research project currently being carried out on Chambo and its fisheries.

35.4   Discussion and conclusion

Problems associated with the data collection methods currently employed in Malawi were presented and discussed at length by Alimoso (1988). It is sufficient to mention that collection of data from the commercial fisheries presents few problems since data from all commercial fishing units are accessible to the Fisheries Department. The number of these units is relatively small and, therefore, not difficult to monitor. Adjustments to the reported data can easily be carried out following any major changes that might be noticed in the fishery. The Fisheries Department is generally satisfied with the quality of the data from these fisheries.

Artisanal fisheries, however, are by their nature difficult to monitor. While the sampling technique used in collecting catch-effort data has, apparently, been successful in some fisheries (e.g. gill net fishery), it does not seem to produce reliable data in other, quite important, fisheries (e.g. beach seine and trap fisheries). The reasons for this are firstly operational and, secondly, related to the design of the sampling technique. The operational problems are mainly related to administrative and financial constraints (Orach Meza, 1991). The main problem with the design of the sampling technique is its inability to take into account particular, and sometimes unique, features of some artisanal fisheries. One design flaw not mentioned in the review of the statistical method by Alimoso (1988) is that the monthly sampling period in a catch assessment survey is by design concentrated between the beginning and the middle of the month. These design and operational problems tend to produce unrealistic estimates in some artisanal fisheries.

Although a number of criticisms have been levelled at catch-effort analyses (Larkin, 1977; Gulland, 1978; Healy, 1984), reasonable and practical alternatives are still lacking. Dynamic pool models cannot be used because data required in fitting the models are either inadequate or unavailable. In addition, the multi-species nature of the fisheries makes it relatively expensive to estimate the various parameters required to fit dynamic pool models. Applying regulations based on catch effort information, together with knowledge of the biology and particular history of the fisheries, has proved to be a viable fisheries management strategy in Malawi.

References

Alimoso, S.B. (1988) An assessment of yields in the Chambo (*Oreochromis* spp.) fishery in southern Lake Malawi. M.Sc. Thesis, University of Wales, Bangor, UK.

Alimoso, S.B. (1986) A review of the present systems of collecting fisheries statistics from Malawi waters. Country Paper, SADDC Fisheries Statistics Workshop, Lasaka, Zambia, 25–29 April, 1988. Govt. of Malawi SADDC Report, Lilongwe, Malawi.

Bazigos, G.P. (1972) Improvement of the Malawi Fisheries Statistical System. Fisheries Dept. Report, Lilongwe.

F.A.O. (1976) An analysis of the various fisheries of Lake Malawi. Promotion of Integrated Fishery Development, Malawi. *FAO/UNDP Tech. Rep* 1 No. FI/DP/MLW71/516

Fox, W.W. (1970) An exponential surplus yield model for optimising by exploited fish populations. *Transactions of the American Fisheries Society* **99**: 80–88.

Gulland, J.A. (1978) Fishery management: New strategies for new conditions. *Trans. Am. Fish. Soc.* **107**: 1–11.

Healy, M.C. (1984) Multiattribute analysis and the concept of optimum yield. *Can. J. Aqua. Sci.* **41**: 1393–1405.

Larkin, P.A. (1977) An epitaph for the concept of maximum sustainable yield. *Trans. Am. Fish. Soc.* **106**: 1–11.

Lewis, D.S.C. (1986) A review of the research conducted on chambo and the chambo fisheries of Lakes Malawi and Malombe, 1859–1985. Malawi Fisheries Department Report, 36 pp.

Lowe, R.H. (1952) Report on the tilapia and other fish and fisheries of Lake Nyasa, 1945–47. Fish. Publ. Colon. Office, London. 1 (2); 126 pp.

Orach-Meza, F.L. (1991) Statistical sampling methods for improving catch assessment in lake fisheries. In *Catch effort sampling strategies: Their application in freshwater fisheries management* (Ed. by I.G. Cowx). Oxford: Fishing News Books, Blackwell Scientific Publications Ltd.

Schaefer, M.B. (1954) Some aspects of the dynamics of populations important to the management of commercial marine fisheries. *Bull. Int-Am. Trop. Tuna Comm.* **1**: 27–56.

Turner, J.L. (1977) Some effects of demersal trawling in Lake Malawi (Lake Nyasa) from 1968 to 1974. *J. Fish. Biol.* **10**: 261–271.

Tweddle, D. and Magasa, J.H. (1989) Assessment of multispecies cichlid fisheries of the Southeast Arm of Lake Malawi, Africa. *J. Cons. Int. Explor. Mer.* **45**: 209–222.

Walker, R.S. (1974) Collection of catch assessment data in tropical fisheries (inland waters). Notes to the Malawi Fisheries Department. 6 p.

Williamson, R.B. (1966) Analysis of ring-net catch data from the South East Arm of Lake Malawi, 1946–1966. Malawi Fisheries Dept, Mimeographed report. 4 pp.

Chapter 36
Catch effort sampling strategies: conclusions and recommendations for management

I.G. COWX *Humberside International Fisheries Institute, University of Hull, Hull, HU6 7RX, UK*

The management of the fisheries of large inland waters is notoriously difficult because of the inability to evaluate the absolute abundance of the fish populations by conventional scientific methods. One approach which has received some attention in recent years is that of catch effort sampling. This technique has been used with varying degrees of success to assess the status of fish populations in both commercial and recreational fisheries and aid policy decision-making with regard to their management. This paper reviews the various catch effort sampling strategies that have been used in the management of commercial and recreational fisheries in both temperate and tropical regions. It assesses the advantages and disadvantages of the different sampling strategies available and how they can be best used to satisfy the management objectives.

36.1   Introduction

Each continent is well endowed with lakes, rivers, impoundments, wetlands and coastal and estuarine waters. However, until recently, arrangements for the rational exploitation of the fisheries resources associated with these waters has received little attention. This is partly due to the inability to evaluate the absolute abundance of the fish populations by conventional scientific methods (EIFAC, 1974). Unfortunately, increasing fishing pressure, often resulting in overfishing, has heightened our recognition of the need to maintain and, in some cases, preserve the fisheries. As a result, significant plans have been initiated by various agencies, governments and institutions to determine the nature and magnitude of the fisheries resources associated with various water bodies and find the best ways of exploiting them in a rational manner, thus perpetuating the availability of the resource. Some of the efforts have been successful but few have managed to lay the foundations upon which future plans could be based.

This requirement to properly manage the fisheries resources has called for continuous research and development into the various methods of evaluating their status and assessing the impact of exploitation. To this end, catch effort sampling strategies, as described in the preceding chapters, provide the opportunity to resolve many of the important technical and management issues in a cost-effective manner. This chapter reviews the concept of catch effort sampling strategies, as highlighted in the preceding chapters, and summarizes the main conclusions and recommendations presented.

36.2 Development of catch effort sampling strategies

Inland fisheries need to be managed for a variety of reasons. The objectives are usually associated with the types of use as well as with socio-economic factors connected with the fishing community. Apart from management goals, which underlie most management practice, a typical, but not exhaustive, list of such objectives is as follows:

(1) Production of food.
(2) Maintenance of stocks for sports fishing.
(3) Maintenance of stocks for recreational fishing.
(4) Supply of ornamental fish.
(5) Control of unwanted organisms (diseases vectors, vegetation, rice borers, etc.).
(6) Sustain employment within a community.
(7) Conservation or other aesthetic values.

In terms of management, all these objectives demand action aimed at the maintenance of the fish stock. However, management of inland fisheries is still far from an exact science because of the difficulty in evaluating stock characteristics. Decisions are often based on poor quality, relative information which is inadequate for the purpose. Catch effort sampling strategies provide a mechanism whereby the quality of the information can be improved, at minimal cost to the fishery manager.

When developing catch effort sampling strategies, a number of design aspects are common to both commercial and recreational inland fisheries, and the different types of water bodies, i.e. lakes, reservoirs and rivers.

Firstly, there is the need for a simple common mechanism which can be applied to locally specific situations by the responsible organizations which will enhance management decision-making to the benefit of the fishery. In other words as much information as possible about the status of the fishery, at any moment in time, is required. This requirement is best illustrated by Fig. 36.1 where the ultimate objective is to assess the current status of the fish stocks and predict future yield. It is important to note that catch effort (fishery dependent) data alone will rarely provide sufficient information about the fishery for sound management. This information needs to be supplemented by fishery independent data (from experimental fishing and routine monitoring) to give an overall perspective of the status of the fishery.

Secondly, management should actively encourage development and improvement in the mechanism of collection and transfer of catch and effort data. This process should involve an active feedback of information from the manager to the user group. Any new or existing catch effort survey programme should attempt to

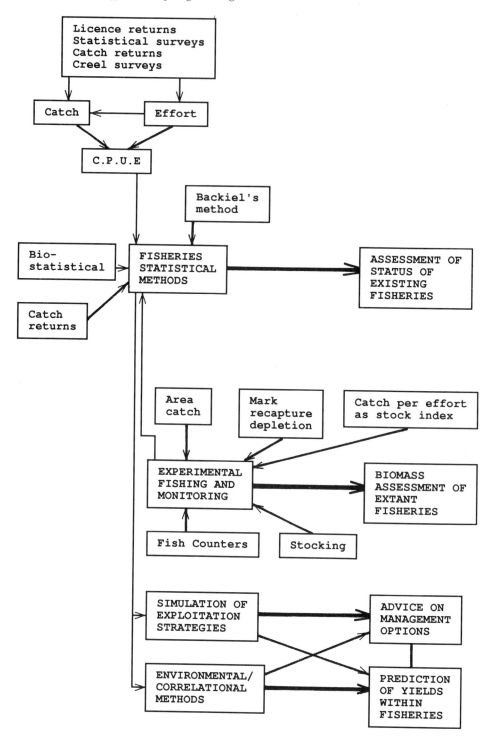

Fig. 36.1 Suggested combination of methods for stock assessments and yield predictions (modified from Welcomme, 1979)

make the data collection and data-processing mechanism simple to ensure the accuracy of the data collected and the maintenance of the system. Several systems have failed because the data processing and interpretation have been neglected (e.g. Lake Volta, L. Braimah *pers. comm.*). A simplified feed-back mechanism to illustrate this point is presented in Fig. 36.2. Adoption of such an approach should improve the reliability of the data collection techniques and also facilitate the dissemination of data on the fish stocks and the fishery.

36.3   Catch effort strategies for commercial fisheries

Collection of catch and effort data for commercial fisheries falls into two categories, those fisheries where the effort is regulated by licences (e.g. River Severn, chapter 1; Lake Annecy, chapter 11; IJsselmeer, chapter 22) and artisanal and subsistence fisheries (e.g. Lake Victoria, chapters 31 and 33; Lake Malawi, chapter 35; Lake Kyoga, chapter 29). The latter are usually associated with developing countries. The solution to the catch effort data collection strategy for these two categories is thus very different.

36.3.1   *Artisanal fisheries*

Until recently, artisanal fisheries of large inland waters have been monitored by direct enumeration or frame surveys. This involves observers visiting a high proportion of the fishing villages or landing sites on a regular or *ad hoc* basis and collecting as much information on the number of fishing boats, fishing gears and fishermen as possible. These data are usually collected by direct enumeration or random sub-sampling. Although this procedure gives an overall picture of the fishery, the data collected are often based on a single sample and may not be truly representative of the pattern over space and time. Thus the accuracy of the information is somewhat suspect. In addition, the manpower and financial resources required to implement such frame surveys on large fisheries are generally excessive and often lead to inadequate coverage of the fishery, particularly when it is diverse with regard to the fishing methods used and species exploited.

These limitations preclude the reliability and applicability of simple frame surveys for direct total enumeration of the status of the fisheries and warrant the use of sub-sampling (stratified sampling) procedures (Bazigos, 1974). Through this sampling design it is possible to take into account the following considerations:

(1)   Many waters are characterized by having multiple species stocks.
(2)   Variability in the fishing methods and fisheries.
(3)   Settlements of fishing communities along the water bodies.
(4)   Variations in the type and size of the fishing gears.
(5)   Variations in the type of fishing vessels.
(6)   Variation in the type of mechanization/propulsion of the fishing vessels.
(7)   Variation in the fishing methods and times.

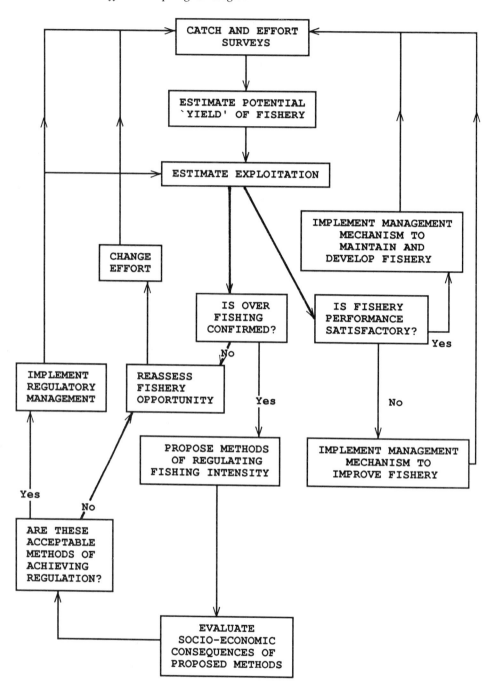

Fig. 36.2 Flow chart for decision-making in the management of inland fisheries using catch effort sampling techniques as the method of evaluation

(8) Variation in the landing times and places.
(9) Diversity in fishing skills.

Therefore, before a random stratified sampling programme can be designed and implemented a considerable amount of information about the fishery is required. If this information is not available it could preclude the use of such a sampling strategy. In addition, it is necessary to accurately define the most appropriate parameter(s) of fishing effort. This has been shown to vary considerably for different fisheries, e.g. number of boats in Lake Kyoga (Orach-Meza, chapter 29) and Lake Malawi (Alimso, chapter 35), number of fishermen in the Niger River (Meredith & Malvestuto, chapter 21), light hours in Lake Victoria (Ligtvoet et al., chapter 31) or traps in Bukit Merah Reservoir (Yap, chapter 32).

To facilitate the design of a suitable sampling technique there is a need to:

(1) Stratify the water body in order to obtain a relatively homogeneous ecological system.
(2) Stratify the fish landing centres by size within the primary strata.
(3) Stratify the fishing vessels by types and fishing methods within the primary strata.

A typical sub-division of the characteristics of a fishery (Lake Kyoga; Orach-Meza, chapter 29) for such statistical purposes is shown in Fig. 36.3.

In this procedure each stratum is assumed to be homogeneous and that the variability of a particular parameter of the fishery is minimal. Thus a precise estimate of any stratum parameter mean can be obtained from a small sample

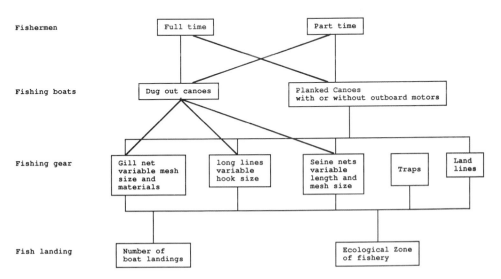

Fig. 36.3 Suggested stratification of the Lake Kyoga fisheries (after Orach-Meza, chapter 29)

within the stratum. The procedures for optimal allocation of the sample size have been formulated by Bazigos (1974) and are illustrated for Lake Kyoga in chapter 29.

These procedures allow the fishery to be stratified to account for all the possible parameters and the relationship between them. Within this procedure, however, accurate data collection of the selected statistical units is imperative. Unless this work is carried out carefully, honestly and thoroughly the returns will give a false impression of the status of the fishery. These data should be supported by fishery independent data which are collected periodically to evaluate the stock size and structure and characteristics of the gear.

36.3.2   *Licensed fisheries*

When the fisheries are regulated by licences, as is usually the case for catching salmonids, there is generally an obligation to make a catch return. The principles of catch effort sampling strategies are similar to those encountered in recreational fisheries with natural recruitment, where the fish are removed, i.e. log books/ catch census recording the catch of all fishermen (see chapters 1, 2, 10, 11, 22, 23 and 24). As with recreational fisheries, the monitoring procedure suffers from the problem of non- or inaccurate returns of catch information (chapters 1, 2, 7, 8, 10 and 22) and a mechanism to adjust for these missing data is therefore required. This procedure has been expounded for recreational fisheries (chapters 7, 8 and 10) but is equally applicable to commercial fisheries.

36.4   **Catch effort strategies for recreational fisheries**

In examining the documentation it was apparent that surveys of recreational fisheries are required to measure the impact of management policy or assess the status of the resource both in terms of stock structure and end user satisfaction. The planning of these surveys are governed by constraints on money and manpower and the degree of accuracy required to meet the management objectives.

Two distinct methods of collecting the information were identified. These were:

(1) Event recall methods such as logbooks, mail questionnaires and phone surveys which require the angler to evaluate the fishing experience some time after the event (chapters 1, 2, 4, 7, 8, 12, 14, 15, 20).
(2) On-site intercept methods such as aerial count methods, access point methods and roving creel methods. They collect information at the time of the event or immediately following it (chapters 5, 6, 12, 19).

Event recall methods are low cost and can have a regional coverage. However, they suffer from non-response and biases associated with recall memory, angling prestige and enthusiasm. To obtain acceptable data the recall time interval should be kept to a minimum (<2 months preferred) and the questions need to be simple. If at all possible the participants should be encouraged to complete the questionnaires at the time of the fishing experience.

On-site intercept methods minimize the biases associated with event recall methods because they allow direct information exchange which can be confirmed by the interviewer. However, their cost is considerably higher and they rely on a stratified survey approach which may introduce biases associated with the estimator used. Although the complexity of the questions in on-site methods can be increased it must not be so great as to interfere with the angler experience.

To ensure accuracy of the data retrieval, several important aspects should be addressed:

(1) Event recall methods must be validated with respect to the accuracy of the responses. Calibration with on-site survey data is one option.
(2) On-site intercept methods must be evaluated with respect to field techniques and design efficiency. New design alternatives must be developed, particularly with respect to obtaining precise estimates of catch rate.
(3) Indices which truly reflect the quality of the fishery from the anglers point of view must be developed and used for justification and evaluation of management programmes.
(4) The value of CPUE as an accurate index of stock abundance for recreational resources must be evaluated, particularly for those fisheries where stock size and high rates of exploitation are critical factors related to efficient functioning and economic value of the fishery (put-and-take systems).

The objectives of the data collection systems in recreational fisheries were considered two fold:

(1) To obtain consistent and understandable information as a basis for discussion on management with regulatory authorities and anglers.
(2) To provide regular, standard format, publications which should be freely available (to the public) on request. It should be simple with easily interpreted diagrams which can be up-dated as necessary.

For exploited, natural stocks, annual assessment of the stock status through catch and effort sampling appears acceptable whilst put-and-take fisheries need weekly monitoring, for angler satisfaction and stock depletion. For unexploited stocks, trends tend to be the main requirement.

All catch and effort data distributions are non-normal (Small, chapter 7), due to fish distribution aggregation and effort variability. However, for monitoring purposes, normalized mean ± variance, plus significance of any differences is satisfactory. In general, indices of abundance are more important, and can be of higher quality, than absolute catch estimates.

Data handling procedures for recreational fisheries should, as with commercial fisheries, be as easy as possible and with fast feedback to the fishery participants.

The use of micro-computers with standard analytical/processing programmes and database format should be used if possible. The output should be simple explaining the need for management data. This will ease acceptance of any regulations and uphold the quality of data (imposed management leads to deterioration and/or bias in data). Regular feedback of the data to the end user is essential.

36.5   Recommendations

On the basis of scientific data provided by the symposium, several recommendations were made. These were:

(1)   Collection of catch and effort data needs to be further enhanced and governments, through their organizations, should endeavour to improve the mechanisms for such collection wherever possible. The gathering of fishery-independent data by routine surveys should be organized in such a way as to meet the information needs of the fisheries management.

(2)   Any new or existing catch effort survey programme should attempt to make the data collection and data-processing system simple to ensure the accuracy of the data collected and the maintenance of the system. The need to support catch effort sampling programmes with fishery independent data was again stressed.

(3)   The literature contains an array of incompatible jargon, definitions and symbols and survey methods used throughout the world. Efforts should be directed toward the standardization and analysis of data collection. The resulting data should be made available in the most appropriate form to aid various user groups.

(4)   Governments should encourage the active participation of professional inland fishermen in the management of fisheries and the freshwater environment.

(5)   A handbook of practical sampling and analytical procedures relating to catch and effort sampling strategies is required. Both commercial and recreational techniques, in temperate and tropical countries, should be covered, either in the same or separate texts.

(6)   Education in fisheries management at the levels of decision makers, user groups and the general public is of major importance. This will ensure all personnel involved in the fishery understand and trust the scientific advice given, and that the implications and likely impacts of management options are understood. This activity is essential if we wish to sustain the resource within biologically safe limits.

Acknowledgements

The author would like to thank all those who contributed to the symposium on catch effort sampling strategies and in particular, Ir B. Steinmetz, Drs S. Malvestuto, W.L.T. van Densen, M.L. Orach-Meza, P. Hickley and M. Pawson, all of whom helped generate the contents of this chapter.

References

Bazigos, G. (1974) The design of fisheries statistical surveys—inland waters. *FAO Fish.. Tech. Pap.* **133**: 122 p

European Inland Fisheries Advisory Committee (EIFAC) (1974) Symposium on methodology for the survey, monitoring and appraisal of fishery resources in lakes and large rivers. EIFAC Tech. Pap. **22**: Rome: FAO Pubs.

Welcomme, R.L. (1979) Fishery management in large rivers. *FAO Fish. Tech. Pap.* **194**: 602 p

Species Index

Site Index

General Index

Books published by
Fishing News Books

Free catalogue available on request from Fishing News Books, Blackwell Scientific Publications Ltd, Osney Mead, Oxford OX2 OEL, England

Abalone farming
Abalone of the world
Advances in fish science and technology
Aquaculture in Taiwan
Aquaculture: principles and practices
Aquaculture training manual
Aquatic weed control
Atlantic salmon: its future
Better angling with simple science
British freshwater fishes
Business management in fisheries and
   aquaculture
Cage aquaculture
Calculations for fishing gear designs
Carp farming
Catch effort sampling strategies
Commercial fishing methods
Control of fish quality
Crab and lobster fishing
Crustacean farming
The crayfish
Culture of bivalve molluscs
Design of small fishing vessels
Developments in electric fishing
Developments in fisheries research in
   Scotland
Echo sounding and sonar for fishing
The economics of salmon aquaculture
The edible crab and its fishery in British
   waters
Eel culture
Engineering, economics and fisheries
   management
European inland water fish: a multilingual
   catalogue
FAO catalogue of fishing gear designs
FAO catalogue of small scale fishing gear
Fibre ropes for fishing gear
Fish and shellfish farming in coastal waters
Fish catching methods of the world
Fisheries oceanography and ecology
Fisheries of Australia
Fisheries sonar
Fisherman's workbook
Fishermen's handbook
Fishery development experiences
Fishing and stock fluctuations
Fishing boats and their equipment
Fishing boats of the world 1
Fishing boats of the world 2
Fishing boats of the world 3
The fishing cadet's handbook
Fishing ports and markets
Fishing with electricity
Fishing with light

Freezing and irradiation of fish
Freshwater fisheries management
Glossary of UK fishing gear terms
Handbook of trout and salmon diseases
A history of marine fish culture in Europe and
   North America
How to make and set nets
Inland aquaculture development handbook
Intensive fish farming
Introduction to fishery by-products
The law of aquaculture: the law relating to the
   farming of fish and shellfish in Great Britain
The lemon sole
A living from lobsters
The mackerel
Making and managing a trout lake
Managerial effectiveness in fisheries and
   aquaculture
Marine fisheries ecosystem
Marine pollution and sea life
Marketing: a practical guide for fish farmers
Marketing in fisheries and aquaculture
Mending of fishing nets
Modern deep sea trawling gear
More Scottish fishing craft
Multilingual dictionary of fish and fish
   products
Navigation primer for fishermen
Net work exercises
Netting materials for fishing gear
Ocean forum
Pair trawling and pair seining
Pelagic and semi-pelagic trawling gear
Pelagic fish: the resource and its exploitation
Penaeid shrimps — their biology and
   management
Planning of aquaculture development
Refrigeration of fishing vessels
Salmon and trout farming in Norway
Salmon farming handbook
Scallop and queen fisheries in the British Isles
Scallop farming
Seafood science and technology
Seine fishing
Squid jigging from small boats
Stability and trim of fishing vessels and other
   small ships
Study of the sea
Textbook of fish culture
Training fishermen at sea
Trends in fish utilization
Trout farming handbook
Trout farming manual
Tuna fishing with pole and line